T0325856

Progress in Physics
Volume 15

A. N. Leznov
M. V. Saveliev

Group-Theoretical Methods for Integration of Nonlinear Dynamical Systems

Translated from the Russian
by D. A. Leites

1992

Birkhäuser Verlag
Basel · Boston · Berlin

Authors' addresses:

Professor A. N. Leznov
Institute for
High Energy Physics
142 284 Protvino, Moscow region
Russia

Professor M. V. Saveliev
Institute for
High Energy Physics
142 284 Protvino, Moscow region
Russia

The present book is the expanded version of the Russian original, published as:
Gruppovye metody integrirovaniya nelineynykh dinamicheskikh sistem,
© Nauka, Moskva, 1988

Deutsche Bibliothek Cataloging-in-Publication Data

Leznov, Andrej N.:
Group theoretical methods for integration of nonlinear
dynamical systems / Andrei N. Leznov ; Mikhail V. Saveliev.
Transl. from the Russ. by D. A. Leites. – Basel ; Boston ;
Berlin : Birkhäuser, 1992
  (Progress in physics ; Vol. 15)
  Einheitssacht.: Gruppovye metody integrirovanija nelinejnych
  dinamičeskich sistem <engl.>
  ISBN 3-7643-2615-8 (Basel ...)
  ISBN 0-8176-2615-8 (Boston)
NE: Savel'ev, Michail V.:; EST; GT

© 1992 for the English edition: Birkhäuser Verlag Basel
Typesetting and Layout: *mathScreen online*, CH–4056 Basel
Printed on acid-free paper in Germany
ISBN 3-7643-2615-8
ISBN 0-8176-2615-8

Contents

Preface

The book reviews a large number of 1- and 2-dimensional equations that describe nonlinear phenomena in various areas of modern theoretical and mathematical physics. It is meant, above all, for physicists who specialize in the field theory and physics of elementary particles and plasma, for mathematicians dealing with nonlinear differential equations, differential geometry, and algebra, and the theory of Lie algebras and groups and their representations, and for students and post-graduates in these fields. We hope that the book will be useful also for experts in hydrodynamics, solid-state physics, nonlinear optics electrophysics, biophysics and physics of the Earth.

The first two chapters of the book present some results from the representation theory of Lie groups and Lie algebras and their counterpart on supermanifolds in a form convenient in what follows. They are addressed to those who are interested in integrable systems but have a scanty vocabulary in the language of representation theory. The experts may refer to the first two chapters only occasionally.

As we wanted to give the reader an opportunity not only to come to grips with the problem on the ideological level but also to integrate her or his own concrete nonlinear equations without reference to the literature, we had to expose in a self-contained way the appropriate parts of the representation theory from a particular point of view.

This has led to some new results in the representation theory itself, to wit: a canonical way for ordering roots of simple finite-dimensional Lie algebras and Lie superalgebras and an asymptotic method in the representation theory of noncompact Lie groups. These results enabled us to get explicit expressions for a number of Lie algebraic and group quantities, e.g., Haar's measure on compact groups and the highest weight vectors of their irreducible representations, the infinitesimal and invariant operators, the generalized Whittaker vectors and Plancherel's measure on semisimple Lie groups and the inter-twining operators, to name a few.

Many of these results are crucial in the study of integrable systems: some of them are recently obtained and are not yet reflected in monographs.

The reference, say, (1.2.3) means formula 3 from § 2 of Chapter 1, while 1.2.3 means subsection 3 from the same section.

References for a particular subsection or section are given generally as a footnote after its title. On the whole, the bibliography is obviously incomplete. In the introductory Chapters 1 and 2 we refer mostly to monographs or textbooks, although original works are cited in some cases. As for the main text (Chapters 3 to 7), since the problems discussed herein are set forth within the framework of a single approach, we did not think it appropriate to cite individual results obtained by other authors using different methods of specialized nature and adopted for particular physical problems.

The absence of an index is justified, we hope, by the detailed list of subjects given in the Table of Contents.

Notational conventions. As usual, $\mathbb{Z}$, $\mathbb{Z}_+$, $\mathbb{N}$, $\mathbb{Q}$, $\mathbb{R}$, $\mathbb{C}$ stand for the sets of integers, nonnegative integers, positive integers, rational, real and complex numbers, respectively; following Bourbaki we usually denote the Lie groups by Latin capitals and the corresponding Lie algebras by small Gothic letters.

Notation $A \equiv B$ is sometimes used instead of "introduce A setting $A = B$".

Let also

$$\mathrm{diag}(a_1, \ldots, a_n) = \begin{pmatrix} a_1 & & 0 \\ & \ddots & \\ 0 & & a_n \end{pmatrix}, \quad 1_n = \mathrm{diag}\,\underbrace{(1, \ldots, 1)}_{n},$$

$$J_{2n} = \begin{pmatrix} 0 & 1_n \\ -1_n & 0 \end{pmatrix}, \quad J_{p,q} = \mathrm{diag}(1_p, -1_q),$$

$$I_{p,q} = \mathrm{diag}(J_{p,q}, J_{p,q}), \quad \theta(a,b) = \begin{cases} 1 & \text{if } a < b \\ 0 & \text{if } a \geq b \end{cases}, \quad \mathrm{sgn}(x) = \begin{cases} 1 & \text{if } x > 0 \\ -1 & \text{if } x < 0 \end{cases}.$$

In what follows $\mathbb{Z}_2$ denotes $\mathbb{Z}/2\mathbb{Z}$, the field of residues modulo 2.

To avoid a misunderstanding sometimes we use for the elements of the algebra $\mathfrak{sl}(2)$ different notations:
$\{H, J_\pm\}$ and $\{h, X_\pm\}$.

Acknowledgments. § 1.7 was rewritten by our translator D.A. Leites. We are deeply thankful to B.A. Arbuzov, V.Ya. Fainberg, I.A. Fedoseev, O.A. Khrustalev, A.A. Kirillov, D.A. Kirzhnitz, D.A. Leites, A.A. Logunov, Yu.I. Manin, V.I. Manko, M.A. Mestvirishvili, S.P. Novikov, A.B. Shabat, V.G. Smirnov, L.D. Soloviev for useful discussions.

December, 1982. Protvino – Moscow.

Introduction *)

This monograph is devoted mostly to the exposition of a group-theoretical method for constructing nonlinear 1- and 2-dimensional integrable dynamical systems and finding explicit solutions of these equations determined by a sufficient number of arbitrary functions to formulate Cauchy or Goursát problems (i.e. initial value or boundary value problems, respectively).

This simple method of obtaining soliton solutions was devised for equations related to infinite-dimensional Lie algebras such as the periodic Toda lattice. Unlike the inverse scattering problem (ISP) method, it does not appeal to the explicit matrix realization of Lax operators.

A quantization procedure in the Heisenberg and Schrödinger pictures for the dynamical systems related to finite-dimensional Lie algebras was also developed.

The first positive results covered in this book were obtained, largely, during the past 10 years; the theme is far from being exhausted. However, the results accumulated already are plentiful and require at least a preliminary systematization. This is exactly the goal we had in mind when setting out to write this book.

The systems of nonlinear equations that arise in the framework of our approach are generated by elements of $\mathbb{Z}$-graded Lie algebras (or Lie superalgebras) via Lax representation in a 2-dimensional space (or $(2, 2N)$-dimensional superspace). Depending on the choice of an adequate algebraic structure they describe a broad class of nonlinear phenomena of various fields of theoretical and mathematical physics: e.g., elementary particle physics (gauge fields and monopole configurations), solid state and plasma physics, electrolite the-

---

*) [1, 2, 5, 9, 10, 12, 18, 22, 25, 26, 29–31, 39, 41–43, 53–55, 58–79, 82, 85–87, 90–92, 96–98, 102, 103, 105–108, 111, 124–129, 130–132, 134–136, 141, 147, 148, 150–152].

ory, nonlinear optics, aerodynamics, cosmology models, radiotechnology and ecology (dynamics of species coexistence).

The common feature of all the studied systems is the existence of a non-trivial group of internal symmetries generated by the Lie-Bäcklund algebra. This enables one to get their explicit solutions in terms of representation of Lie groups and their Lie algebras. The representation theory plays an essential role in all the exactly integrable dynamical systems, seemingly without exceptions.

The group-theoretical methods constitute an unalienable part of the mathematical apparatus of the modern mathematical physics. The lucidity and relative simplicity of these methods, and especially their adequacy to the problems considered, which have various symmetries related to the main laws of physical processes, distinguish it among the other methods.

It was only during the past decades, however, that the importance of the representation theory for the study of integrable systems has been understood (perhaps, not yet sufficiently) and the necessary ideas have been expressed in a constructive form.

As a confirmation of an underestimation of the role of the study of the representations of the Lie group and algebra theory for integration of differential equations that had been taken place we refer to N. Bourbaki, who wrote: "These investigations influenced but a little the general theory of differential equations..."

A concrete relation between the exactly integrable dynamical systems and the representation theory had been established and effectively used rather recently. However, the idea itself goes back to pioneering works by Sophus Lie, who foresaw the leading role of the group-theoretical methods as a powerful means for integrating differential systems, especially nonlinear ones. According to him the transformation groups of differential equations play the same role as do the Galois groups of algebraic equations. It turns out that the possibility of constructing their solutions in a given representation and with certain generality and the description of the group of internal symmetries itself are determined to a large extent by the Lie algebra of the group.

This set of problems, as applied to dynamical systems, was investigated by Liouville who established a general criterion for their complete integrability. It requires a certain number (equal to the rank of the system) of functionally independent global integrals of motion in involution.

It is important to notice that even in the 1-dimensional case the knowledge of the form of these integrals does not always allow one to explicitly integrate the corresponding system in the usual sense, i.e. describe in a closed form its evolution from the initial data.

The 2-dimensional case is similar: the Cauchy problem often has an explicit solution while the explicit expressions for the dynamical variables of the system may be expressed in terms of asymptotic (or free) fields, i.e., the system's dynamical characteristics in the infinite past or future. This calls for elucidation, which we will provide.

Usually, the Hamiltonian of a system is represented as the sum of a kinetic term (a free Hamiltonian) and an interaction term with a "small" parameter λ. The complete solution of the quantum problem for Heisenberg operators is then expressed in the form of functionals of the operators of the free field. A similar situation occurs also in the classical region, where the complete solution of a nonlinear system is expressed in terms of solutions of linear (free) equations with $\lambda = 0$, i.e. via the asymptotic values of the fields in the group-theoretical sense.

For exactly integrable systems the Heisenberg operators of a certain kind are polynomials in λ and naturally retain this property after the passage to the classical limit. This latter fact may be related to the late 19th century investigations of Bäcklund, who studied a number of nonlinear equations with the help of the transformations that reduce the nonlinear equations to linear ones.

On the other hand, from the representation theory viewpoint this fact may be interpreted in terms of Inönü-Wigner's contraction or its generalization with contraction parameter λ which coincides with the coupling constant λ. In the limit as $\lambda \to 0$ the initial system turns into a linear system of free equations.

Since the integrals of such deformations coincide with integrals of motion, it is possible to perform a complete integration of the corresponding system and construct its solution with the help of the perturbation theory if there are sufficiently many of the integrals of the deformations. This group-theoretical foundation of the relation of the Heisenberg fields of exactly solvable models with their asymptotic values enables one to construct them with the well-developed machinery of quantum field theory. According to this method the asymptotic fields are related with Heisenberg's fields by a unitary transformation realized by a "halved" Möller's $S(t, -\infty)$-matrix. (Recall that in 1- and 2-dimensional cases there is no Haag's trivialization of the corresponding models; therefore the corresponding operator is well-defined and can be constructed.)

The halved S-matrix is determined by a perturbation theory power series in λ and leads to explicit expressions for the dynamical quantities of a certain form, which for exactly solvable cases are polynomials in λ.

A similar situation takes place in the classical limit where the role of S-matrix is played by a function performing the corresponding Bäcklund transformation in both 1- and 2-dimensional cases.

Therefore two completely different approaches for constructing exactly integrable nonlinear dynamical systems are at our disposal:

- a method for classical systems based on Lax-type representation;
- a typical perturbation theory technique for both classical and quantum system.

We wish to emphasize that the perturbation theory is used here not to obtain some approximate results but to get exact solutions of the corresponding equations in a polynomial form.

As has been mentioned above, many of the 1- and 2-dimensional systems considered here, both in the classical and quantum sectors, pertain directly to concrete problems of theoretical physics. They describe physical phenomena in a "real" (actual) 3- or 4-dimensional space under certain additional invariance conditions such as, say, spherical symmetry of stationary configuration (in 1-dimensional problems) or cylindrical symmetry (in 2-dimensional problems).

Besides, there are intrinsically 2-dimensional objects, e.g. 2-dimensional surfaces whose description leads to exactly integrable systems. Notice that this field of investigations contributes to a classical differential geometry with a concise classification of embedded surfaces.

From the physical point of view the investigation of 1- and 2-dimensional completely integrable dynamical systems seems to be important not only in a methodological or model-theoretical way, but also due to a possibility to extend the methods developed here to spaces of higher dimension in a hope to find criteria for complete integrability, first of all, for the real 4-dimensional case.

A priori we cannot exclude the following possibility. In higher dimensions the systems under discussion also possess a group of internal symmetries; however, the group is insufficient for their complete integrability. This situation is like that of gauge theories in the Euclidean space $\mathbb{R}^4$, where 2-dimensional equations describing cylindrically symmetric self-dual Yang-Mills fields are completely integrable, whereas without these symmetry constraints it is possible to construct only instanton (parametrized) solutions in Atiyah-Drinfeld-Hitchin-Manin's approach. Neither these solutions nor the soliton solutions for periodic Toda lattice, evolution KdV-type equations, etc., guarantee the complete solution of the considered problem in the sense of functional dependence of a solution on a needed number of arbitrary functions determined on characteristics. They are a subclass of general solutions singled out by certain boundary conditions and depending on the corresponding number of numerical parameters.

Problems of this type first appeared in studies of isospectral deformations for some non-linear problems of mathematical physics. The ISP method was developed provided appropriate transformations are reversible (see, for example, [1, 39, 102, 103, 105, 130, 134, 135]). It produced a special subclass of soliton-like solutions for some non-linear wave equations such as the

Korteweg-de Vries equation and its modifications, Kadomtsev-Petviashvili equation, the non-linear Schrödinger equation or sine-Gordon equations, etc. This method is essentially a non-linear generalization of Fourier analysis and can be regarded as non-local linearization of original non-linear wave equations associated with the given linear eigenvalue problem via the integrability condition for a pair of partial differential equations.

In what follows the equations possessing solutions obtained by an ISP method or an equivalent one will be (conditionally) called *completely integrable*. The term *exactly integrable* will be retained for the systems whose solutions are expressed in quadratures and defined by a set of arbitrary functions necessary to solve the corresponding Cauchy or Goursat problem.

The examples accumulated so far lead us to the following conjecture on a relation between the properties of a group of internal symmetries and integrability criteria: *a system is exactly integrable if the Lie-Bäcklund algebra is solvable and finite-dimensional; a system is completely integrable if it is solvable and of finite growth.* In particular, if the algebra in the last case has finite-dimensional representations then the system is completely integrable.

Roughly speaking, the method developed in this book is related to the ISP method as follows. In the above-mentioned two cases of exactly and completely integrable systems the spectral parameter in the Lax representation either can or cannot be excluded by transformations from the group of internal symmetries. If the spectral parameter can be eliminated then the ISP method fails but our method works.

If the spectral parameter cannot be excluded then the ISP method leads to a non-trivial spectrum of soliton-type solutions while our method (in its initial form) enables us to get a solution of the Goursát problem for the corresponding system only in the form of an infinite though absolutely convergent series with recursively defined terms; the problem of singling out soliton-type solutions from the general ones remains open. Thus these two approaches are complementary. In Chapter 6, however, we give a method for obtaining the soliton solutions without the concrete matrix realization necessary in the ISP method.

A number of nonlinear differential equations of the type considered in this book admit a direct geometric interpretation. In particular, Gauss, Peterson-Codazzi and Ricci equations can be reformulated and their solutions can be used to express the components of the metric tensor, torsion, curvature and the 2nd fundamental form of a minimal 2-dimensional surface. In general this interpretation is related with the internal geometry of surfaces in the Euclidean, pseudoeuclidean or affine spaces (the minimal surfaces and surfaces of constant curvature). The simplest of these equations, e.g. Liouville, sine-Gordon and Lund-Regge, first appeared precisely in the problems of differential geometry.

Chapter 1
Background of the theory of Lie algebras and Lie groups and their representations

The first two Chapters of this book should not be regarded as a consistent description of fundamentals of the theory of Lie groups and their Lie algebras. Rather it is a list of branches of the representation theory of the Lie algebras and Lie groups and of the results needed to understand the main text which describes a group method for integrating a broad class of non-linear equations of theoretical and mathematical physics. We refer the reader to excellent monographs in this area of mathematics for a deeper and more detailed knowledge (see, e.g., [12, 27, 35, 36, 51, 95, 99, 101, 112–115]).

The first two chapters of the book are recommended to those who are interested in specific results of integration of non-linear dynamical systems but have zero or almost zero "vocabulary" of the theory of Lie algebras and Lie groups or their superalgebra counterparts. At the same time, since we intended to let the reader not only understand the philosophy of the problem under consideration but also learn to integrate specific non-linear equations (without turning to any additional literature), we had to describe appropriate areas of the theory of Lie algebras and Lie groups with this purpose in mind.

In addition, as the integration method itself was developed, a need often arose to investigate specifically some parts of the Lie theory, which resulted in a new procedure for ordering roots of a simple finite-dimensional Lie algebra and in an asymptotic method in the representation theory of noncompact Lie groups [61]. This produced explicit expressions for many ingredients of the theory pertaining to Lie groups and Lie algebras such as the Haar measure on compact groups and highest vectors of irreducible representations, infinitesimal and invariant operators, Whittaker vectors and the Plancherel measure of semi-simple Lie groups, and intertwining operators among others. Many of the results obtained recently and not described yet in the literature (except in the original articles) are used extensively in the study of the non-linear dynamical systems.

§ 1.1 Lie algebras and Lie groups

1.1.1 Basic definitions *). In abstract terms, a *group* G is a set of elements g_α and a rule under which to any two of the elements, say g_α and g_β, a third element corresponds, symbolically,

$$g_\alpha g_\beta = g_{(\alpha,\beta)}. \qquad (*)$$

The group composition law obeys the condition of associativity:

$$(g_\alpha g_\beta)g_\gamma = g_\alpha(g_\beta g_\gamma).$$

The elements of the group should contain a unit element g_0 such that

$$g_0 g_\alpha = g_\alpha g_0 = g_\alpha$$

for any element g_α from G. In addition, for each g_α there exists an inverse element $g_{-\alpha}$ which satisfies

$$g_\alpha g_{-\alpha} = g_{-\alpha} g_\alpha = g_0.$$

The subscript α that numbers elements of a group may be discrete as well as continuous and may have a finite or infinite range of values. Accordingly, groups may be discrete (like permutation groups that have a finite number of elements) or continuous (like a group of rotations of three-dimensional space). In what follows, we will be interested mainly in continuous groups which were introduced in the 1890s under some additional conditions by a Norwegian mathematician Sophus Lie and were named after him. It is worth mentioning here that the Lie groups were introduced precisely in order to provide the task of integration of partial differential equations.

An element of a (finite-dimensional) Lie group G is given by a set of continuous parameters α_a with the subscript a running through the values 1 to N. The integer N is called the dimension of the group. Here the group composition law is none other than the (non-commutative, generally) summation rule of group parameters which associates to a pair of continuous parameters α_a and β_a a third parameter, $\gamma = \gamma(\alpha,\beta)$. In general the term $\gamma(\alpha,\beta)$ does not equal $\gamma(\beta,\alpha)$. If $\gamma(\alpha,\beta) = \gamma(\beta,\alpha)$ for all α and β, the group is called *Abelian* or *commutative*. If the basic group relation (*) is explicitly solvable, we speak about a realization of the group, i.e., about a representation of the elements g_α of the group G, for example, by linear operators on a particular linear space. (For the matrix realization, $g_\alpha \equiv g(\alpha)$ is a matrix of a certain form, whose dependence on the set of parameters α_a is fixed. The matrix multiplication of g_α and g_β produces another matrix of the same form that depends on the set of parameters (α,β). The representation can also be realized on a

---

*) [7, 12, 13, 27, 35, 43, 51, 95, 99, 101, 112–115]

space of functions of a certain form that depend on the appropriate number of independent variables. Then the function transformed under the action of the group element g_α satifies the same additional conditions; two successive transformations of this type with the parameters α and β are equivalent to the action of the element $g_{(\alpha,\beta)}$.)

Parametrize a group element, say g_α, close to zero: $\alpha = 0 + \alpha^0\varepsilon$, ($\alpha_a = \alpha_a^0 \cdot \varepsilon$), with an infinitely small transformation parameter ε. Using the main relation (*) and ignoring higher powers of ε we get:

$$g_{(\alpha^0\cdot\varepsilon,\beta)} = g_{\alpha^0\cdot\varepsilon}g_\beta = \left(1 + \sum_{a=1}^{N} \mathfrak{F}_a \cdot \alpha_a^0 \cdot \varepsilon\right) g_\beta, \qquad (1.1)$$

i.e., $\delta g = \sum_a \mathfrak{F}_a \cdot \alpha_a^0 \cdot g \cdot \varepsilon$, where the (tangent) operators $\mathfrak{F}_a$ act in the space of this group representation as operators of left shifts along G and generate the Lie algebra of G. The composition law in terms of the Lie algebra of G is formulated as a system of commutative relations for its generators. This system can be obtained from (1.1) by putting $\beta = 0 + \beta^0\varepsilon'$ and considering the difference of two elements $g_{(\alpha,\beta)}$ and $g_{(\beta,\alpha)}$ which are close to the unit, namely:

$$g_{(\alpha^0\varepsilon,\beta^0\varepsilon')} - g_{(\beta^0\varepsilon',\alpha^0\varepsilon)} = \sum_{a,b}[\mathfrak{F}_a, \mathfrak{F}_b]\alpha_a^0\beta_b^0\varepsilon\varepsilon'.$$

The Taylor series expansion of the Lie group element $g_{(\alpha,\beta)}$ (i.e., the expansion of the composition function) near zero values of α and β takes the following form up to terms of order $\varepsilon\varepsilon'$:

$$g_{(\alpha,\beta)} \approx g_0 + \sum_d \left(\alpha_d^0\varepsilon + \beta_d^0\varepsilon' + \sum_{a,b}\alpha_a^0\beta_b^0\varepsilon\varepsilon' D_{ab}^d\right)\mathfrak{F}_d,$$

where

$$D_{ab}^d \equiv \partial^2(\alpha,\beta)_d/\partial\alpha_a\partial\beta_b|_{(\alpha,\beta)=(0,0)}.$$

Taking into account the previous formula, one finds

$$[\mathfrak{F}_a, \mathfrak{F}_b] = \sum_d c_{ab}^d \mathfrak{F}_d. \qquad (1.2)$$

Here c_{ab}^d ($\equiv D_{ab}^d - D_{ba}^d$) are referred to as *structure constants*.

By definition, the structure constants satisfy the conditions:

$$c_{ab}^d = -c_{ba}^d \;(antisymmetricity), \qquad (1.3)$$

$$c_{ab}^d c_{mn}^b + c_{mb}^d c_{na}^b + c_{nb}^d c_{am}^b = 0 \;(Jacobi\ identity), \qquad (1.4)$$

where the first condition is obvious while the second one is a direct consequence of the associativity of the composition law for the group G. The Lie algebras of commutative groups ($c_{ab}^d \equiv 0$) are also called *commutative*.

Summing up the above conditions satisfied by the operators $\mathfrak{F}$, we can give now a general definition of a Lie algebra $\mathfrak{g}$ as a vector space $\mathfrak{g}$ over a field Φ, where the bilinear law of multiplication of the elements from $\mathfrak{g}$, $(F, F') \rightarrow [F, F']$, satisfies the conditions:

$$[F, F] = 0, \tag{1.5}$$

$$[[F, F']F''] + [[F', F'']F] + [[F'', F]F'] = 0. \tag{1.6}$$

Clearly, (1.5) implies anticommutativity of the bilinear operation $[F, F'] = -[F', F]$. In what follows, we regard the field Φ (over which Lie algebras and vector spaces are considered) as a field of zero characteristic, which, if not specified otherwise, is the field of real numbers $\mathbb{R}$ or complex numbers $\mathbb{C}$ (the algebra $\mathfrak{g}$ is called in this case real or complex, respectively). The structure constants for finite-dimensional Lie algebras in coordinate representation $(F = \sum_{a=1}^{N} F_a f_a)$, where the basis elements F_a in the space $\mathfrak{g}$ are subject to commutation relations as in (1.2), satisfy (1.3) and (1.4) due to conditions (1.5) and (1.6).

Complete classification of finite-dimensional Lie algebras reduces to finding all possible sets of structure constants that satisfy the above conditions (and define completely the composition law in a vicinity of the corresponding Lie group). In other words, the description of all types of Lie algebras is equivalent to finding all solutions of the Jacobi identity on a class of tensors of the third rank which have one contravariant index and two covariant indices and which are antisymmetric with respect to the covariant indices. This problem has not yet been completely solved.

We introduce some notions of Lie algebra theory which are needed in order to understand the text below. We define a subalgebra $\mathcal{A}$, an ideal $\mathcal{B}$, and the center $\mathcal{C}$ of a Lie algebra $\mathfrak{g}$ as its vector subspaces that satisfy the relations:

$$[\mathcal{A}, \mathcal{A}] \subset \mathcal{A}, \quad [\mathcal{B}, \mathfrak{g}] \subset \mathcal{B} \text{ and } [\mathcal{C}, \mathfrak{g}] = 0,$$

where $[\mathfrak{g}', \mathfrak{g}'']$ denotes the linear span of all vectors $[F', F'']$, where $F' \in \mathfrak{g}'$, $F'' \in \mathfrak{g}''$. Clearly, an ideal is a subalgebra and the center is a commutative ideal.

If the Lie algebra $\mathfrak{g}$ is expandable as a vector space into a direct sum of vector subspaces $\mathfrak{g}_1$ and $\mathfrak{g}_2$ with $[\mathfrak{g}_i, \mathfrak{g}_i] \subset \mathfrak{g}_i$, where $i = 1, 2$, and $[\mathfrak{g}_1, \mathfrak{g}_2] = 0$ ($[\mathfrak{g}_1, \mathfrak{g}_2] \subset \mathfrak{g}_2$), then $\mathfrak{g}$ is called the "direct (semidirect) sum" of the algebras $\mathfrak{g}_1$ and $\mathfrak{g}_2$, i.e. $\mathfrak{g} = \mathfrak{g}_1 \oplus \mathfrak{g}_2$ ($\mathfrak{g} = \mathfrak{g}_2 \dotplus \mathfrak{g}_1$).

The mapping σ of the Lie algebras $\mathfrak{g} \rightarrow \mathfrak{g}'$ over the field Φ is referred to as a *homomorphism* if, for any $F_a \in \mathfrak{g}$, $c_a \in \Phi$,

$$\sigma\left(\sum_a c_a F_a\right) = \sum_a c_a \sigma(F_a), \quad \sigma([F_a, F_b]) = [\sigma(F_a), \sigma(F_b)];$$

it is an *isomorphism* if it is a one-to-one homomorphism, and an *automorphism* if it is an isomorphism of $\mathfrak{g}$ with itself.

According to general Lie-Engel theorems, there exists a correspondence between Lie algebras and Lie groups: a Lie algebra defines a Lie group up to a local isomorphism, i.e. a map between vicinities of group units, which transfers the unit element into the unit and preserves the composition law in some neighbourhoods of unit elements. So the classification of Lie groups is tantamount in a sense to the classification of their Lie algebras. Given a Lie group, its Lie algebra always exists; however, the procedure of "integration" or "exponentiation", i.e. transition from a Lie algebra to a Lie group encounters some obstructions and is not always possible. Moreover, it is not true in general that every linear representation of $\mathfrak{g}$ corresponds to a linear representation of the Lie group G, or that to any Lie subalgebra $\mathfrak{g}'$ of $\mathfrak{g}$ there corresponds a closed subgroup G' of G with Lie algebra $\mathfrak{g}' = \text{Lie}(G')$.

1.1.2 Contractions and deformations. Suppose a Lie algebra $\mathfrak{g}$ is given in terms of its structure constants c_{ij}^k with respect to a basis $\{X_i, 1 \le i \le N\}$. After selecting an other basis of $\mathfrak{g}$, $X_i = \sum_k a_i^k Y_k$, we pass to structure constants $\tilde{c}_{ij}^k$ with respect to the new basis. Let $\tilde{a} = (a_i^k)^{-1}$. Then

$$[Y_i, Y_j] = \left[\sum_m \tilde{a}_i^m X_m, \sum_n \tilde{a}_j^n X_n\right] = \sum_{m,n,l,k} \tilde{a}_i^m \tilde{a}_j^n c_{mn}^l a_l^k Y_k,$$

$$c_{ij}^k \mapsto \sum_{m,n,l} \tilde{a}_i^m \tilde{a}_j^n c_{mn}^l a_l^k = \tilde{c}_{ij}^k.$$

Suppose now that the transformation given by the matrix $a = (a_i^k)$ depends on a parameter t, is invertible for $t \ne 0$ and is singular for $t = 0$. Suppose that for some bases X and Y the transformation $a(t)$ is singular at $t = 0$, but the constants $\tilde{c}_{ij}^k(0)$ are still well defined as $t \to 0$. The described operation is called the *Inönü-Wigner contraction*; to the contracted algebra the contracted group corresponds.

One can certainly consider the iterated combinations of contractions and also view t as a vector. There is an operation in a sense inverse to contraction; it is called deformation; the study of deformations is a part of cohomology theory, a powerful tool of modern mathematics and theoretical physics, see e.g. [161].

1.1.3 Functional algebras. In what follows we will need the notion of functional algebra introduced by Lie and Engel (they called it the "object functional group" but since it is a Lie algebra, our term seems to be more appropriate). This is the Lie algebra structure with respect to the Poisson bracket in the space $F(T^*G)$ of functions T^*G for a Lie group G, cf. [51]. Note that $T^*G \simeq G \times \mathfrak{g}^*$, where $\mathfrak{g} = \text{Lie } G$.

The functional algebra appears, for instance, when we consider a quantum system whose dynamical variables constitute a Lie algebra $\mathfrak{g}$ with Casimir operators of $\mathfrak{g}$ being the integrals of motion. Passing to the classical limit, we replace $U(\mathfrak{g})$, the universal envelopping of $\mathfrak{g}$, by $\mathfrak{g}^f = F(T^*G)$, with the bracket

$$\{F_a, F_b\} = \sum_\alpha \left[\frac{\partial F_a}{\partial p_\alpha} \frac{\partial F_b}{\partial q_\alpha} - \frac{\partial F_a}{\partial q_\alpha} \frac{\partial F_b}{\partial p_\alpha} \right], \tag{1.7}$$

where p_α are coordinates on $\mathfrak{g}$ and q_α are coordinates on G.

Note that the Hamiltonian (and Lagrangian) of the classical dynamical system does not explicitly depend on cyclic coordinates and the conjugate momenta are constants.

Before we give a general definition and a list of basic properties of a functional algebra, let us explain the formulation of a problem whose solution leads to it.

Consider a quantum system whose dynamical variables satisfy commutation relations of a semisimple Lie algebra $\mathfrak{g}$, while the integrals of motion are (Casimir) operators invariant with respect to $\mathfrak{g}$ and constructed from its elements (see § 1.4.3). Transition to the classical system involves replacement of commutators $[F_a, F_b]$ in $\mathfrak{g}$ by the appropriate Poisson brackets $\{F_a, F_b\}$ and the algebra $\mathfrak{g}$ itself by the functional algebra $\mathfrak{g}^F$ whose elements F_a are functions $F_a(x; p)$ given on the phase space of $2N$ variables x_α and p_β, $1 \leq \alpha, \beta \leq N$ (generalized positions and momenta satisfying $\{x_\alpha, x_\beta\} = \{p_\alpha, p_\beta\} = 0$, $\{p_\alpha, x_\beta\} = \delta_{\alpha\beta}$). Here the Poisson bracket is given by (1.7).

Let us explain the above once again. The elements F_a of Lie algebra, as was mentioned in sect. 1.1.1 above, can be realized by shift operators on an appropriate group, which are expressed linearly via the derivatives with respect to the group parameters α_a, $1 \leq a \leq N$. When passing to a functional algebra, the space of group parameters splits in two: the variables α_a act as generalized positions while variables $\hat{\alpha}_a \equiv \partial/\partial\alpha_a$ play the role of their conjugate momenta in a $2N$-dimensional space. Here there are r cyclic coordinates α_i, $1 \leq i \leq r$ (r is the "rank" of $\mathfrak{g}$ to be defined later) which do not participate in the play while their conjugate momenta $\hat{\alpha}_i$ commute (with respect to the Poisson bracket) with all the elements of $\mathfrak{g}^F$. The invariant operators of $\mathfrak{g}^F$ and the eigenvalues of these operators for $\mathfrak{g}$ are expressed as polynomials of these momenta; one of these operators (the quadratic operator) coincides with the Hamiltonian of the system. Thus the minimal phase space of states of the classical system is set by $2(N - r)$ generalized positions and momenta $\mathcal{P}_\alpha$ and Q_α, $r + 1 \leq \alpha \leq N$, and in addition there are r "constant" momenta

$\mathcal{P}_i$, $1 \le i \le r$, such that

$$\{Q_\alpha, Q_\beta\} = 0, \ \{\mathcal{P}_a, \mathcal{P}_b\} = 0, \ \text{and} \ \{\mathcal{P}_a, Q_\beta\} = \delta_{a\beta}. \tag{1.8}$$

In the general case, $\mathfrak{g}^F$ is defined as a Lie algebra whose elements are functions $F_a(x; p)$ on the space of $2N$ variables x_α and p_α, $1 \le \alpha, \beta \le N$, with a bilinear multiplication law in the form of Poisson brackets,

$$\{F_a, F_b\} = F_{ab}(F). \tag{1.9}$$

Here the composition function $F_{ab}(F_1, \ldots, F_N)$ $(= -F_{ba})$ satisfies the conditions imposed by the Jacobi identity on its functional dependence on the elements F_a, due to the identity

$$\{\Phi_1, \Phi_2\} = \sum_{a,b} \frac{\partial \Phi_1}{\partial F_a} \frac{\partial \Phi_2}{\partial F_b} \{F_a, F_b\}, \ \Phi = \Phi(F_1, \ldots, F_N). \tag{1.10}$$

Now, it is easy to demonstrate that the functions $F_{ab} = \sum_c C_{ab}^c F_c$ (where C_{ab}^c are constants satisfying (1.3) and (1.4)) define a functional algebra with relations of type (1.2) where the commutator is replaced by the Poisson bracket. It is also clear that the bracket of any two functions F_{ab} identically satisfies the equations

$$F_{ab} \frac{\partial F_{cd}}{\partial F_a} + F_{ac} \frac{\partial F_{db}}{\partial F_a} + F_{ad} \frac{\partial F_{bc}}{\partial F_a} = 0. \tag{1.11}$$

Conversely, if $F_{ab}(= -F_{ba})$ are given functions of $F_1, \ldots, F_N$ for which (1.11) are true, then the functions F_a are expressed in terms of $2N$ variables x_α and p_α, $1 \le \alpha \le N$, and generate $\mathfrak{g}^F$; moreover, $F_{ab} = \{F_a, F_b\}$.

Following Eisenhart [115], the description of $\mathfrak{g}^F$ can be reduced to the construction of a sequence of embeddings $\mathfrak{g}_{(1)} \subset \mathfrak{g}_{(2)} \subset \ldots \subset \mathfrak{g}_{(k)} \to \mathfrak{g}^F$ with the following rule of recurrence: $\mathfrak{g}_{(i+1)} = \mathfrak{g}_{(i)} \cup \{\mathfrak{g}_{(i)}, \mathfrak{g}_{(i)}\}$, $1 \le i \le k - 1$, where the linear span of the elements $\{F, F'\}$ with $F, F' \in \mathfrak{g}_{(i)}$ is denoted by $\{\mathfrak{g}_{(i)}, \mathfrak{g}_{(i)}\}$. Here $\mathfrak{g}_{(1)}$ is generated by s_1 independent functions $F_1, \ldots, F_{s_1}$ and $\mathfrak{g}_{(i+1)}$ is the set generated by s_{i+1} independent functions obtained by adjoining to $\mathfrak{g}_{(i)}$ those brackets $\{F, F'\}$ of the elements of $\mathfrak{g}_{(i)}$ which are not functions of the original ones from $\mathfrak{g}_{(i)}$. The result is N independent functions F_a, $s_1 \le s_2 \le \ldots \le s_k \equiv N$ which form the Lie algebra $\mathfrak{g}^F$.

The identity of type (1.10) can be shown to imply that the requirement that a function $\Phi(F_1, \ldots, F_N)$ commute with all the elements of $\mathfrak{g}^F$ ($\{\Phi, F_a\} = 0$ for all F_a from $\mathfrak{g}^F$) is equivalent to the following full system of equations

$$\sum_b \{F_a, F_b\} \frac{\partial \Phi}{\partial F_b} = 0.$$

From here one can get the number r of invariants constructed from the elements of the Lie algebra $\mathfrak{g}^F$ and in involution with Φ.

According to the main theorem of Lie-Engel on the functional algebra, there exists a basis formed by the elements $\mathcal{P}_a$, $1 \leq a \leq N$, and Q_α, $r + 1 \leq \alpha \leq N$, which satisfies the relations (1.8) [115].

§ 1.2 $\mathbb{Z}$-graded Lie algebras and their classification *)

1.2.1 Definitions. We introduce an important notion of grading of Lie algebras $\mathfrak{g}$. This notion is particularly useful and constructive both for describing their structure and classification and for solving the embedding problem of subalgebras $\mathfrak{g}'$ of $\mathfrak{g}$ into $\mathfrak{g}$. We call a presentation of the Lie algebra $\mathfrak{g}$ as a direct sum of subspaces (which are often assumed to be finite-dimensional, $\dim \mathfrak{g}_a \equiv d_a < \infty$) indexed by elements of a commutative group G,

$$\mathfrak{g} = \bigoplus_{a \in G} \mathfrak{g}_a, \tag{2.1}$$

a *G-grading* provided the following condition is satisfied:

$$[\mathfrak{g}_a, \mathfrak{g}_b] \subset \mathfrak{g}_{a+b}, \quad a, b \in G. \tag{2.2}$$

In other words, (2.2) means that a nonzero commutator of any element from the subspace $\mathfrak{g}_b$ with any element from $\mathfrak{g}_a$ lies in the subspace $\mathfrak{g}_{a+b}$. The Lie algebras which have such a grading are called graded Lie algebras. In particular, any Lie algebra $\mathfrak{g}$ may be regarded in general as a graded Lie algebra comprising only the subspace $\mathfrak{g}_0$ in (2.1). In what follows, we consider gradings with integer indices, i.e. $\mathbb{Z}$-gradings.

The basic definitions of the theory of Lie algebras can be introduced for graded Lie algebras as well.

In accordance with relation (2.2), the subspace $\mathfrak{g}_0$ is closed with respect to the bracket in $\mathfrak{g}$, and is a subalgebra of $\mathfrak{g}$. The relation $[\mathfrak{g}_0, \mathfrak{g}_a] \subset \mathfrak{g}_a$ induces the linear representation T_a of the Lie algebra $\mathfrak{g}_0$ in the space $\mathfrak{g}_a$, $T_a(F^0)F^a = [F_0, F_a]$, $F^0 \in \mathfrak{g}_0$, $F^a \in \mathfrak{g}_a$. (In what follows, we call a graded Lie algebra $\mathfrak{g}$ *irreducible* if the representation T_{-1} is irreducible.)

If a Lie algebra has a grading that contains only subspaces with non-negative indices, $\mathfrak{g} = \sum_{a>0} \oplus \mathfrak{g}_a$ (or non-positive indices, $\mathfrak{g} = \sum_{a \leq 0} \oplus \mathfrak{g}_a$), then the subspace $\mathfrak{g}' \equiv \sum_{a>0} \oplus \mathfrak{g}_a$ (resp. $\mathfrak{g}' \equiv \sum_{a<0} \oplus \mathfrak{g}_a$) is an ideal in $\mathfrak{g}$; $[\mathfrak{g}', \mathfrak{g}] \subset \mathfrak{g}'$ according to (2.2). Clearly, this ideal is commutative if $\mathfrak{g} = \mathfrak{g}_0 \oplus \mathfrak{g}_1$ ($\mathfrak{g} = \mathfrak{g}_{-1} \oplus \mathfrak{g}_0$).

For a broad class of gradings and for all the gradings used below, it is possible to define a notion of order p_a of the subspace $\mathfrak{g}_a$ in the expansion (2.1) with respect to a grading operator $\hat{\mathfrak{h}}$,

$$[\hat{\mathfrak{h}}, F^a] = 2p_a F^a, \text{ so that } p_{a+1} > p_a, \ p_0 \equiv 0, \tag{2.3}$$

---

*) [7, 12, 16, 17, 27, 32–35, 44–49, 51, 60, 61, 83, 93–95, 99, 101, 113–115]

for any element F^a from $\mathfrak{g}_a$. As will be seen further on, this operator might either be from the algebra $\mathfrak{g}$ or be an exterior one. The subspace $\mathfrak{g}_0$ whose elements are of order 0 with respect to $\hat{\mathfrak{h}}$ will be called "the invariance subalgebra".

The above definition of a graded Lie algebra is usually supplemented by the requirement that the algebra as a whole be generated by what is called its *local part* $\hat{\mathfrak{g}} \equiv \mathfrak{g}_{-1} \oplus \mathfrak{g}_0 \oplus \mathfrak{g}_1$. In coordinates, the generators X_a^0 ($\in \mathfrak{g}_0$, $1 \le a \le d_0$) and $X_\alpha^\pm$ ($\in \mathfrak{g}_{\pm 1}$, $1 \le \alpha < d_{\pm 1}$) of the local part obey the commutation relations:

$$[X_a^0, X_b^0] = \sum_d B_{ab}^d X_d^0, \quad [X_a^0, X_\alpha^\pm] = \sum_\beta C_{a\alpha}^{\pm\beta} X_\beta^\pm ,$$

$$[X_\alpha^+, X_\beta^-] = \sum_a A_{\alpha\beta}^a X_a^0 \qquad (B_{ab}^d = -B_{ba}^d). \tag{2.4}$$

Here the structure constants A, B and C determine completely the structure of the maximal Lie algebra $\mathfrak{g}$ (with a given local part) and satisfy the following conditions as a result of the Jacobi identity (1.6):

$$[B_a, B_b] = -\sum_d B_{ab}^d B_d, \quad [C_a^\pm, C_b^\pm] = -\sum_d B_{ab}^d C_d^\pm ,$$

$$C_a^+ A_b + A_b C_a^- = \sum_d B_{ad}^b A_d. \tag{2.5}$$

Here A_a, B_a and $C_a^\pm$ are defined by their matrix elements $(A_a)_{\alpha\beta} \equiv A_{\alpha\beta}^a$, $(B_a)_{cd} \equiv B_{ac}^d$ and $(C_a^\pm)_{\alpha\beta} \equiv C_{a\alpha}^{\pm\beta}$. In particular, $A_a \equiv 0$, $C_a^\pm \equiv 0$ for the trivial grading $\mathfrak{g} = \mathfrak{g}_0$.

1.2.2 Semisimple, nilpotent and solvable Lie algebras. The Levi-Malcev theorem.

The paramount fundamental problem of the structural theory of Lie algebras is the classification of all non-isomorphic Lie algebras. This classification can be reduced in a sense to two complementary types — solvable and semisimple algebras — due to the Levi-Malcev theorem. Solvable Lie algebras are graded with subspaces of non-negative (or non-positive) degree (indices in (2.1)), with a commutative $\mathfrak{g}_0$. If, moreover, in the expansion (2.1) $\mathfrak{g}_0 = 0$, such solvable Lie algebras are called nilpotent. Clearly, any subalgebra of a solvable Lie algebra is solvable and any solvable Lie algebra contains a commutative ideal. The maximal solvable ideal of a Lie algebra containing any other solvable ideal is called its *radical* $\mathcal{R}$.

A Lie algebra $\mathfrak{g}$ is called *semisimple* if it contains no nontrivial commutative ideals; when in addition, $\mathfrak{g}$ has no ideals except itself and $\{0\}$, it is called *simple*. The semisimple Lie algebra $\mathfrak{g}$ has center $\{0\}$ and can be presented as a direct sum of simple ideals. Further on we will need the notion of the Cartan subalgebra $\mathfrak{h}$ of the semisimple algebra $\mathfrak{g}$ which is defined as a maximal Abelian subalgebra in $\mathfrak{g}$ whose linear transformation operators ad $\mathfrak{h}$ ($\mathfrak{g} \to [\mathfrak{h}, \mathfrak{g}] \equiv \mathrm{ad}_\mathfrak{g}\mathfrak{h}$) can be diagonalized. The dimension of the algebra $\mathfrak{h}$ is referred to as the *rank* of $\mathfrak{g}$.

Now we are prepared to formulate the *Levi-Malcev theorem* which states that any Lie algebra $\mathfrak{g}$ decomposes (as a linear *space*) into the direct sum of its radical $\mathcal{R}$ and a semisimple algebra $\mathfrak{g}_s$,

$$\mathfrak{g} = \mathcal{R} \oplus \mathfrak{g}_s, \qquad (2.6)$$

where $\mathfrak{g}_s$ is a maximal semisimple subalgebra in the algebra $\mathfrak{g}$ and is defined up to an automorphism. Here, due to the commutation relations

$$[\mathfrak{g}_s, \mathfrak{g}_s] \subset \mathfrak{g}_s, \ [\mathcal{R}, \mathcal{R}] \subset \mathcal{R}, \ [\mathcal{R}, \mathfrak{g}_s] \subset \mathcal{R}, \qquad (2.7)$$

the *algebra* $\mathfrak{g}$ is a *semidirect* sum $\mathcal{R} \dotplus \mathfrak{g}_s$.

All the above definitions of the theory of Lie algebras are applicable almost completely to Lie groups. In particular, connected Lie groups are called semisimple, simple, nilpotent or solvable if their respective Lie algebras are so. In complete analogy with the notion of direct and semidirect sum of algebras, definitions of direct and semidirect product of groups are introduced according to the invariance properties of group-multipliers with respect to the internal automorphisms, namely:

for $gg'g^{-1} \subset G_1$ and $hh'h^{-1} \subset G_2$, $G = G_1 \oplus G_2$ if $ghg^{-1} \equiv h$,

and $G = G_2 \rtimes G_1$ if $ghg^{-1} \in G_2$ and $\neq h$;

$\forall g, g' \in G_1, \ h, h' \in G_2$.

The Levi-Malcev theorem for a connected Lie group G establishes its local isomorphism to the semidirect product of the connected subgroups R and G_s of a group G, which correspond to the subalgebras $\mathcal{R}$ and $\mathfrak{g}_s$ of the Lie algebra $\mathfrak{g}$ of the group G in the decomposition (2.6)

$$G = R \rtimes G_s.$$

This representation is unambiguous for a simply connected Lie group G; the subgroups R and G_s are then also simply connected.

Thus, the Levi-Malcev theorem reduces the problem of classification of all Lie algebras to that of the summands of the decomposition (2.6). We note that not much advancement has been made yet in that direction for solvable Lie algebras.

The classification of all semisimple Lie algebras reduces to that of simple Lie algebras since according to the Cartan theorem any semisimple Lie algebra is represented uniquely in the form of a direct sum of pairwise orthogonal simple subalgebras. For simple Lie algebras it is possible to solve the system of Jacobi equations (1.4) for structure constants and thus describe all of these algebras. Late in the 19th century, Killing and Cartan classified complex simple Lie algebras and less than a quarter of a century later Cartan listed all of their real forms. The latter and the complex simple Lie algebras considered as real algebras exhaust all the real simple Lie algebras.

All these remarkable results were obtained for finite-dimensional simple Lie algebras. At the same time, these algebras are a particular case of an essentially wider class of simple Lie algebras brought together by the requirement that their growth should be finite. This concept is an important qualitative characteristic of **Z**-graded Lie algebras permitting to classify such simple algebras and is defined as follows.

The *growth* $GR(\mathfrak{g})$ of a graded Lie algebra $\mathfrak{g}$ is expressed via the dimension of the subspaces in the decomposition (2.1) by the formula:

$$GR(\mathfrak{g}) = \varlimsup_{a \to \infty} \frac{\log\left(\sum\limits_{b=-a}^{a} d_b\right)}{\log a}.$$

Accordingly, there can be three typical cases.

1) Starting with some $a = a_{max}$ ($a = a_{min}$), the subspaces $\mathfrak{g}_{\pm a}$ with $a > a_{max}$ ($a < a_{min}$) become zero, i.e. the algebra $\mathfrak{g}$ is finite-dimensional and its growth is zero.

2) In the decomposition (2.1) for any value of the grading index the corresponding subspace is nonzero, i.e. the algebra $\mathfrak{g}$ is infinite-dimensional but the limit in the definition of $GR(\mathfrak{g})$ exists and is equal to a finite number.

3) The expansion (2.1) contains an infinite number of subspaces $\mathfrak{g}_a$, i.e. the algebra $\mathfrak{g}$ is infinite-dimensional, and its growth $GR(\mathfrak{g})$ is equal to infinity.

The first two possibilities correspond to algebras of finite growth. For simple graded Lie algebras over **C**, case 1 involves "ordinary" finite-dimensional simple complex Lie algebras from Cartan's list while the case 2 already leads to a different type of Lie algebra of finite (nonzero) growth. Less than 20 years ago considerable progress was made in their description [44]. For such algebras, the dimensions of the subspaces $\mathfrak{g}_a$ grow as a power function $|a|^m$ and, accordingly, $GR(\mathfrak{g}) = m + 1$. Clearly, these algebras are exceptional in a sense since the dimensions of the subspaces $\mathfrak{g}_a$ of generic graded Lie algebras grow exponentially with a.

To conclude the introductory description of simple Lie algebras above, we note that after the pioneering work of Killing and Cartan a great number of various methods were proposed to solve the classification problem in the finite-dimensional case both for the complex algebras (Weyl, Van der Waerden, Dynkin, etc.) and for real forms (Gantmakher, and others).

Extraordinarily, there exists another, until recently unknown, deduction of the classification theorem based on the requirement of solvability in quadratures of the system of second-order partial differential equations with exponential non-linearities (for more details see Chapter 6).

1.2.3 Simple Lie algebras of finite growth: Classification and Dynkin-Coxeter diagrams. The method for constructing integrable systems discussed in this book suits any $\mathbb{Z}$-graded Lie algebra.

However, the most interesting equations are related to the so to say self-symmetric, $\mathbb{Z}$-graded simple Lie algebras of finite growth (or their nontrivial central extensions). Three groups of these algebras over $\mathbb{C}$ are distinguished by V. Kac [44]:

1) simple finite-dimensional Lie algebras (no central extensions);

2) loop algebras whose target algebra is simple and finite-dimensional (the only nontrivial central extensions are called *affine Kac-Moody* algebras, cf. [93], [94]);

3) $W = der\mathbb{C}\ [t^{-1}, t]$, the *Witt algebra* (its only nontrivial central extension Vir is called the *Virasoro algebra*).

The Lie algebras of types 1) and 2) can be described uniformly in a wider class of algebras as follows. They have a basis $X_{\pm 1}, \ldots, X_{\pm r}$ satisfying relations which are easier to describe with the help of additional generators $h_i = [X_{+i}, X_{-i}]$:

$$[h_i, X_{\pm j}] = \pm k_{ji} X_{\pm j}, \ [X_{+i}, X_{-j}] = \delta_{ij} h_i, \quad [h_i, h_j] = 0, \qquad (2.8)$$

where $k = (k_{ij})$ is called a *generalized Cartan matrix*.

The Cartan matrix is defined up to a left multiplication by an invertible diagonal matrix; therefore we assume that (k_{ij}) is normalized so that $k_{ii} = 2$ for all i (if we allow $k_{ii} = 0$ then thanks to the Jacobi identity and (2.8) the i-th row and i-th column are zeros).

If (k_{ij}) is invertible, it is positive definite. Otherwise it has one zero eigenvalue and the numbers over the nodes of the Dynkin diagram defined in what follows indicate the coefficients of linear dependence of the columns of the Cartan matrix. Invertible matrices correspond to finite-dimensional Lie algebras.

Setting $degX_{\pm i} = \pm 1$ for $1 \leq i \leq r$ we define what is called *canonical* or *principal $\mathbb{Z}$-grading* in these algebras, i.e. any such algebra $\mathfrak{g}$ takes the form

$$\underset{i \in \mathbb{Z}}{\oplus} \mathfrak{g}_i \text{ with } [\mathfrak{g}_i, \mathfrak{g}_j] \subset \mathfrak{g}_{i+j}.$$

The growth of all these algebras is finite. Under the assumption that $\mathfrak{g}$ is generated by $\mathfrak{g}_{\pm 1}$ such algebras were classified by V. Kac around 1967: they comprise the above-mentioned classes 1) and 2) and Lie algebras of polynomial vector fields.[*]

---

[*] Recently (1990) O. Mathieu announced that he proved Kac' conjecture that simple $\mathbb{Z}$-graded Lie algebras of finite growth are exhausted by these and W.

The notion of (perhaps twisted) *loop algebras* appears as follows. For a finite-dimensional $\mathfrak{g}$ set $\tilde{\mathfrak{g}} = \mathfrak{g} \otimes \mathbb{C}[t^{-1}, t]$. If φ is an automorphism of $\mathfrak{g}$ of finite order m, set

$$\mathfrak{g}_\varphi^{(m)} = \underset{m \in \mathbb{Z}, k \in \mathbb{Z}/m\mathbb{Z}}{\oplus} \mathfrak{g}_i t^{mj+k}, \quad \text{where}$$

$$\mathfrak{g}_j = \{g \in \mathfrak{g} : \varphi(g) = \exp(2\pi i j/m)g\} \quad (i = \sqrt{-1}). \tag{2.9}$$

It turns out that nonisomorphic algebras correspond to outer automorphisms of $\mathfrak{g}$, which in turn are in $1-1$ correspondence with automorphisms of the Dynkin diagram of $\mathfrak{g}$ defined at the end of this subsection. (The term "loop" is due to the fact that t may be viewed as $\exp(ia)$, where a is the angle parameter on the circle S; then $\mathfrak{g}^{(1)}$ is the space of Fourier-transforms of functions on S with values in $\mathfrak{g}$, i.e. maps $S \to \mathfrak{g}$, or loops.)

An algebra generated by $X_{\pm i}$, $1 \le i \le r$, for a non-invertible Cartan matrix is a central extension of $\mathfrak{g}_\varphi^{(m)}$ with 1-dimensional center. Such central extensions are called (affine) Kac-Moody algebras.

On the algebra of type 1) and 2) there are *ad*-invariant nondegenerate bilinear forms denoted by $B(.,.)$ and defined by the formula:

$$B(X, Y) = res B(ad\ X\ ad\ Y). \tag{2.10}$$

If $\det(k_{ij}) = 0$, as happens for loop algebras, the center of $\mathfrak{g}_\varphi^{(m)}$ is generated by an element of degree 0 (with respect to any grading). This element is a multiple of $\sum_{1 \le i \le r} c_i h_i$, where $\sum_{1 \le j \le r} k_{ij} c_j = 0$.

For a finite-dimensional Lie algebra in the principal grading there exists an operator $\hat{\mathfrak{h}}$ in the algebra $\mathfrak{g}$ which is defined by the conditions:

$$[\hat{\mathfrak{h}}, X_{\pm j}] = \pm 2X_{\pm j}, \ [\hat{\mathfrak{h}}, h_j] = 0, \quad 1 \le j \le r,$$

and thus each subspace $\mathfrak{g}_{\pm a}$, $a > 0$, is generated by all nonzero commutators of the generators $X_{\pm j}$ of multiplicity a. Here $p_a \equiv a$ and $\hat{\mathfrak{h}} \equiv h = \sum_j k_j h_j$, where $k_i \equiv 2\sum_{1 \le j \le r}(k^{-1})_{ij}$ for an invertible matrix k. Let us denote by $\mathfrak{g}(k)$ the Lie algebra constructed as above with the help of k. When $\det k = 0$ this algebra $\mathfrak{g}(k)$ can be embedded into the Lie algebra $\mathfrak{g}(k')$, $\det k' \ne 0$, so that all canonical generators of $\mathfrak{g}(k)$ are included in the system of canonical generators of $\mathfrak{g}(k')$. For example, $k' = \begin{Vmatrix} k & -n\mathbf{1} \\ -2\mathbf{1} & 2\mathbf{1} \end{Vmatrix}$, where $n \in \mathbb{Z}_+$ is not an eigenvalue of k.

An element of the subspace $\mathfrak{g}_a$ (a linear combination of a-multiple commutators of the operators X_{+j}) is considered to be nonzero and so belongs to $\mathfrak{g}$ if its commutator with at least one element of $\mathfrak{g}_{-1}$ is nonzero. Here the elements of $[\mathfrak{g}_a, \mathfrak{g}_{-1}] \subset \mathfrak{g}_{a-1}$ are easy to calculate explicitly using (2.8).

It is not difficult to prove that the center of $\mathfrak{g}(k)$ consists of the elements $\sum c_j h_j$, where the constants c_j are defined by the system $\sum_{1 \le j \le r} k_{ij} c_j = 0$. Clearly, there is no center if $\det k \ne 0$.

On the algebra $\mathfrak{g}(k)$, define a symmetric bilinear form $(\cdot, \cdot)$ whose invariance is ensured by the requirement that the equality

$$([F, F'], F'') = (F, [F', F'']) \tag{2.11}$$

be satisfied for any F, F' $F'' \in \mathfrak{g}(k)$. To introduce such a form on a graded Lie algebra generated by its local part $\hat{\mathfrak{g}}$, define the form on $\hat{\mathfrak{g}}$ and extend it to the whole algebra $\mathfrak{g}$. In other words, the invariant form is defined first on the generators of $\hat{\mathfrak{g}}$: $\{X_a^0, X_\alpha^\pm; 1 \le a \le d_0, 1 \le \alpha \le d_{\pm 1}\}$ so that $(\mathfrak{g}_i, \mathfrak{g}_j) = 0$ for $i + j \ne 0$, $-1 \le i, j \le 1$. Then we try to extend $(\cdot, \cdot)$ to an invariant form on $\mathfrak{g}$ so as $(\mathfrak{g}_a, \mathfrak{g}_b) = 0$ for $a + b \ne 0$. Thanks to (2.4) and (2.11) we have

$$\sum_a A_{\alpha\beta}^b (X_a^0, X_b^0) = - \sum_\nu C_{\alpha\beta}^{-\nu} (X_\alpha^+, X_\nu^-). \tag{2.12}$$

Thus, if a form is set for elements of subspaces $\mathfrak{g}_{\pm 1}$, the invariance condition (2.12) fixes the form on $\mathfrak{g}_0$. In the canonical grading, i.e. when $\mathfrak{g} = \{X_{\pm j}, h_j, 1 \le j \le r\}$, and assuming that

$$(X_{+i}, X_{-j}) = \delta_{ij} v_i, \tag{2.13}$$

where v_i are elements of a diagonal $r \times r$ matrix V which symmetrizes the matrix k, i.e. $Vk = k^T V$, it is easy to see that the condition

$$(h_i, h_j) = v_i k_{ij} \tag{2.14}$$

is sufficient for the form $(\cdot, \cdot)$ on $\hat{\mathfrak{g}}(k)$ to be invariant. The value of the form on a pair of arbitrary vectors from the subspaces $\mathfrak{g}_{\pm a}$ is calculated using relations (2.8) and the normalization condition (2.13), taking (2.11) into account.

So to single out the Lie algebras we are interested in from among the whole set of Lie algebras $\mathfrak{g}(k)$, we must specify the algebraic construction under consideration by imposing the conditions of simplicity (in the form of the generalized Cartan criterion of non-degeneracy of the form) and finiteness of growth. This results in rigid constraints on the matrix k. Namely, all its elements outside the main diagonal are non-positive integers with $k_{ij} = 0$ implying $k_{ji} = 0$. In addition, the equality

$$k_{i_1 i_2} k_{i_2 i_3} \cdots k_{i_s i_1} = k_{i_2 i_1} k_{i_3 i_2} \cdots k_{i_1 i_s} \tag{2.15}$$

is valid for any natural $i_1, \dots, i_s$ not exceeding r.

Furthermore, as can be seen from (2.8), under the replacement $h_i \to \lambda h_i$, $X_{-i} \to \lambda X_{-i}$, $\lambda \in \Phi$, the i-th column of the matrix k is multiplied by λ. Therefore, the generators h_i, $X_{\pm i}$ can be chosen so that in the defining relations (2.8) all the diagonal elements of k are equal to, say, 2. Note also that the conditions (2.15) and $k_{ij} = 0 \Leftrightarrow k_{ji} = 0$ for $i \ne j$, make it possible to define the form $(\cdot, \cdot)$ unambiguously on $\hat{\mathfrak{g}}(k)$ (and consequently on $\mathfrak{g}(k)$ with the help of (2.13) and (2.14)) and ensure the relations

$$(adX_{+i})^{-k_{ji}+1}X_{+j} = (adX_{-i})^{-k_{ji}+1}X_{-j} = 0, \quad i \neq j \tag{2.16}$$

It is also natural to impose an additional condition of indecomposability on the matrix k since otherwise (i.e. when it becomes block-diagonal $k = \left\| \begin{matrix} k_1 & 0 \\ 0 & k_2 \end{matrix} \right\|$) the algebra $\mathfrak{g}(k)$ is the direct sum of the algebras, $\mathfrak{g}(k_1) \oplus \mathfrak{g}(k_2)$.

For the case when $\det k = 0$, the elimination of the center of the algebra $\mathfrak{g}(k)$ leads to the condition

$$\sum_{i=1}^{r} c_i h_i = 0, \tag{2.17}$$

where the coefficients c_i are determined from the system

$$\sum_{j=1} k_{ij} c_j = 0. \tag{2.18}$$

Finally, the finiteness of the growth of the algebra $\mathfrak{g}(k)$ is ensured by non-negative definiteness of the symmetric matrix Vk which satisfies the above listed requirements and, in accordance with the results of Coxeter on the discrete groups generated by reflections, leads to rigidly defined matrices k (up to multiplication on the right by a diagonal matrix with arbitrary nonzero elements) which are called generalized Cartan matrices. As can be seen from the above reasoning, the generalized Cartan matrix completely determines the structure of the subspaces $\mathfrak{g}_{\pm a}$ and thus that of the whole algebra $\mathfrak{g}(k)$.

In what follows, when loop algebras are to be distinguished from simple finite-dimensional Lie algebras $\mathfrak{g}$ they will be denoted by $\tilde{\mathfrak{g}}$.

Notice particularly some characteristic differences between the Cartan matrix k of an algebra $\mathfrak{g}$ and the Cartan matrix $\tilde{k}$ of $\tilde{\mathfrak{g}}$. For $\mathfrak{g}$ the determinant of k is nonzero, k is positively definite, the eigenvalues of k and all the elements of k^{-1} are positive. The absolute values of elements of k do not exceed 3.

The generalized (affine) Cartan matrix $\tilde{k}$ of the algebras $\tilde{\mathfrak{g}}$ are not invertible, have all eigenvalues positive except one of zero value, and the absolute value of their elements does not exceed 4.

In spite of the differences, a close interconnection between the algebras $\mathfrak{g}$ and $\tilde{\mathfrak{g}}$ can be traced by reviewing one of the possible methods (described in § 1.3) for direct construction of the subspaces $\hat{\tilde{\mathfrak{g}}}$ starting from $\hat{\mathfrak{g}}$.

The *Dynkin-Coxeter diagrams* provide us with a compact, one-to-one way of expressing the generalized Cartan matrices (see Table 1.2.3).

Each diagram is a set of r points on a plane, each two of the points (i-th and j-th) being connected by $k_{ij}k_{ji}$ segments. The arrows on the segments

Finite-dimensional case			
$A_{r\geq 1}$	$\underset{1}{\circ}-\underset{1}{\circ}-\cdots-\underset{1}{\circ}-\underset{1}{\circ}$	G_2	$\underset{3}{\circ}\Leftarrow\underset{2}{\circ}$
$B_{r\geq 3}$	$\underset{1}{\circ}-\underset{2}{\circ}-\cdots-\underset{2}{\circ}\Rightarrow\underset{2}{\circ}$	F_4	$\underset{2}{\circ}-\underset{4}{\circ}\Leftarrow\underset{3}{\circ}-\underset{2}{\circ}$
B_2	$\circ\Rightarrow\circ$	E_6	$\underset{1}{\circ}-\underset{2}{\circ}-\underset{3}{\circ}-\underset{2}{\circ}-\underset{1}{\circ}$ with $\underset{2}{\circ}$ above the $\underset{3}{\circ}$ node
$C_{r\geq 2}$	$\underset{2}{\circ}-\underset{2}{\circ}-\cdots-\underset{2}{\circ}\Leftarrow\underset{1}{\circ}$	E_7	$\underset{1}{\circ}-\underset{2}{\circ}-\underset{3}{\circ}-\underset{4}{\circ}-\underset{3}{\circ}-\underset{2}{\circ}$ with $\underset{2}{\circ}$ above the $\underset{4}{\circ}$ node
$D_{r\geq 3}$	$\underset{1}{\circ}-\underset{2}{\circ}-\cdots-\underset{2}{\circ}$ with $\underset{1}{\circ}$ and $\underset{1}{\circ}$ branching	E_8	$\underset{2}{\circ}-\underset{3}{\circ}-\underset{4}{\circ}-\underset{5}{\circ}-\underset{6}{\circ}-\underset{4}{\circ}-\underset{2}{\circ}$ with $\underset{3}{\circ}$ above the $\underset{6}{\circ}$ node
Infinite-dimensional case			
$A^{(1)}_{r\geq 2}$	triangle with apex $\underset{1}{\circ}$, base $\underset{1}{\circ}-\underset{1}{\circ}-\cdots-\underset{1}{\circ}-\underset{1}{\circ}$	$G^{(1)}_2$	$\underset{3}{\circ}\Leftarrow\underset{2}{\circ}-\underset{1}{\circ}$
$A^{(1)}_1$	$\circ\Leftrightarrow\circ$	$F^{(1)}_4$	$\underset{2}{\circ}-\underset{4}{\circ}\Leftarrow\underset{3}{\circ}-\underset{2}{\circ}-\underset{1}{\circ}$
$B^{(1)}_{r\geq 3}$	$\underset{2}{\circ}-\underset{2}{\circ}-\cdots-\underset{2}{\circ}\Rightarrow\underset{2}{\circ}$ with $\underset{1}{\circ}$ and $\underset{1}{\circ}$ branching at left	$E^{(1)}_6$	$\underset{1}{\circ}-\underset{2}{\circ}-\underset{3}{\circ}-\underset{2}{\circ}-\underset{1}{\circ}$ with $\underset{2}{\circ}$ and $\underset{1}{\circ}$ above
$C^{(1)}_{r\geq 2}$	$\underset{1}{\circ}\Rightarrow\underset{2}{\circ}-\cdots-\underset{2}{\circ}\Leftarrow\underset{1}{\circ}$		
$D^{(1)}_{r\geq 4}$	$\underset{2}{\circ}-\underset{2}{\circ}-\cdots-\underset{2}{\circ}$ with $\underset{1}{\circ},\underset{1}{\circ}$ branching at left and $\underset{1}{\circ},\underset{1}{\circ}$ at right	$E^{(1)}_7$	$\underset{1}{\circ}-\underset{2}{\circ}-\underset{3}{\circ}-\underset{4}{\circ}-\underset{3}{\circ}-\underset{2}{\circ}-\underset{1}{\circ}$ with $\underset{2}{\circ}$ above the $\underset{4}{\circ}$ node
$A^{(2)}_{2r,\ r\geq 2}$	$\underset{2}{\circ}\Leftarrow\underset{2}{\circ}-\cdots-\underset{2}{\circ}\Leftarrow\underset{1}{\circ}$		
$A^{(2)}_2$	$\underset{2}{\circ}\Lleftarrow\underset{1}{\circ}$	$E^{(1)}_8$	$\underset{1}{\circ}-\underset{2}{\circ}-\underset{3}{\circ}-\underset{4}{\circ}-\underset{5}{\circ}-\underset{6}{\circ}-\underset{4}{\circ}-\underset{2}{\circ}$ with $\underset{3}{\circ}$ above the $\underset{6}{\circ}$ node
$D^{(2)}_{r+1,\ r\geq 2}$	$\underset{1}{\circ}\Leftarrow\underset{1}{\circ}-\cdots-\underset{1}{\circ}\Rightarrow\underset{1}{\circ}$	$E^{(2)}_6$	$\underset{1}{\circ}-\underset{2}{\circ}\Rightarrow\underset{3}{\circ}-\underset{2}{\circ}-\underset{1}{\circ}$
$A^{(2)}_{2r-1,\ r\geq 3}$	$\underset{2}{\circ}-\underset{2}{\circ}-\cdots-\underset{2}{\circ}\Leftarrow\underset{1}{\circ}$ with $\underset{1}{\circ}$ and $\underset{1}{\circ}$ branching at left	$D^{(3)}_4$	$\underset{1}{\circ}\Rrightarrow\underset{2}{\circ}-\underset{1}{\circ}$

Table 1.2.3 Dynkin-Coxeter diagrams for simple Lie algebras of finite growth

indicate the i-th point if $|k_{ji}| > |k_{ij}|$. A number c_i is assigned to each point on the diagram. It coincides with the coefficient c_i in (2.17) for algebras $\tilde{\mathfrak{g}}$. For finite-dimensional algebras the number c_i is equal to the multiplicity of the generator $X_{\pm i}$ entering the multiple commutator that expresses $X_{\pm m}$.

In other words, according to the terminology to be introduced in the next subsection, the numbers c_i coincide in the finite-dimensional case with coefficients of expansion of the maximal root m with respect to the simple roots π_j, $m = \sum_{1 \le j \le r} c_j \pi_j$.

Throughout the book we use conventional (and sometimes confusing) notations for Lie algebras and their root systems:

$sl(n+1)$, the Lie algebra of traceless $(n+1) \times (n+1)$ matrices, is sometimes
   denoted as A_n;

$0(2n+1)$ is sometimes denoted as $B_n = \{x \in sl(2n+1); x^t = -x\}$;

$sp(2n)$ is sometimes denoted as $C_n = \{x \in sl(2n) : x^t J_{2n} + J_{2n} x = 0\}$;

$0(2n)$ is sometimes denoted as $D_n = \{x \in sl(2n) : x^t = -x\}$.

The most confusing is the case of A_1 which is also $sl(2) \simeq sp(2) \simeq 0(3)$. In order to avoid confusion, when we write A_1 we mean, unless otherwise specified, the realization as $sl(2)$ with the basis

$$X_- = \begin{pmatrix} 0 & 0 \\ 1 & 0 \end{pmatrix}, \quad h = \begin{pmatrix} 1 & 0 \\ 0 & -1 \end{pmatrix}, \quad X_+ = \begin{pmatrix} 0 & 1 \\ 0 & 0 \end{pmatrix}.$$

Clearly, $C_1 \simeq B_1 \simeq A_1$, $C_2 \simeq B_2$, $D_2 \simeq A_1 \oplus A_1$, $D_3 \simeq A_3$ and D_1 is a commutative Lie algebra. The restrictions in the Table rule out nonisomorphic simple algebras.

The sole Lie algebra, W, constituting class 3) is naturally **Z**-graded: set $deg\ t = 1$, so that the degree of $X_i = t^{i+1} d/dt$ is i. The algebra W is generated by $X_{\pm 1}$ and $X_{\pm 2}$. Its central extension is called Vir, short of Virasoro algebra, and is given by the formula (where c is the generator of the center and the division by 12 is performed as a historical tribute to some physical applications of Vir):

$$[X_i, X_j] = (j - i)X_{i+j} + (i^3 - i)\delta_{ij} c/12. \tag{2.19}$$

The definitions introduced earlier were quite enough to describe briefly the results of the classification theory of Lie algebras. However, we need a more universal and constructive language of invariant root systems for a more detailed discussion of this problem, and even more so for studying the structure of root subspaces $\mathfrak{g}_a$ of the Lie algebras $\mathfrak{g}$ and for describing of the corresponding Lie groups and their representations.

1.2.4 Root systems and the Weyl group. The structure of Lie algebras we use is described best with the help of a maximal torus, i.e. a maximal abelian subalgebra whose adjoint representation is diagonalizable. (For finite-dimensional algebras, maximal tori coincide with Cartan subalgebras; in general the notion of Cartan subalgebra is not very useful.) Having selected a maximal torus in a simple Lie algebra $\mathfrak{g}$ of the above list (all the maximal tori are conjugate with respect to an automorphism of the algebra), we encounter a remarkable fact: the nonzero weights of the adjoint representation of the torus in $\mathfrak{g}$, called *roots* (of $\mathfrak{g}$), can be divided into two groups: positive and negative ones; the positive roots not presentable as sums of other positive roots are called *simple* roots and if $\pi_1, \ldots, \pi_r$ are simple positive roots, then every root is of the form:

$$\pm \sum n_i \pi_i \text{ where } n_i \in \mathbb{Z}_+.$$

Let $\mathfrak{t} \subset \mathfrak{g}$ be a maximal toral subalgebra or, briefly, torus. We have been discussing everything over $\mathbb{C}$ but now select a basis of $\mathfrak{t}$ and consider $\mathbb{Q}$-span of this basis, $\mathfrak{t}_\mathbb{Q}$. Define the Weyl group $W(\mathfrak{g})$ as the one generated by reflection r_i of $\mathfrak{t}_\mathbb{Q}$ with respect to the hyperplanes singled out by the π_i:

$$r_i(\alpha) = \alpha - 2\frac{(\pi_i, \alpha)}{(\pi_i, \pi_i)}\alpha_i. \qquad (2.20)$$

Since there is a nondegenerate symmetric bilinear form B on $\mathfrak{g}$, the restriction of B onto $\mathfrak{t}$ determines a nondegenerate symmetric form on $\mathfrak{t}_\mathbb{Q}$, and we have made use of this fact in (2.20).

As was mentioned above, the structure of the Lie algebra $\mathfrak{g}(k)$ is determined completely by the commutation relations (2.8) for the generators $X_{\pm j}$ called its *Chevalley generators* and h_j, $1 \le j \le r$, of the algebra $\hat{\mathfrak{g}}(k)$ and by the form of its Cartan matrix k. All other elements of the algebra $\mathfrak{g}(k)$ are obtained by successive commutation of the generators $X_{\pm j}$, $1 \le j \le r$; this commutation does not vanish at the intermediate stages (which is easy to observe in direct calculations taking into account the relations (2.16)). At the same time, the canonical grading (*deg* $X_{\pm j} = \pm 1$ for all j) generally classifies the subspaces $\mathfrak{g}_{\pm a}$, $a > 1$, only by the multiplicity of entry of any of the operators $X_{\pm j}$ into a multiple commutator and does not distinguish elements inside of $\mathfrak{g}_{\pm a}$.

For this purpose, the "one-dimensional grading" (with respect to the order p_a) should be replaced by a multidimensional one allowing for specific features of entry of the operators $X_{\pm J}$ with different $1 \le j \le r$ into a multiple commutator. This grading[*]) $\mathfrak{g} = \oplus_\alpha \mathfrak{g}^\alpha$ is defined by vector-functions (*weights*) α determined on the subspace $\mathfrak{g}_0$ by the relations $[h, X_\alpha] = \alpha(h)X_\alpha$ for any $h \in \mathfrak{g}_0$. The components of these functions are $\alpha_i(h_j) = k_{ij}$ according to (2.8) for $X_\alpha \equiv X_j$, $1 \le i, j \le r$. The nonzero linear functions α are called *roots* of $\mathfrak{g}$ with respect to $\mathfrak{g}_0$ if $\mathfrak{g}^\alpha \ne \{0\}$, and $\mathfrak{g}^\alpha$ is called a *root space* corresponding to α. The set R of all roots will be referred to as the root system of the algebra $\mathfrak{g}(k)$. Since $\mathfrak{g}_0$ is a maximal abelian subalgebra of $\mathfrak{g}$, the space $\mathfrak{g}^0$

---

[*]) The algebraic closedness of the main field guarantees the existence of this root expansion.

corresponding to the zero root $\alpha = 0$ coincides with $\mathfrak{g}_0$. The Jacobi identity implies that $[\mathfrak{g}^\alpha, \mathfrak{g}^\beta] \subset \mathfrak{g}^{\alpha+\beta}$ for any roots α, β and $\alpha + \beta$.

The roots α_i (also denoted by π_i), $1 \leq i \leq r$, are simple ones and the elements $X_{\pm i}$ and h_i are called the root vectors corresponding to simple roots and the generators of $\mathfrak{g}^0$, respectively. It can be seen from the construction that any root α is a linear combination of simple roots with all coefficients non-negative (or non-positive) integers, $\alpha = \sum_{j=1}^r t_j^\alpha \pi_j$; the value $|\sum_j t_j^\alpha|$ is called the *height* of the root α. Here the root space $\mathfrak{g}^\alpha$, $\alpha \gtrless 0$ is spanned by all possible vectors:

$$X_\alpha \equiv \left[\ldots [X_{\pm\pi_{i_1}}, X_{\pm\pi_{i_2}}] \ldots X_{\pm\pi_{i_s}}\right] \text{ with } \alpha = \pm \sum_{1 \leq q \leq s} \pi_{i_q}.$$

The roots α are called *positive* ($\alpha > 0$) or *negative* ($\alpha < 0$) for a nonzero root vector X_α and the set of all positive (negative) roots is denoted by R^+ (R^-). In other words, the root α is called positive, $\alpha > 0$, if the first nonzero coefficient t_j^α is positive.

It is easy to demonstrate that the Cartan matrix can be directly expressed via a symmetric bilinear form on the vector space $\mathcal{A}$ (over the field $\mathbb{Q}$ of rational numbers) spanned by simple roots. This is done by formula

$$k_{ij} = 2(\pi_j, \pi_i)/(\pi_j, \pi_j). \tag{2.21}$$

Hence the following defining relations take place in the root space of the algebra $\mathfrak{g}(k) = \mathfrak{g}^0 \oplus \sum_{\alpha \in R} \oplus \mathfrak{g}^\alpha$ which is a generalization of the Cartan-Weyl basis for a finite-dimensional case:

$$[X_\alpha, X_\beta] = \begin{cases} N_{\alpha\beta} X_{\alpha+\beta}, & \alpha + \beta \in R; \\ 0 & \alpha + \beta \neq 0, \ \alpha + \beta \notin R; \\ h_\alpha & \alpha + \beta = 0; \end{cases} \tag{2.22}$$

$$[h_i, h_j] = 0; \ [h_i, X_{\pm\alpha}] = \pm\alpha(h_i)X_{\pm\alpha};$$

$$\alpha_j(h_i) \equiv k_{ji}; \ (\mathfrak{g}^\alpha, \mathfrak{g}^\beta) = 0 \text{ if } \alpha + \beta \neq 0;$$

where $N_{\alpha\beta} (\equiv -N_{\beta\alpha})$ are nonzero constants. These constants depend on the choice of the root vectors X_α. In particular, in the finite-dimensional case there exists a Chevalley normalization for which $|N_{\alpha\beta}| = p + 1$, where p is the maximal number for which $\beta - p\alpha \in R$ and $N_{\alpha\beta} = -N_{-\alpha,-\beta}$. Besides, there are the following simple but rather useful relations:

$$(\gamma, \gamma)^{-1} N_{\alpha\beta} = (\alpha, \alpha)^{-1} N_{\beta\gamma} = (\beta, \beta)^{-1} N_{\gamma\alpha} \text{ if } \alpha + \beta + \gamma = 0;$$

$$(\gamma + \delta, \gamma + \delta)^{-1} N_{\alpha\beta} N_{\gamma\delta} + (\alpha + \delta, \alpha + \delta)^{-1} N_{\beta\gamma} N_{\alpha\delta} +$$

$$+ (\beta + \delta, \beta + \delta)^{-1} N_{\gamma\alpha} N_{\beta\delta} = 0 \text{ if }$$

$$\alpha + \beta + \gamma + \delta = 0; \tag{2.23}$$

and the sum of any two roots is not zero.

An important property of the Weyl group is that the root system R of the Lie algebra $\mathfrak{g}(k)$ and the bilinear symmetric form on $\mathfrak{t}_Q$ described above are $W(\mathfrak{g})$-invariant.

This allows us to establish the characteristic difference between the root systems of finite-dimensional and infinite-dimensional simple Lie algebras, the latter being marked again with a tilde in what follows. For an algebra $\mathfrak{g}(k)$, any root α from R can be transformed into a simple root under the action of an element w from $W(\mathfrak{g})$. However, this does not happen in the case of $\tilde{\mathfrak{g}}(\tilde{k})$ since, in addition to roots corresponding to the submatrix k (which is the Cartan matrix of $\mathfrak{g}(k)$),there are roots like $\varepsilon \equiv \sum_{j=1}^{r} c_j \pi_j$ (recall that $\sum_j \tilde{k}_{ij} c_j = 0$). This root cannot be mapped to a simple one by any transformation from $W(\tilde{\mathfrak{g}})$. A characteristic feature of such roots is that $(\varepsilon, \pi_i) = 0$, $1 \le i \le r$ and $(\varepsilon, \varepsilon) \le 0$; accordingly, such roots are called *imaginary* (in contrast to "ordinary" *real* roots). The presence of an imaginary root entails in turn the appearance of both an infinite number of real roots and of an infinite series of imaginary roots $l\varepsilon$, where l is any integer. At the same time, all real roots α of the algebra $\tilde{\mathfrak{g}}(\tilde{k})$ possess characteristic features of root systems of finite-dimensional algebras, namely: $l\alpha \in R(\tilde{k})$ only for $l = 0, \pm 1$; $\dim \mathfrak{g}^\alpha = 1$; $\beta + l\alpha \in R(\tilde{k})$ for $\alpha, \beta \in R(\tilde{k})$ if and only if $-p \le l \le q$; and $p - q = 2(\alpha, \beta)/(\alpha, \alpha)$.

1.2.5 A parameterization and ordering of roots of simple finite-dimensional Lie algebras. Bearing in mind practical applications of general results in the next chapters for the concrete types of finite-dimensional complex Lie algebras $\mathfrak{g}$, we give in this subsection some reference information about their root systems as may be required for the purpose. In the process we introduce some definitions characteristic for the finite-dimensional case.

As was noted above, each root from $R^+(R^-)$ can be decomposed with respect to a system of simple roots that forms a basis in the space $\mathcal{A}$. (Note that each simple root π_i occurs in the system R^+ exactly $2 \sum_j (k^{-1})_{ji}$ times.) The corresponding expressions for classical series A_r, B_r, C_r, and D_r, and for the exceptional Lie algebras G_2, F_4 and E_6, E_7, E_8 are given in Table 1.2.5$_1$. The simple roots of these Lie algebras in turn can be expressed via the elements $e_1, e_2, \ldots, e_p$ of an orthonormal basis of the p-dimensional Euclidean space, $p \ge r$.

Compiling Table 1.2.5$_1$ raises a question: how should one arrange the roots of R^+ relative to each other? There are different ways of ordering the roots, e.g., the lexicographic ordering where we assume that $\alpha > \beta$ if $\alpha - \beta > 0$, if $\alpha - \beta$ is a positive root. This order does not control in any way the relative arrangement of the roots for which $\alpha - \beta$ is not a root, e.g., roots of the same height.

An analysis of the structure of the positive root system of simple Lie algebras (see Table 1.2.5$_1$) implies the existence of a linear ordering such that each nonsimple root is positioned between its summands. An ordering of the roots of $\mathfrak{g}$ (not unique in general) fixed this way is called a *normal* or $\sum^{+}$-ordering. It has the following important property: the set of positive (negative) roots greater than any given root, say α, in $\sum^{+}$ and of negative (positive) roots smaller than $-\alpha$ form the root system of a subalgebra of $\mathfrak{g}$. (Indeed, as evidently follows from the definition of the $\sum^{+}$-ordering, the sum $\beta + \gamma \in R^{+}$ of any two positive roots β and γ which are smaller than a root α is positioned between β and γ and consequently is smaller than α.)

This ordering plays a distinguished role among other root arrangements (not only because it fixes them most rigidly). It has some remarkable features. In particular, as will be shown later, it leads to a universal parameterization of elements of any simple compact Lie group $\mathcal{K}$ in terms of the generalized Euler angles. This parametrization makes it possible to obtain expressions for the invariant Haar measure on $\mathcal{K}$ which are rather easily factorizable in terms of these angles, highest vectors of the irreducible representations of these groups, etc. (Note that just this ordering of the roots leads to the known factorized form of the universal R-matrix of the corresponding integrable quantum systems.)

Examples of universal orderings are (we have identfied the roots with the corresponding root vectors)

$sl(n)$	$sp(2n)$
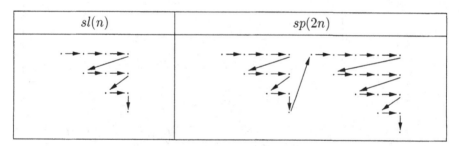	

$O(2n)$	$O(2n+1)$
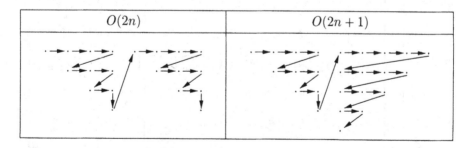	

We used the realizations:

$$sp = \left\{ \begin{pmatrix} A & B \\ C & -A^t \end{pmatrix} \text{ with } B = B^t, \ C = C^t \right\}$$

Table 1.2.5₁ Positive roots of finite-dimensional complex simple Lie algebras

$\mathfrak{g}$	R^+	m	π_i
$A_{r(\ge 1)}$	$e_i - e_j = \sum_{i \le k \le j-1} \pi_k,\ 1 \le i < j \le r+1$	$\sum_{1 \le j \le r} \pi_j \equiv \pi$	$\pi_i = e_i - e_{i+1},\ 1 \le i \le r$
$B_{r(\ge 2)}$	$e_i = \sum_{i \le k \le r} \pi_k,\ 1 \le i \le r;$ $e_i - e_j = \sum_{i \le k \le j-1} \pi_k,\ 1 \le i < j \le r;$ $e_i + e_i = \sum_{i \le k \le j-1} \pi_k + 2\sum_{j \le k \le r} \pi_k,\ 1 \le i < j \le r$	$2\pi - \pi_1$	$\pi_i = e_i - e_{i+1},\ 1 \le i \le r-1;$ $\pi_r = e_r$
$C_{r(\ge 2)}$	$2e_i = 2\sum_{i \le k \le r-1} \pi_k + \pi_r,\ 1 \le i \le r;$ $e_i - e_j = \sum_{i \le k \le j-1} \pi_k,\ 1 \le i < j \le r;$ $e_i + e_j = \sum_{i \le k \le j-1} \pi_k + 2\sum_{j \le k \le r-1} \pi_k + \pi_r,\ 1 \le i < j \le r$	$2\pi - \pi_r$	$\pi_i = e_i - e_{i+1},\ 1 \le i \le r-1;$ $\pi_r = 2e_r$
$D_{r(\ge 3)}$	$e_i - e_j = \sum_{i \le k \le j-1} \pi_k,\ 1 \le i < j \le r;$ $e_i + e_r = \sum_{i \le k \le r-2} \pi_k + \pi_r,\ 1 \le i \le r-1;$ $e_i + e_j = \sum_{i \le k \le j-1} \pi_k + 2\sum_{j \le k \le r-2} \pi_k + \pi_{r-1} + \pi_r,\ 1 \le i < j \le r-1$	$2\pi - \pi_1 - \pi_{r-1} - \pi_r$	$\pi_i = e_i - e_{i+1},\ 1 \le i \le r-1;$ $\pi_r = e_{r-1} + e_r$

G_2	$\pi_1,\ \pi_1+\pi_2,\ 2\pi_1+\pi_2,\ 3\pi_1+\pi_2,\ 3\pi_1+2\pi_2,\ \pi_2$	$3\pi_1+2\pi_2\equiv(32)$	$\pi_1=e_1-e_2;$ $\pi_2=-2e_1+e_2+e_3$
F_4	$e_i,\ 1\leq i\leq 4;$ $e_i\pm e_j,\ 1\leq i<j\leq 4;$ $^1\!/\!_2(e_1\pm e_2\pm e_3\pm e_4)$	$2\pi_1+3\pi_2+4\pi_3+\\ +2\pi_4\equiv(2342)$	$\pi_1=e_2-e_3;$ $\pi_2=e_3-e_4;$ $\pi_3=e_4;$ $\pi_4={}^1\!/\!_2(e_1-e_2-e_3-e_4)$
E_6	$\pm e_i+e_j,\ 1\leq i<j\leq 5;$ $^1\!/\!_2\left(e_8-e_7-e_6+\sum_{i=1}^{5}(-1)^{\nu(i)}e_i\right),$ $\sum_{1\leq i\leq 5}\nu(i)$ is even	(122321)	$\pi_1={}^1\!/\!_2(e_1+e_8)-{}^1\!/\!_2\displaystyle\sum_{2\leq i\leq 7}e_i;$ $\pi_2=e_1+e_2;$ $\pi_j=e_{j-1}-e_{j-2},\ 3\leq j\leq 6$
E_7	$\pm e_i+e_j,\ 1\leq i<j\leq 6;$ $e_8-e_7;$ $^1\!/\!_2\left(e_8-e_7+\sum_{1\leq i\leq 6}(-1)^{\nu(i)}e_i\right),$ $\sum_{1\leq i\leq 6}\nu(i)$ is odd	(2234321)	$\pi_1={}^1\!/\!_2(e_1+e_8)-{}^1\!/\!_2\displaystyle\sum_{2\leq i\leq 7}e_i;$ $\pi_2=e_1+e_2;$ $\pi_3=e_3-e_1;$ $\pi_j=e_{j-1}-e_{j-2},\ 4\leq j\leq 7$
E_8	$\pm e_i+e_j,\ i<j;$ $^1\!/\!_2\left(e_8+\sum_{1\leq i\leq 7}(-1)^{\nu(i)}e_i\right),$ $\sum_{1\leq i\leq 7}\nu(i)$ is even	(23465432)	$\pi_1={}^1\!/\!_2(e_1+e_8)-{}^1\!/\!_2\displaystyle\sum_{2\leq i\leq 7}e_i;$ $\pi_2=e_1+e_2;$ $\pi_j=e_{j-1}-e_{j-2}\ \ 3\leq j\leq 8$

$$0(2n) = \left\{ \begin{pmatrix} A & B \\ C & -A^t \end{pmatrix} \text{ with } B = -B^t, \ C = C^t \right\}$$

$$0(2n+1) = \left\{ \begin{pmatrix} A & X & B \\ Y & 0 & -X^t \\ C & -Y^t & -A^t \end{pmatrix} \begin{array}{l} \text{with } B \text{ and } C \text{ skew-} \\ \text{symmetric with respect} \\ \text{to the side diagonal} \end{array} \right\}.$$

Then positive root vectors correspond to the upper triangular part of $sl(n)$ complemented for sp and 0 with the upper triangular part of matrices A and any of the symmetric halves of matrices B (and with X for $0(2n+1)$). Depicted above are orderings for $n = 5$.

The possible weights of irreducible finite-dimensional representations of a simple finite-dimensional Lie algebra constitute a lattice in $t_\mathbb{Q}$, a system of positive weights (with respect to the lexicographic ordering).

Among all positive (negative) roots of a simple Lie algebra $\mathfrak{g}$ of rank r there exists a set θ^+ of fundamental highest roots $s_1, \ldots, s_r$ the sum of each of which with any simple root (if the sum itself is a root) coincides with a root, again from θ^+ greater than the original one, i.e. $s_i + \pi_j \geq s_{i+1}$, $1 \leq i$, $j \leq r$ and, consequently, $s_i + \alpha \geq s_{i+1}$ for any $\alpha \in \sum^+$. Clearly, $s_r + \alpha \notin R^+$ for any root α from R^+. This root $s_r (\equiv m)$ is called *maximal*; m being equal to $\sum_j c_j \pi_j$, where the numbers c_j are given in Dynkin's diagrams (see Table 1.2.3) for all algebras $\mathfrak{g}$. (Recall that we have already used above that $[X_{+m}, X_\alpha] = 0$ for any $X_\alpha \in \mathfrak{g}^\alpha$, $\alpha > 0$.) One of the possible systems of linearly independent highest roots for all types of Lie algebras $\mathfrak{g}$ is given in Table 1.2.5$_2$.

Later on we will also need the notion of fundamental weights $\tilde{\omega}_i$, $1 \leq i \leq r$ which for finite-dimensional case are defined as $\tilde{\omega}_i \equiv \sum_j k_{ij}^{-1} \pi_j$ and have the heights $\sum_j k_{ij}^{-1} \equiv \frac{1}{2} k_i$, respectively.

To each simple root π_i (or fundamental weight $\tilde{\omega}_i$) of an algebra $\mathfrak{g}$ assign an indeterminate y_i and define the function $\alpha(y)$ by the expression $\alpha(y) = \sum_j t_j^\alpha y_j$, where the set t_j^α corresponds to the expansion of the root α.

1.2.6 The real forms of complex simple Lie algebras. A clear explanation of the fact that real forms correspond to involutive antilinear automorphisms of complex Lie algebras and a transparent description of them via Satake diagrams are given in [162]. We will just recall the basics.

A Lie algebra $\mathfrak{h}$ over $\mathbb{R}$ is called a *real form* of a Lie algebra $\mathfrak{g}$ over $\mathbb{C}$ if $\mathfrak{h} \oplus_\mathbb{R} \mathbb{C} \simeq \mathfrak{g}$. All real forms are in $1 - 1$ correspondence with antilinear automorphisms $\sigma \in Aut \ \mathfrak{g}$ such that $\sigma^2 = id$. In particular, selecting a basis of $\mathfrak{g}$ over $\mathbb{R}$, i.e. having expressed every $g \in \mathfrak{g}$ in the form $g = g_r + i g_{im}$, we define the conjugation automorphism setting $g^\dagger = g_r - i g_{im}$ (here $i = \sqrt{-1}$).

The elements g such that $g = g^\dagger$ are called *noncompact*, while those that satisfy $g = -g^\dagger$ are called *compact* (because they generate noncompact or compact groups, respectively).

$A_{r(\geq 1)}$	$\pi_1,\ \pi_1+\pi_2,\ldots,\pi \equiv \displaystyle\sum_{1\leq j\leq r}\pi_j,$
$B_{r(\geq 1)}$	$\pi,\ \pi+\pi_r,\ \pi+\pi_{r-1}+\pi_r,\ldots,2\pi-\pi_1$
$C_{r(\geq 1)}$	$\pi,\ \pi+\pi_{r-1},\ \pi+\pi_{r-1}+\pi_{r-2},\ldots,2\pi-\pi_r$
$D_{r(\geq 3)}$	$\pi-\pi_{r-1},\ \pi-\pi_r,\ \pi,\ \pi+\pi_{r-2},\ \pi+\pi_{r-3}+\pi_{r-2},\ldots$ $\ldots,2\pi-\pi_{r-1}-\pi_r-\pi_1$
G_2	(13), (23)
F_4	(1232), (1242), (1243), (2243)
E_6	(101212), (112211), (111212), (112212), (112312), (112322)
E_7	(1112322), (1212322), (1212313), (1212323), (1212423), (1213423), (1223423)
E_8	(12324524), (12323534), (12324534), (12324634), (12324635), (12424635), (13424635), (23424635)

Table 1.2.5$_2$ The sequence of highest roots arranged in order of increasing height

Note. To simplify the notation for the highest roots of the exceptional algebras we write down only the coefficients of expansion of the highest roots with respect to simple roots.

Let

$$X = \sum_{\alpha>0} a_\alpha^+ X_\alpha + \sum_{1\leq j\leq r} a_j^0 h_j + \sum_{\alpha<0} a_\alpha^- X_\alpha$$

be an element of $\mathfrak{g}$; hence a_α, a_j^0 are complex numbers. Let σ be an automorphism of $\mathfrak{g}$ of order 2; then all real forms of $\mathfrak{g}$ are obtained as solutions of the equation:

$$\sigma = (X(a)^\dagger) = -X(a).$$

In particular, for $\sigma = id$ we have

$$\bar{a}_\alpha^\pm = -a_\alpha^\mp, \bar{a}_j^0 = -a_j^0,$$

which corresponds to a compact form of a Lie algebra. The automorphism

$$\sigma(X_{\pm\alpha}) = X_{\mp\alpha},\ \sigma(h_j) = -h_j$$

corresponds to the so-called *natural* form of a Lie algebra; the corresponding conditions are

$$\bar{a}_\alpha^\pm = a_\alpha^\pm,\ \bar{a}_j^0 = a_j^0$$

There are three types of roots for a simple $\mathfrak{g}$:

1) $\sigma(X_{+\alpha}) = X_{-\beta}$ for $\alpha \neq \beta$. Such α is called "of multiplicity 2"; four generators correspond to it as follows:

compact	non-compact
$X_\alpha - X_{-\alpha} + \sigma(X_\alpha - X_{-\alpha})$, $i(X_\alpha + X_{-\alpha} + \sigma(X_\alpha + X_{-\alpha}))$	$i(X_\alpha - X_{-\alpha} + \sigma(X_\alpha - X_{-\alpha}))$ $X_\alpha + X_{-\alpha} + \sigma(X_\alpha + X_{-\alpha})$

2) $\sigma(X_{+\alpha}) = \varepsilon X_{-\alpha}$ for $\varepsilon = \pm 1$. Such a root α is called "of multiplicity 1"; two generators correspond to it:

ε	compact	non-compact
1	$i(X_\alpha + X_{-\alpha})$,	$i(X_\alpha - X_{-\alpha})$
-1	$X_\alpha + X_{-\alpha}$	$X_\alpha + X_{-\alpha}$

3) $\sigma(X_\alpha) = X_\alpha$. Such a root α is called "of multiplicity 0"; such roots generate the σ-invariant subalgebra. See Tables 1.2.6 and [162] for more details.

Table 1.2.6 Characteristics of noncompact Lie groups

$G = \mathcal{K}\mathcal{A}\mathcal{K}$	rank r_G	dim N_G	$\sigma(X)$	$\mathcal{K}$	$r_{\mathcal{K}}$
$L(n,C)\,(A)$	$2n$	$2n^2$	$-$	$U(n)$	n
$L(n,R)\,(A_{\mathrm{I}})$	n	n^2	$\sigma(X) = X^*$	$O(n)$	$\left[\frac{n}{2}\right]$
$U^*(2n)\,(A_{\mathrm{II}})$	$2n$	$4n^2$	$\sigma(X) = J_n X^* J_n^{-1}$	$SP(2n)$	n
$U(p,q)\,(A_{\mathrm{III}})$ $p \geq q$, $p+q=n$	n	n^2	$\sigma(X) = $ $= -J_{pq} X^\dagger J_{pq}$	$U(p) \times U(q)$	n
$O(n,C)\,(B,D)$	$2\left[\frac{n}{2}\right]$	$n(n-1)$	$-$	$O(n)$	$\left[\frac{n}{2}\right]$
$O^*(2n)\,(D_{\mathrm{III}})$	n	$n(2n-1)$	$\sigma(X) = J_n X^* J_n^{-1}$	$U(n)$	n
$O(p,q)\,(BD_{\mathrm{I}})$ $p \geq q$, $p+q=n$	$\left[\frac{n}{2}\right]$	$\dfrac{n(n-1)}{2}$	$\sigma(X) = J_{pq} X^* J_{pq}$	$O(p) \times O(q)$	$\left[\frac{n}{2}\right]$
$Sp(2n,C)\,(C)$	$2n$	$2n(1+2n)$	$-$	$Sp(2n)$	n
$Sp(2n,R)\,(CI)$	n	$n(1+2n)$	$\sigma(X) = X^*$	$U(n)$	n
$Sp(2p,2q)\,(CII)$ $p \geq q$, $p+q=n$	n	$n(1+2n)$	$\sigma(X) = $ $= -I_{pq} X^\dagger I_{pq}$	$Sp(2p) \times$ $\times Sp(2q)$	n

$G = \mathcal{KAK}$	$N_{\mathcal{K}}$	$\mathcal{A} = T(\tau)$ $-\infty < \tau < \infty$	$r_{\mathcal{A}}$
$L(n,C)\,(A)$	n^2	$\exp\tau$	n
$L(n,R)\,(A_{\mathrm{I}})$	$\dfrac{n(n-1)}{2}$	$\exp\tau$	n
$U^*(2n)\,(A_{\mathrm{II}})$	$n(1+2n)$	$\begin{pmatrix} \exp\tau & 0 \\ 0 & \exp\tau \end{pmatrix}$	n
$U(p,q)\,(A_{\mathrm{III}})$ $p \geq q,\, p+q=n$	p^2+q^2	$\begin{pmatrix} \mathrm{ch}\,\tau & \mathrm{sh}\,\tau \\ \mathrm{sh}\,\tau & \mathrm{ch}\,\tau \end{pmatrix}$	q
$O(n,C)\,(B,D)$	$\dfrac{n(n-1)}{2}$	$\begin{pmatrix} \mathrm{ch}\,\tau & i\,\mathrm{sh}\,\tau \\ -i\,\mathrm{sh}\,\tau & \mathrm{ch}\,\tau \end{pmatrix}$	$\left[\dfrac{n}{2}\right]$
$O^*(2n)\,(D_{III})$	n^2	$\begin{pmatrix} \mathrm{ch}\,\tau & i\,\mathrm{sh}\,\tau & 0 & 0 \\ -i\,\mathrm{sh}\,\tau & \mathrm{ch}\,\tau & 0 & 0 \\ 0 & 0 & \mathrm{ch}\,\tau & -i\,\mathrm{sh}\,\tau \\ 0 & 0 & i\,\mathrm{sh}\,\tau & \mathrm{ch}\,\tau \end{pmatrix}$	$\left[\dfrac{n}{2}\right]$
$O(p,q)\,(BD_{\mathrm{I}})$ $p \geq q,\, p+q=n$	$\dfrac{p(p-1)}{2}+$ $+\dfrac{q(q-1)}{2}$	$\begin{pmatrix} \mathrm{ch}\,\tau & \mathrm{sh}\,\tau \\ \mathrm{sh}\,\tau & \mathrm{ch}\,\tau \end{pmatrix}$	q
$Sp(2n,C)\,(C)$	$n(1+2n)$	$\begin{pmatrix} \exp\tau & 0 \\ 0 & \exp(-\tau) \end{pmatrix}$	n
$Sp(2n,R)\,(CI)$	n^2	$\begin{pmatrix} \exp\tau & 0 \\ 0 & \exp(-\tau) \end{pmatrix}$	n
$Sp(2p,2q)\,(CII)$ $p \geq q,\, p+q=n$	$p(2p+1)+$ $+q(2q+1)$	$\begin{pmatrix} \mathrm{ch}\,\tau & \mathrm{sh}\,\tau & 0 & 0 \\ \mathrm{sh}\,\tau & \mathrm{ch}\,\tau & 0 & 0 \\ 0 & 0 & \mathrm{ch}\,\tau & \mathrm{sh}\,\tau \\ 0 & 0 & \mathrm{sh}\,\tau & \mathrm{ch}\,\tau \end{pmatrix}$	q

$G = \mathcal{KAK}$	$\mathcal{S}$	$r_{\mathcal{S}}$	$N_{\mathcal{S}}$
$L(n,C)\,(A)$	$\underbrace{U(1) \times \ldots \times U(1)}_{n}$	n	n
$L(n,R)\,(A_{\mathrm{I}})$	$\underbrace{\mathbb{Z}_2 \times \ldots \times \mathbb{Z}_2}_{n}$	$-$	$-$
$U^*(2n)\,(A_{\mathrm{II}})$	$\underbrace{SU(2) \times \ldots \times SU(2)}_{n}$	n	$3n$
$U(p,q)\,(A_{\mathrm{III}})$ $p \geq q,$ $p+q=n$	$U(p-q) \times \underbrace{U(1) \times \ldots \times U(1)}_{q}$	p	$(p-q)^2 + q$

$O(n,C)\,(B,D)$	$\underbrace{U(1) \times \ldots \times (1)}_{\left[\frac{n}{2}\right]}$	$\left[\frac{n}{2}\right]$	$\left[\frac{n}{2}\right]$
$O^*(2n)\,(D_{\mathrm{III}})$	$\underbrace{SU(2) \times \ldots \times SU(2)}_{k,\,n=2k}$	k	$3k$
	$\underbrace{SU(2) \times \ldots \times SU(2)}_{k,\,n=2k-1} \times U(1)$	$k+1$	$3k+1$
$O(p,q)\,(BD_{\mathrm{I}})$ $p \geq q,$ $p+q=n$	$O(p-q) \times \underbrace{\mathbb{Z}_2 \times \ldots \times \mathbb{Z}}_{q}$	$\left[\dfrac{p-q}{2}\right]$	$\dfrac{(p-q)(p-q-1)}{2}$
$Sp(2n,C)\,(C)$	$\underbrace{U(1) \times \ldots \times U(1)}_{n}$	n	n
$Sp(2n,R)\,(CI)$	$\underbrace{\mathbb{Z}_2 \times \ldots \times \mathbb{Z}_2}_{n}$	$-$	$-$
$Sp(2p,2q)\,(CII)$ $p \geq q,$ $p+q=n$	$\underbrace{Sp(2p-2q) \times {} \times Sp(2) \times \ldots \times Sp(2)}_{q}$	p	$(p-q) \times {} \times (2p-2q+1) + 3q$

§ 1.3 *sl*(2)-subalgebras of Lie algebras

1.3.1 Embeddings of *sl*(2) into Lie algebras. The description of all nonequivalent embeddings of semisimple algebras $\mathfrak{g}'$ into a finite-dimensional simple
Lie algebra $\mathfrak{g}$ is completely solved in [32]–[34], where, in particular, there is
a list of all $sl(2)$-subalgebras. Thanks to Gantmakher's theorem [16], [17] for
any embedding $\mathfrak{g}' \subset \mathfrak{g}$ we can select a Cartan subalgebra $\mathfrak{h} \subset \mathfrak{g}$ so that the
elements of the Cartan subalgebra $\mathfrak{h}' \subset \mathfrak{g}'$ would belong to $\mathfrak{h}$.

Consider in detail the case $\mathfrak{g}' = A_1 \simeq sl(2)$ with generators H, J_+, J_-
selected so that $[H, J_\pm] = \pm 2 J_\pm$, $[J_+, J_-] = H$. It can be shown that all the
non-equivalent (with respect to an inner automorphism of $\mathfrak{g}$) embeddings of
$sl(2)$ into a semisimple finite-dimensional Lie algebra $\mathfrak{g}$ are determined by
a vector $\vec{k}_{(h)}$ whose components are expressed in terms of the coefficients of
the decomposition of the Cartan element $H = \sum_{1 \leq j \leq r} k^i_{(h)} h_i$ with respect to
the generators of $\mathfrak{h} \subset \mathfrak{g}$, where $k^i_{(h)} \equiv (H, h^i)$.

With the help of H we can introduce a grading $\mathfrak{g} = \oplus \mathfrak{g}_a$, where a is
a weight on H, i.e. H plays the role of the grading operator. (Notice that
for infinite-dimensional simple Lie algebras the grading operator does not
generally belong to the algebra itself.)

An embedding of $sl(2)$ into $\mathfrak{g}$, $\tilde{\mathfrak{g}}$ or Vir determines a decomposition of $\mathfrak{g}$ (resp. $\tilde{\mathfrak{g}}$ or Vir) into irreducible $sl(2)$-modules. The elements F_m^{l,ν_l} of these modules are numbered by the value of the angular momentum l (the highest weight of the representation), its projection m and the index ν_l equal to the multiplicity of l. The spectrum of different values of l, i.e. $\{l\} \equiv \{l_1, \ldots, l_s\}$ is fixed for every embedding and in what follows we will assume that the numbers l_i are always even (recall that we have realized A_1 as $sl(2)$; had we realized it as $O(3)$ we would assume the l_i to be integers), as is the case for all semisimple $\mathfrak{g}$. Obviously, $\mathfrak{g}_0$ contains all the elements of the multiplets with $m = 0$ and $\dim \mathfrak{g}_0 = d_0 = \sum_i \nu_{l_i}$ for $l_i \in \{l\}$. In $\mathfrak{g}_0$, distinguish the A_1-invariant subalgebra $\mathfrak{g}_0^0$, i.e. $[J_\pm, \mathfrak{g}_0^0] = 0$ whose elements are scalars with respect to A_1. Set $\mathfrak{g}_0^f = \mathfrak{g}_0/\mathfrak{g}_0^0$ and describe its elements as elements of A_1-modules of highest weight l (clearly $l > 0$):

$$[J_+, [J_-, F_0^{l,\nu_l}]] = l(l+1)F_0^{l,\nu_l}.$$

This relation is invariant with respect to $\mathfrak{g}_0^0$ and therefore the identification of the elements of $\mathfrak{g}$ with respect to ν_l can be performed only with respect to irreducible representations of $\mathfrak{g}_0^0$. Obviously, to describe $\mathfrak{g}$ completely as $sl(2)$-module it suffices to define its elements corresponding to $m = 0$. Then by applying $J_\pm$ to these elements we can obtain all the other elements of $\mathfrak{g}$ since we have assumed that all the l_i are even.

Therefore, we can perform the grading (2.1) in $\mathfrak{g}$ with respect to the eigenvalues m of $H \in A_1$. In what follows we take $m/2$ as the grading index; it is more convenient than m.

In particular, $\mathfrak{g}_{\pm 1}$ is generated by the elements of the form $[J_\pm, F_0^{l,\nu_l}]$. Notice that the dimensions $d_{\pm 1}$ coincide for such a construction with $d_0^f = \dim \mathfrak{g}_0^f$.

As applied to simple Lie algebras, this approach enables us to describe all the embeddings $A_1 \subset \mathfrak{g}$. Every embedding is unique up to the equivalence given by the decomposition of H with respect to generators of $\mathfrak{h} \subset \mathfrak{g}$. The constants c_j of the embedding are in turn defined by the following construction. A system of positive simple roots (also called a π-system) of $\mathfrak{g}$ should be supplemented by the minimal root yielding an extended π-system; the corresponding root vectors satisfy the following relations for $1 \leq I, J \leq r+1$:

$$[h_I, h_J] = 0, \quad [h_I, X_{\pm J}] = \pm \tilde{k}_{JI} X_{\pm J}, \quad [X_I, X_{-J}] = \delta_{IJ} h_I. \qquad (3.1)$$

Here $\tilde{k}$ is the extended Cartan matrix $\tilde{k}_{ij} \equiv k_{ij}$, $1 \leq i, j \leq r$; $X_{\pm(r+1)} \equiv X_{\mp m}$. Next, in the π-system, pick a set, $p^{(s)} \equiv \{p_1, \ldots, p_s; s \leq r\}$ of roots corresponding to a semisimple subalgebra $\mathfrak{g}^{(s)} \subset \mathfrak{g}$ of rank $r_s \leq r$. Then the desired element $H = \sum_{1 \leq i \leq s} c_i h_{p_i}$ is found from the condition

$$[H, X_{p_i}] = 2X_{p_i}.$$

Therefore such an embedding $A_1 \subset \mathfrak{g}$ for which $[H, X_i] = 2X_i$, $1 \leq i \leq r$, plays a distinguished role; the corresponding subalgebra A_1 is called the *principal* one. Notice that for this embedding, which clearly corresponds to the canonical grading of $\mathfrak{g}$, the spectrum $\{l\}$ coincides with the values of the exponents of $\mathfrak{g}$ and $\nu_l = 1$ except for D_{2r} when $l_r = l_{2r} = 2r - 1$.

Indeed, the description of practically all the embeddings $A_1 \subset \mathfrak{g}$ reduces to consideration of principal subalgebras in all the algebras $\mathfrak{g}^{(s)}$. (The exceptions are the $[(r - 2)/2]$ embeddings for D_r and the $[(r' - 3)/2]$ embeddings for $E_{r'}$, where $r' = 6, 7, 8$, see [32]–[34].) The constants c_j are, clearly, expressed via the elements of $\tilde{k}$ with the help of relations (3.1).

It is not difficult to generalize the above scheme to adjust it to embeddings with odd highest weights of irreducible A_1-modules.

1.3.2 Infinite-dimensional graded Lie algebras corresponding to embeddings of sl(2) into simple finite-dimensional Lie algebras.

To every embedding of $sl(2) = A_1$ into a simple finite-dimensional Lie algebra $\mathfrak{g}$ we can assign an infinite-dimensional graded Lie algebra $\tilde{\mathfrak{g}}$ which is of finite growth. In doing so we extend the subspaces $\mathfrak{g}_0$, $\mathfrak{g}_{\pm 1}$ of $\mathfrak{g}$ with basis $\{X^0\}$, $\{X^{\pm}\}$ (see (2.4)) by adding a certain number of additional generators Y^0 and $Y^{\pm}$ so that the finite-dimensional representations of $\mathfrak{g}$ and those of the infinite-dimensional algebra to be constructed are the same. To this end consider the highest and lowest vectors $F_{\pm l}^{l, \nu_l}$ of irreducible A_1-submodules of $\mathfrak{g}$ for a fixed embedding with $l_{\max} \equiv L$, i.e. $F_{\pm L}^{l, \nu_l} \equiv F_{\nu}^{\pm}$, $1 \leq \nu \leq \nu_l$. They satisfy

$$[F_{\nu}^-, F_{\nu'}^+] = \sum_i R_{\nu\nu'}^i V_i^0, \quad [X_a^0, F_\nu^{\pm}] = \sum_{\nu'} Q_{a\nu}^{\pm\nu'} F_{\nu'}^{\pm}, \qquad (3.2)$$

where V_i^0 are linearly independent combinations of elements from $\mathfrak{g}_0$, and $V_i^0 = \sum_a b_{ia} X_a^0$. Extend $\mathfrak{g}_{\pm 1}$ to $\tilde{\mathfrak{g}}_{\pm 1}$ by adding the elements $Y_\nu^{\pm}$, $1 \leq \nu \leq \nu_L$, i.e. by setting $\tilde{\mathfrak{g}}_{\pm 1} = \{X_\alpha^{\pm}, Y_\nu^{\pm}\}$ so that

$$[X_\alpha^+, Y_\nu^-] = 0, \quad [X_\alpha^-, Y_\nu^+] = 0, \quad [Y_\nu^+, Y_{\nu'}^-] = \sum_i R_{\nu\nu'}^i (Y_i^0 + V_i^0). \qquad (3.3)$$

Add the elements Y_i^+ obtained via (3.3) to $\mathfrak{g}_0$. The constants $R_{\nu\nu'}^i$ are defined in accordance with (3.2).

In all the constructions to follow let us make sure that the extended local part of $\tilde{\mathfrak{g}}$ admits finite-dimensional representations. For such representations $Y_\nu^{\pm} = F_\nu^{\mp}$ and all the additional generators of $\tilde{\mathfrak{g}}_0$ (i.e., $Y_i^0 \in \tilde{\mathfrak{g}}_0/\mathfrak{g}_0$) vanish. By (3.3) these conditions are satisfied; clearly thanks to (3.2) they also hold for $Y_\nu^{\pm} = F_\nu^{\mp}$. Furthermore, set

$$[X_a^0, Y_\nu^{\pm}] = \sum_{\nu'} Q_{a\nu}^{\mp\nu'} Y_{\nu'}^{\pm}, \qquad (3.4)$$

with $Q_{a\nu}^{\pm\nu'}$ given by (3.2). Relations similar to (2.5) follow from the Jacobi identity for $\hat{\mathfrak{g}}$ and distinguish nonzero elements Y_i^0. Therefore it follows from (2.5) that the Y_i^0 belong to the center of $\tilde{\mathfrak{g}}_0$ i.e.

$$[Y_i^0, X_\alpha^\pm] = [Y_i^0, Y_\nu^\pm] = [Y_i^0, X_a^0] = [Y_i^0, Y_j^0] = 0, \qquad (3.5)$$

and hence the Y_i^0 belong to the center of the whole $\tilde{\mathfrak{g}}$. One can recover the whole $\tilde{\mathfrak{g}}$ by introducing the invariant symmetric bilinear form on $\hat{\mathfrak{g}}$ and its subsequent continuation onto $\tilde{\mathfrak{g}}$, cf. 1.2.3. Thus, the commutation relations (2.4) and (3.2)–(3.5) determine the structure constants of the infinite-dimensional Lie algebra $\tilde{\mathfrak{g}}$ with local part $\tilde{\mathfrak{g}}_{-1} \oplus \tilde{\mathfrak{g}}_0 \oplus \tilde{\mathfrak{g}}_{+1}$ and play the same role as the relation (2.4) for the finite-dimensional Lie algebra $\mathfrak{g}$ whose extension is $\tilde{\mathfrak{g}}$. By definition, $\tilde{\mathfrak{g}}$ possesses the same finite-dimensional representations as $\mathfrak{g}$ and therefore (3.2)–(3.5) are well-defined.

If A_1 is the principal subalgebra in a finite-dimensional simple Lie algebra $\mathfrak{g}$, let us complete the local part (2.8) of $\mathfrak{g}$ via the above scheme with Y^-, Y^+ and $Y^0 \equiv [Y^+, Y^-]$.

The generators of the local part $\tilde{\mathfrak{g}}_{-1} \oplus \tilde{\mathfrak{g}}_0 \oplus \tilde{\mathfrak{g}}_1$ of $\tilde{\mathfrak{g}}$ satisfy (3.1) and

$$\sum_{1 \leq I \leq r+1} \lambda_I h_I$$

belongs to the center of $\tilde{\mathfrak{g}}$, where $\lambda \equiv \{\lambda_I\}$ is the eigenvector of the Cartan matrix $\tilde{k}$ corresponding to the eigenvalue 0 and $\sum_{1 \leq J \leq r+1} \tilde{k}_{IJ} \lambda_J = 0$. This construction leads to one of posssible types of affine Kac-Moody algebras.

Notice that since in a finite-dimensional representation of $\tilde{\mathfrak{g}} = \oplus_{a \in \mathbf{Z}} \tilde{\mathfrak{g}}_a$ no $\tilde{\mathfrak{g}}_a$ can contain more elements than dim $\mathfrak{g}$, the constructed $\tilde{\mathfrak{g}}$ is of finite growth. In any case the functions with values in $\tilde{\mathfrak{g}}$ (e.g. the operators of the Lax-type representations, see Chapter 3) can be realized by finite-dimensional matrices. This fact suffices, as is known, to find soliton-like solutions of the corresponding dynamical systems via ISP method.

1.3.3 Explicit realization of simple finite-dimensional Lie algebras for the principal embedding of $\mathfrak{sl}(2)$.

For the principal embedding the number of irreducible A_1-submodules of $\mathfrak{g}$ equals $r = \text{rank}\,\mathfrak{g}$ and the invariance subalgebra $\mathfrak{g}_0$ is abelian, i.e. $\mathfrak{g}_0 = \oplus_{1 \leq j \leq r} \mathfrak{gl}(1)$.

We will sometimes call the principal embedding minimal, having in mind that for other embeddings the number of irreducible A_1-components is greater than rank $\mathfrak{g}$.

Let the principal subalgebra A_1 be generated by $H = \sum_j k_j h_j$, $J_\pm = \sum_j k_j^{1/2} X_{\pm j}$, where $k_i \equiv 2 \sum_j (k^{-1})_{ij}$ are the heights of fundamental weights (see 3.2) and $\mathfrak{g}$ splits into the set of irreducible subrepresentations which can be studied taking into account the obvious relation $[H, X_{\pm\alpha}] = \pm m(\alpha) X_{\pm\alpha}$,

$\mathfrak{g}$	spectrum of l (exponents)	k_α	$s_\alpha \tilde{H}_\alpha^1$	$\tilde{H}_\alpha^l,\ l \neq 1$
A_r	m $1 \le m \le r$	$k_\alpha = \alpha(r - \alpha + 1)$	1	$\tilde{H}_1^l = 1;\ \tilde{H}_\alpha^l = \Delta_{\alpha-1} \Big/ \prod_{1 \le \beta \le \alpha-1} k_\beta,\ 2 \le \alpha \le r$
B_r	$2m - 1$ $1 \le m \le r$	$k_\alpha = \alpha(2r - \alpha + 1),$ $1 \le \alpha \le r - 1;$ $k_r = r(r+1)/2$	$1, 1, \ldots, 1, 2$	$\tilde{H}_\alpha^l = \Delta_{\alpha-1} \Big/ \prod_{1 \le \beta \le \alpha-1} k_\beta,\ 1 \le \alpha \le r - 1;\ \tilde{H}_r^l = \frac{1}{2}\Delta_{r-1} \Big/ \prod_{1 \le \beta \le r-1} k_\beta$
C_r	$2m - 1$ $1 \le m \le r$	$k_\alpha = \alpha(2r - \alpha)$	$1, 1, \ldots, 1, \frac{1}{2}$	$\tilde{H}_\alpha^l = \Delta_{\alpha-1} \Big/ \prod_{1 \le \beta \le \alpha-1} k_\beta$
D_r	$2m - 1,$ $1 \le m \le r - 1,$ $r - 1$	$k_\alpha(2r - \alpha - 1)$ $1 \le \alpha \le r - 2;$ $k_{r-1} = k_r = \dfrac{r(r - 1)}{2}$	1	$l \neq r - 1\ (2k_r - l(l+1) \equiv r(r-1) - l(l+1) \neq 0)$ $\tilde{H}_\alpha^l = \Delta_{\alpha-1} \Big/ \prod_{1 \le \beta \le \alpha-1} k_\beta,\ 1 \le \alpha \le r - 2;$ $\tilde{H}_\alpha^l = \Delta_{r-3} \Big/ \left([2k_r + y_l] \prod_{1 \le \beta \le r-3} k_\beta\right),\ \alpha = r - 1, r$ $l = r - 1,\ r$ odd $\tilde{H}_\alpha^{r-1} = 0,\ 1 \le \alpha \le r - 2;\ \tilde{H}_r^{r-1} = -\tilde{H}_{r-1}^{r-1} = 1$ $l = r - 1,\ r$ even. The moment $r - 1$ is of multiplicity 2 and the second vector with $l = r - 1$ is $\tilde{\tilde{H}}_\alpha^{r-1} = \Delta_{\alpha-1} \Big/ \prod_{1 \le \beta \le \alpha-1} k_\beta,\ 1 \le \alpha \le r - 2;$ $\tilde{\tilde{H}}_{r-1}^{r-1} = -\Delta_{r-4} \Big/ \prod_{1 \le \beta \le r-4} k_\beta;\ \tilde{\tilde{H}}_r^{r-1} = 0.$

E_6	$1,4,5,7,8,11$	$16,22,30,$ $42,30,16$	1	$\tilde{H}_1^l = 1,\ \tilde{H}_2^l = \dfrac{k_2}{2k_2+y_l},\ \dfrac{y_l^2+2(k_1+k_3)y_l+3k_1k_3}{k_1k_3},$ $\tilde{H}_3^l = \dfrac{2k_1+y_l}{k_1},\ \tilde{H}_4^l = \dfrac{y_l^2+2(k_1+k_3)y_l+3k_1k_3}{k_1k_3},$ $\tilde{H}_5^l = \dfrac{\Delta_4}{k_5},\ \tilde{H}_6^l = \dfrac{\Delta_5}{k_6}.$
E_7	$1,5,7,9,$ $11,13,17$	$34,49,66,96$ $75,52,27$	1	$\tilde{H}_\alpha^l,\ 1\le\alpha\le4,$ are determined by the corresponding formulas for E_6; $\tilde{H}_5^l = \dfrac{\Delta_4}{k_4},\ \tilde{H}_6^l = \dfrac{\Delta_5}{k_4k_5},\ \tilde{H}_7^l = \dfrac{\Delta_6}{k_4k_5k_6}$
E_8	$1,7,11,13,$ $17,19,23,29$	$2\cdot(46,68,91,$ $135,110,84,$ $57,29)$	1	$\tilde{H}_\alpha^l,\ 1\le\alpha\le4,$ are determined by the corresponding formulas for E_6; $\tilde{H}_5^l = \dfrac{\Delta_4}{k_4},\ \tilde{H}_6^l = \dfrac{\Delta_5}{k_4k_5},\ \tilde{H}_7^l = \dfrac{\Delta_6}{k_4k_5k_6},$ $\tilde{H}_8^l = \dfrac{\Delta_7}{k_4k_5k_6k_7}$
F_4	$1,5,7,11$	$2\cdot(11,21,15,8)$	$\dfrac{1}{2},\dfrac{1}{2},1,1$	$\tilde{H}_\alpha^l = \Delta_{\alpha-1}\Big/\displaystyle\prod_{\beta=1}^{\alpha-1} k_\beta$
G_2	$1,5$	$2\cdot(3,5)$	$3,1$	$\tilde{H}_1^l = 1,\ \tilde{H}_2^l = \dfrac{2k_1+y_l}{3k_1}$

Note: Δ_α are principal minors of order α of the matrix $\mathcal{R} \equiv P + y_l 1$, where $y_l \equiv -l(l+1)$, $1_{\alpha\beta} \equiv \delta_{\alpha\beta}$.

where $m(\alpha) = ht(X_{\pm\alpha})$ is the height of $X_{\pm\alpha}$, i.e. the number of simple roots (multiplicities counted) constituting α. Notice that the maximal l coincides with the height of the maximal root vector of $\mathfrak{g}$.

To recover an irreducible representation of A_1 completely, it suffices to know at least one of its elements since all others can constructed applying J_+ or J_-.

The Casimir operator of A_1 (see § 1.5) equals $l(l+1) \times id$ on the irreducuble representation with highest weight l, i.e.,

$$[J_+[J_-, F_0^l]] = l(l+1)F_0^l \text{ or } \sum_j k_{ij} H_i^{(1)} H_j^{(l)} = l(l+1)H_i^{(l)},$$

$$\text{where } F_0^l \equiv \sum_j H_j^{(l)} h_j; \quad H \equiv F_0^1, H_j^{(1)} \equiv k_j. \tag{3.6}$$

Thus, the multiplet structure of the embedding is completely determined by the eigenvalues of the matrix $P_{ij} \equiv k_i k_{ij}$ entering (3.6). This in turn enables us to express the exponents of the Weyl group directly via the Cartan matrix.

The matrix P, like k, is not generally symmetric and its eigenvectors can be normalized with the help of a diagonal matrix S such that $SP = P^T S$, $s_i \equiv s_{ii} = v_i k_i^{-1}$, see (2.13). It is convenient to normalize the eigenvectors $H_i^{(l)}$ of P by the relation

$$\sum_i H_i^{(l)} s_i H_i^{(l')} = [l(l+1)]^{-2} \delta_{ll'}.$$

The vectors of the orthonormal basis are defined by the equation $\tilde{H}_i^l \equiv s_i^{1/2} l(l+1) H_i^{(l)}$ and we have

$$k_{ij} = k_i^{-1}(s_i/s_j)^{-1/2} \sum_{\{l\}} l(l+1) \tilde{H}_i^l \tilde{H}_j^l.$$

The remaining elements of the irreducible representation with highest weight l are obtained from F_0^l applying $J_\pm$ m times:

$$F_{\pm m}^l = [(l-m)!/(l+m)!]^{1/2} \underbrace{[J_\pm[J_\pm[\ldots[J_\pm F_0^l]\ldots]]]}_{m}. \tag{3.7}$$

In the basis (3.7) the commutation relations are of the form

$$[F_m^{l_1}, F_n^{l_2}] = \sum_{\{l\}} a(l_1, l_2, L) c(l_1, l_2, L; m, n) F_{m+n}^L, \tag{3.8}$$

where $c(l_1, l_2, L; m, n)$ are Clebsch-Gordan coefficients for $SL(2; \mathbb{C})$ and $a(l_1, l_2, L)$ is the set of structure constants which do not depend on the indices of basic vectors; the sum in (3.8) runs over the whole multiplet spectrum of $\mathfrak{g}$.

To find $a(l_1, l_2, L)$ explicitly, set $m = 1$ and $n = 0$ in (3.8) yielding

$$[F_1,{}^{l_1}, F_0^{l_2}] = [l_1(l_1+1)]^{-1/2}[[J_+ F_0^{l_1}]F_0^{l_2}]$$
$$= \sum_L a(l_1, l_2, L) c(l_1, l_2, L; 1, 0)[J_+, F_0^L] \cdot [L(L+1)]^{-1/2},$$

which implies

$$[l_1(l_1+1)]^{1/2} l_2(l_2+1) \tilde{H}_m^{l_1} \tilde{H}_m^{l_2} k_m^{-1}$$
$$= \sum_L [L(L+1)]^{1/2} a(l_1, l_2, L) c(l_1, l_2, L; 1, 0) \tilde{H}_m^L.$$

Making use of the normalization of $\tilde{H}_m^l$ we find that

$$a(l_1, l_2, L) = [l_1(l_1+1)/L(L+1)]^{1/2} l_2(l_2+1)$$
$$\times \sum_m \tilde{H}_m^{l_1} \tilde{H}_m^{l_2} k_m^{-1} s_m \tilde{H}_m^L / c(l_1, l_2, L; 1, 0) \sum_n \tilde{H}_n^L s_n \tilde{H}_n^L. \tag{3.9}$$

Therefore the formulas (3.8) with structure functions (3.9) solve the problem of constructing the commutation relations of a simple finite-dimensional Lie algebra with basis (3.7). The quantities H_m^l, k_m, s_m and the spectrum of l (directly determined by the Cartan matrix of the algebra) which are necessary for concrete calculations in basis (3.7) are given for all simple algebras in Table 1.3, cf. [16, 17, 32–34, 44, 52, 113], where proofs can also be found.

Examples illustrating an application of the results of this section are given in Chapters 3 and 4.

§ 1.4 The structure of representations *)

Though in this section we consider finite-dimensional Lie groups and Lie algebras, many of the constructions and formulations we give are valid in infinite-dimensional case.

1.4.1 Terminology. A *representation of a Lie algebra* $\mathfrak{g}$ in a linear space V is a Lie algebra homomorphism $\mathfrak{g} \to \mathfrak{gl}(V)$. Similarly, a *representation of a Lie group* G in V is a smooth function T on G with values in $GL(V)$ such that

$$T(g_1)T(g_2) = T(g_1 g_2). \tag{4.1}$$

This relation implies $T(g^{-1}) = T^{-1}(g)$ and $T(g_0) = \mathbf{1}$, where $\mathbf{1}$ is the identity transformation of V. The space V and its dimension are called the

---

*) [7, 13, 19, 20, 28, 35, 36, 51, 95, 101, 107, 112]

space and the *dimension of the representation T* and *V* is sometimes called a *module over* $\mathfrak{g}$ or *G*, respectively. Representations *T* and *T'* are said to be *equivalent* if there exists an invertible operator *B* (called an *intertwining operator*) mapping the space of *T* onto the space of *T'* and satisfying

$$T'(g) = BT(g)B^{-1}. \tag{4.2}$$

Over $\mathbb{C}$, the space of *T* may be endowed with an invariant Hermitian form, i.e. a form $(.\,,.)$ such that

$$(f_1, f_2) = \overline{(f_2, f_1)};$$

$$(T(g)f_1, T(g)f_2) = (f_1, f_2) \text{ for all } g \in G \text{ and all } f_1, f_2 \in V; \tag{4.3}$$

$$(f, f) > 0 \text{ if } f \neq 0.$$

In this case $T^\dagger(g)T(g) = \mathbf{1}$ and the representation is called *unitary*. Clearly, if *T* is unitary then its *Hermitian conjugate* defined by $\tilde{T}(g) \equiv (T(g^{-1}))^\dagger$ is equivalent to *T*. The problem of distinguishing all the unitary representations is rather complicated for arbitrary Lie groups. For compact Lie groups it is not difficult to show that any finite-dimensional representation over $\mathbb{C}$ is unitarizable.

The structure of the spaces of representations of Lie groups may be rather involved and its study requires introducing the notion of *topological reducibility* of representations. The simplest representations are the *irreducible* ones, defined as follows. A subspace *V'* of *V* is called *invariant* if for all the operators of the representation $T(g)V' \subset V'$ for any $g \in G$. Obviously, 0 and *V* itself are invariant. Accordingly, a representation *T* is called *irreducible* if its space does not contain irreducible subspaces different from the above-mentioned two trivial examples; otherwise *T* is called *reducible*. It can be shown that in the space of any representation there is at least one irreducible subrepresentation. If an invariant subspace *V'* contains an invariant complement, i.e. $V = V' \oplus V''$ then *T* is uniquely determined by its subrepresentations *T'* and *T''* in *V'* and *V''* respectively and we say that *T* is the *direct sum* of *T'* and *T''*. A representation is called *completely reducible* if it is representable as the direct sum of irreducible ones. Notice that all unitary representations are completely reducible.

The notion of *operator irreducibility* is less restrictive than that of topological irreducibility, although the two kinds of irreducibility coincide for the finite-dimensional representations. A representation is said to be *operator irreducible* if any intertwining operator is a scalar one. The operator irreducibility of finite-dimensional representations is the statement of Schur's lemma.

In a number of problems it is convenient to pass from operator solutions of the functional equation (4.1) to the usual scalar functions. Such are the matrix elements of T, namely the functions

$$D(g) \equiv (T(g)v_1, v_2), \qquad v_1, v_2 \in V. \tag{4.4}$$

In particular, selecting a basis $\{v_i\}$ in the representation space we get $T(g)v_j = \sum_i d_{ij}(g)v_i$, where $d_{ij}(g) = (T(g)v_j, v_i)$. Then the equation (4.1) implies the relations

$$t_{ij}(g_1 g_2) = \sum_k t_{ik}(g_1)t_{kj}(g_2). \tag{4.5}$$

1.4.2 The adjoint representation. An arbitrary finite-dimensional Lie algebra $\mathfrak{g}$ possesses at least one finite-dimensional representation, namely, the *adjoint* one. To construct matrix elements for the adjoint representation of a Lie group G, consider the space $\mathfrak{g}^*$ dual to $\mathfrak{g}$ with the dual basis, i.e. $(X^a, X_b) = \delta_{ab}$. Since X_a can be considered as numerical $N \times N$-matrices, the same applies to X^a and we can define the scalar product setting $(X^a, X_b) = tr(X^a X_b)$. In this notation the matrix elements of an arbitrary transformation g take the form

$$\langle a|g|b \rangle = (X^a, gX_b g^{-1}) = tr(X^a gX_b g^{-1}), \tag{4.6}$$

and as in (4.5) we have

$$\langle a|g_1 g_2|b \rangle = \sum_c \langle a|g_1|c \rangle \langle c|g_2|b \rangle = tr(X^a g_1 g_2 X_b g_2^{-1} g_1^{-1}). \tag{4.7}$$

Here the basis elements in the the representation space and its dual are denoted with Dirac's symbols "bra" $\langle a|$ and "ket" $|b \rangle$ (constituting the word bracket modulo a humane editing). Dirac introduced these symbols for quantum mechanical vectors of state labeled by the set $\{a\}$ of eigenvalues of commuting operators corresponding to simultaneously measurable observables. Clearly, the scalar product $\langle a|b \rangle$ is linear in $|b \rangle$ and anti-linear in $|a \rangle$, i.e. $\langle a|b \rangle = \overline{\langle b|a \rangle}$ and $\langle a|a \rangle > 0$ for $a \neq 0$; the dual of $\hat{X}|b \rangle$ is $\langle b|\hat{X}^\dagger$, where $\hat{X}$ is an operator.

The matrix elements of the adjoint representation of a Lie group G play an important role in many branches of the representation theory. As we will see, they bridge infinitesimal operators of left and right shifts on G and the weight function of an invariant (Haar)measure on G is expressed in terms of them. For semisimple Lie groups they are instrumental in the description of the highest weights of irreducible representations which completely determine the representation space.

1.4.3 The regular representation and Casimir operators. Let us consider representations of a Lie group G by left (L) and right (R) shifts in the space of functions on G:

$$L(g_0)f(g) = f(g_0^{-1}g), \ R(g_0)f(g) = f(gg_0); \ g, g_0 \in G. \qquad (4.8)$$

It can be shown that any irreducible representation of G is equivalent to a subrepresentation of this, the *regular* representation. This realization of G leads to the following definition of left- or right-invariant measures, respectively:

$$\int f(g)d\mu^L(g) = \int f(g_0^{-1}g)d\mu^L(g) \quad \text{or} \quad \int f(g)d\mu^R(g) = \int f(gg_0)d\mu^R(g) \qquad (4.9)$$

for any $g_0 \in G$. If there is an invariant measure $d\mu(g)$ on G then we can take for the representation space the space of functions, square integrable with respect to this measure, with the invariant inner product

$$(f_1, f_2) = \int_G \bar{f}_1(g)f_2(g)d\mu(g). \qquad (4.10)$$

In this case L and R are called *left-* and *right-regular* representations of G, respectively. Thanks to (4.3) they are unitary, as is not difficult to see. Notice that for a number of groups, in particular for all compact groups, all of the irreducible representations are contained in the regular one. The irreducible unitary components of the regular representation are called *unitary representations* of the *principal series*.

Infinitesimal shift operators on G are given by linear derivatives with respect to group parameters α_a by the formula

$$F_a^L = -\sum_b L_{ab}(g)\partial/\partial\alpha_b, \ F_a^R = \sum_b R_{ab}(g)\partial/\partial\alpha_b, \ g \in G, \qquad (4.11)$$

where (L_{ab}) and (R_{ab}) are invertible matrix-functions in g such that any F^L commutes with any F^R. Since (L_{ab}) and (R_{ab}) are invertible,

$$F_a^L = -\sum_b \omega_{ab}(g)F_b^R, \ F_a^R = -\sum_b \omega_{ab}^{-1}(g)F_b^L, \qquad (4.12)$$

where $\omega \equiv LR^{-1}$.

The weight function $\mu(g)$ of the (left-) right-invariant measure on G is given by the formula

$$\mu^L(g) = \det L_{ab}^{-1}(g), \ \mu^R(g) = \det R_{ab}^{-1}(g), \qquad (4.13)$$

and the coincidence of these measure implies

$$|\det(L_{ab})| = |\det(R_{ab})|, \ |\det(RL^{-1})| = 1. \qquad (4.14)$$

In what follows, to construct explicit expressions for the infinitesimal shift operators on a Lie group we will start from the relations

$$F^R g = g\mathcal{F}, \quad F^L g = -\mathcal{F}g, \tag{4.15}$$

where for the tangent operators $\mathcal{F}$ we may take matrices in an appropriate representation.

Introduce Casimir operators c_i, $1 \le i \le r = rkG$ (Casimir was the first who calculate these operators for semisimple Lie algebras). They generate the center of the associative algebra of all the differential operators on G generated by left (or right) shifts on G. By Schur's lemma on irreducible representations these operators act as scalars. In Gelfand's construction the commutative associative algebra of Casimir operators of $\mathfrak{g}$ is isomorphic to the algebra of $\mathfrak{g}$-invariant polynomials on $\mathfrak{g}$ or, as Chevalley showed, to the algebra of $W(\mathfrak{g})$-invariant polynomials on the Cartan subalgebra $\mathfrak{t}$. In Chapter 2 we will describe the algebra of Casimir operators for reductive groups. Notice that in a number of physical applications of the representation theory it is important to know the spectrum of eigenvalues and find eigenfunctions of Casimir operators.

1.4.4 Bases in the space of representation. Irreducible representations (not necessarily finite-dimensional ones) of complex semisimple Lie algebras are completely described if the space V of the representation contains an element $\underline{\xi}^{\{L\}}$ satisfying

$$X_j \underline{\xi}^{\{L\}} = 0, \quad h_j \underline{\xi}^{\{L\}} = L_j \underline{\xi}^{\{L\}}, \quad 1 \le j \le rk\mathfrak{g}, \tag{4.16}$$

and if V is $\mathfrak{t}$-diagonalizable, i.e. $V = \oplus_{a \in \mathfrak{f}^*} V^a$, where $V^a = \{v \in V : hv = av$ for any $h \in \mathfrak{t}\}$ and $\mathfrak{t}$ is a Cartan subalgebra. The vector $\underline{\xi}^{\{L\}}(\equiv |L\rangle)$ is called the *highest (weight) vector* and the L_i, where $L_i(\equiv L(h_i))$ are called the *numerical labels* of the representation V. The element $|L\rangle$ is uniquely determined from (4.16) up to scalar factor. Obviously, all of the root vectors X_α, $\alpha \in R^+(\mathfrak{g})$, annihilate $|L\rangle$ whereas the operators $\hat{F}^-$ constructed from $X_{-\alpha}$ acting on $|L\rangle$ generate the whole space V. Every weight l on the Cartan subalgebra $\mathfrak{t}$ can be expressed as

$$l_i = L_i - \sum_j k_{ij} n_j, \quad n_j \in \mathbb{Z}_+. \tag{4.17}$$

Thus, the highest weight or the highest vector completely determine an irreducible representation up to equivalence.

The necessary and sufficient condition for finite dimensionality of a representation with highest weight $\{L\}$ is $L_i \in \mathbb{Z}_+$ for all i. If the representation space is of finite dimension then there exists a lowest element $\bar{\xi}^{\{L\}}$ such that

$$X_{-j} \bar{\xi}^{\{L\}} = 0, \quad h_j \bar{\xi}^{\{L\}} = -L_j \bar{\xi}^{\{L\}}, \quad 1 \le j \le r. \tag{4.18}$$

(Obviously, in the above construction we could have started not from the highest weight but from the lowest one.)

1.4.5 Fundamental representations. Among the irreducible representations of simple Lie algebras (groups) of rank r the *fundamental* representations, i.e. representations with highest weight $\{1\}_i \equiv \{0, \ldots, 0, 1, 0, \ldots, 0\}$ (unit on the i-th place) are distinguished. Indeed, any irreducible representation with highest weight L is constructed from the fundamental ones via the formula $L(h_i) \equiv \sum_j L_j \{1\}_i(h_j)$.

In what follows we will make use of the following properties of the fundamental representations valid both for finite- and infinite-dimensional Lie algebras:

$$X_j|i\rangle = 0; \quad h_j|i\rangle = \delta_{ij}|i\rangle \text{ for } 1 \leq j \leq r; X_{-j}|i\rangle = 0 \text{ for } i \neq j. \qquad (4.19)$$

The first two of these properties follow from (4.18) if we take into account that $\{1\}_i(h_j) = \delta_{ij}$ for the components of $\{1\}_i(h)$. The validity of the third condition follows easily from the calculations of the norm of $X_{-j}|i\rangle$:

$$\langle i|X_j X_{-j}|i\rangle = \langle i|h_j|i\rangle = \delta_{ij}\langle i|i\rangle. \qquad (4.20)$$

The second important property of fundamental representations concerns the scalar products of the elements obtained by the action of negative weight operators on the highest vector. Namely, all such products are non-negative:

$$\langle i|X_{+j_1} X_{+j_2} \ldots X_{+j_m} X_{-i_m} X_{-i_{m-1}} \ldots X_{-i_1}|i\rangle \geq 0. \qquad (4.21)$$

For them an explicit expression can be obtained. It is determined completely by the Cartan matrix of the corresponding Lie algebra. Indeed, conditions (4.19) imply that this matrix element is nonzero only for $i_1 = j_1 = i$ and $i_s = j_{\omega(s)}$, $1 \leq s \leq m$, where ω is any permutation of numbers $1, \ldots, m$. Denote by S_l the minimal element from $\{1, \ldots, m\}$ such that $i_{S_l} \equiv j_{\omega(S_l)} = j_l$. Let I_m be the set $j_{\omega(1)}, \ldots, j_{\omega(m)}$, and $I_m^{(S_l)}$ the subset of I_m containing the values $\omega(S_l + 1), \ldots, \omega(m)$ but not containing $\omega(S_n)$ for $n \leq l$. Then, using (2.8), we get by induction

$$\langle i|X_{+j_m} \ldots X_{+j_1} X_{-j_{\omega(1)}} \ldots X_{-j_{\omega(m)}}|i\rangle$$

$$= \prod_{1 \leq l \leq m} \left(\delta_{ij_l} - \sum_{s \in I_m^{(S_l)}} k_{j_{\omega(s)} j_l} + \delta_{ij_l} \delta_{S_l m - 1} \right), \qquad (4.22_1)$$

or, equivalently,

$$\langle i|X_{+j_1} \ldots X_{+j_m} X_{-j_{\omega(m)}} \ldots X_{-j_{\omega(1)}}|i\rangle \equiv D\{j, \omega\}$$

$$= \delta_{ij_1} \prod_{2 \leq l \leq m} \left[\delta_{ij_l}(1 + \delta_{j_l}) - k_{ij_l} - \sum_{2 \leq s \leq l-1} k_{j_s j_l} \theta(j_{\omega(l)} - j_{\omega(s)}) \right], \qquad (4.22_2)$$

where $\delta_{j_l} \equiv \begin{cases} 1 & \text{if } \sum\limits_{2 \leq s \leq l-1} k_{j_s j_l} \theta(j_{\omega(l)} - j_{\omega(s)}) = 0, \\ 0 & \text{otherwise.} \end{cases}$

§ 1.5   A parametrization of simple Lie groups *)

An element g of a Lie group G is sometimes represented as the exponent of an element of its Lie algebra $\mathfrak{g}$:

$$g = \exp \sum_a (\boldsymbol{a}_a \cdot \boldsymbol{F}_a), \tag{5.1}$$

where parentheses denote the scalar product of the vectors from the subspace $\mathfrak{g}_a$ from (2.1) by the corresponding parameters of G. The elements $F_a^s \in \mathfrak{g}_a$ play the role of the tangent vectors to the curves $\exp(a_a^s \cdot F_a^s)$ at $a_a^s = 0$ on the corresponding one-parameter subgroups. Notice that such an exponential map $\mathfrak{g} \to G$ does not generally cover G. For example, while it is onto for $GL(n, \mathbb{C})$ and compact connected Lie groups, for $GL(n, \mathbb{R})$ it is not onto.

In what follows elements of G will be called *regular* if they can be presented in the form (5.1) and *singular* otherwise. Making use of the Cambell-Baker-Hausdorff formula (CBH), which expresses exponents of the sum of operators in terms of exponents of summands and their brackets, one can sometimes write the representation (5.1) explicitly. In particular, if $\mathfrak{g}$ has a subalgebra $\mathfrak{g}_0$, we have $g = g_0 g_{(-)}$, where g_0 is an exponent of elements from $\mathfrak{g}_0$ and $g_{(-)}$ is an exponent of elements from $\mathfrak{g}/\mathfrak{g}_0$. This implies that for a $\mathbb{Z}$-graded Lie algebra $\mathfrak{g}$ a regular element of its group can be represented in the following form:

$$g = g_+ g_0 g_- \text{ or } g = g'_- g'_0 g'_+. \tag{5.2}$$

Such a representation is obtained in two steps. First we take the subalgebra of $\mathfrak{g}$ corresponding to $\oplus_{a<0} \mathfrak{g}_a$ in (2.1) and then $\oplus_{a \geq 0} \mathfrak{g}_a$ is subdivided into $\mathfrak{g}_0$ and $\oplus_{a>0} \mathfrak{g}_a$. As a result we get the forms (5.2) which we will refer to as *Gauss decomposition* corresponding to the chosen grading. Obviously, we can also consider representations

$$g = g_+ g_- g_0, \quad g = g'_- g'_+ g'_0. \tag{5.3}$$

Clearly, the parameters of elements in (5.3) differ from the parameters in (5.2) which implies the useful identity

$$g_+^{-1} g'_- = g_- (g_0 g'^{-1}_0) g'^{-1}_+. \tag{5.4}$$

Thanks to the Levi-Malcev theorem, an element $g \in G$ factors into the product of elements from its maximal semisimple (S) and solvable (R) subgroups. If G is simply connected then the representation

$$g = r \cdot s, \ r \in R, \ s \in S \tag{5.5}$$

is unique and holds for any $g \in G$. For semisimple finite-dimensional Lie groups there are several useful decompositions.

---

*) [7, 13, 35, 51, 60, 61, 113]

A *Gauss decomposition* of an arbitrary regular element g of a connected semisimple complex Lie group G is of the form

$$g = z_+ \cdot a \cdot z_- \text{ or } g = z'_- \cdot a' \cdot z'_+, \tag{5.6}$$

where a, a' belongs to a Cartan subgroup H and $z_\pm, z'_\pm$ to the maximal simply connected nilpotent subgroups $Z^\pm$ spanned by positive (resp. negative) root vectors. The set of singular elements is closed and its dimension is less than that of G; the elements $z_\pm, z'_\pm$ and a, a' from (5.6) are continuous functions on G and their parameters are complex numbers.

A decomposition of the form (5.6) holds also for normal real forms of G. In this case the parameters of $z_\pm, z'_\pm$ and a, a' are real and satisfy certain restrictions imposed by the automorphism which distinguishes this real form. Decomposition (5.6) can be directly formulated for a reductive connected complex Lie group whose Lie algebra is the complexification of a compact algebra. The connected compact semisimple Lie groups themselves do not admit a Gauss decomposition since they do not contain solvable subgroups of the form Z^+H or HZ^-. For noncompact semisimple Lie groups G over $\mathbb{R}$ we can rewrite (5.6) for regular elements in the form

$$g = z_+ \cdot a_k \cdot z_- \text{ or } g = z'_- \cdot a'_k \cdot z'_+, \tag{5.7}$$

where a_k, a'_k belong to the direct product of a simply connected abelian subgroup and the maximal compact subgroup K and $z_\pm, z'_\pm$ belong to simply connected nilpotent subgroups of G.

The invariant measure on G factors with respect to the Gauss decomposition as:

$$d\mu(g) = d\mu(z_+)d\mu(z_-)d\mu(a). \tag{5.8}$$

The *Iwasawa decomposition* for an arbitrary element of a connected semisimple complex Lie group G represents any element of G uniquely in the form

$$g = k \cdot a \cdot z_\pm \text{ or } g = z'_\pm \cdot a' \cdot k', \tag{5.9}$$

where k, k' belong to the maximal compact subgroup $K \subset G$ and $a \cdot z_\pm, z'_\pm a'$ to a maximal simply connected solvable one, i.e., a, a' belong to the maximal abelian noncompact subgroup $A \subset G$. This decomposition can be generalized to all connected reductive complex Lie groups and arbitrary connected real Lie groups. The invariant measure on G factorizes with respect to the Iwasawa decomposition as:

$$d\mu(g) = d\mu(k)d\mu(a)d\mu(z_\pm). \tag{5.10}$$

In the notation of (5.9) the Cartan decomposition of a regular element takes the form

$$g = k \cdot a \cdot k'. \tag{5.11}$$

To make this representation unique the parameters of elements k' and k that belong to the centralizer of A in K should be added to parameters of k and k' respectively. The invariant measure on G factors with respect to Cartan decomposition as:

$$d\mu(g) = d\mu(k)d\mu(k')d\mu(a). \tag{5.12}$$

In the concrete calculations connected with the above factorizations one should know the structure of the corresponding nilpotent and compact subgroups and be able to parametrize the elements conveniently. For semisimple Lie groups there exists a universal parametrization which enables us to explicitly take into account the dependence of an arbitrary element on all its parameters. The effectiveness and natural advantages of this parametrization as compared with other ones are due to especially nice properties of the $\sum^+$-ordering of roots of simple Lie algebras introduced in 2.5.

Let us represent an arbitrary element of a complex simple Lie group G as the product of elements of its A_1-triples G_α generated by root vectors $X_\alpha, X_{-\alpha}$ and $h_\alpha = [X_\alpha, X_{-\alpha}]$, i.e.:

$$g = \prod_{\alpha \in R^+} g_\alpha, \text{ where } g_\alpha \in G_\alpha. \tag{5.13}$$

Then

$$g_\alpha = \exp h_\alpha \Phi_\alpha \cdot \exp(X_\alpha + X_{-\alpha})\Theta_\alpha \cdot \exp h_\alpha \Psi_\alpha \tag{5.14}$$

and the number of parameters $\{\Phi, \Theta, \Psi\}$ from G entering (5.13) is equal to $3N$ (where $N = \dim R^+$) which is in general greater than the total number $2N + r$ of parameters of G (here $r = rkG$). However, making use of the commutation relations (2.18) we can move all the elements of one-parameter subgroups of G_α related to Ψ_α to the right-hand side of (5.13). We get

$$g = \prod_{\alpha \in R^+} \exp h_\alpha \Phi_\alpha \exp(X_\alpha + X_{-\alpha})\Theta_\alpha \prod_{1 \le j \le r} \exp h_j \Psi_j. \tag{5.15}$$

Assuming the position of factors in (5.15) to be compatible with $\sum^+$-ordering and denoting this circumstance by the symbol

$$\prod_{\alpha > 0}^{\Sigma^+}$$

we get the *universal parametrization* of the group. In this parametrization an element k of an arbitrary compact subgroup K can be represented in the form

$$k = \prod_{\alpha > 0}^{\Sigma^+} \exp i h_\alpha \frac{\Phi_\alpha}{(\alpha, \alpha)} \exp i(X_\alpha + X_{-\alpha})\frac{\theta_\alpha}{\sqrt{2(\alpha, \alpha)}} \prod_{1 \le j \le r} \exp i h_j \frac{\Psi_j}{2}, \tag{5.16}$$

$$0 \le \theta_\alpha < \pi, \quad 0 \le \Phi_\alpha < 2\pi, \quad 0 \le \Psi_j < 4\pi,$$

where normalizing factors (α, α) are introduced for further convenience. The CBH formula applied to root vectors of a simple Lie algebra yields

$$\exp\left[i\frac{X_\alpha + X_{-\alpha}}{\sqrt{2(\alpha,\alpha)}}\theta_\alpha\right] = \exp\left[i\sqrt{\frac{2}{(\alpha,\alpha)}}X_{-\alpha}\,\mathrm{tg}\,\frac{\theta_\alpha}{2}\right]$$

$$\times \exp\left[\frac{2}{(\alpha,\alpha)}h_\alpha\ln\cos\frac{\theta_\alpha}{2}\right]\exp\left[i\sqrt{\frac{2}{(\alpha,\alpha)}}X_\alpha\,\mathrm{tg}\,\frac{\theta_\alpha}{2}\right],$$

(5.17)

and it is directly verifiable that the invariant measure $d\mu(k)$ on K takes in the parametrization (5.16) the form

$$d\mu(k) = \prod_{\alpha>0}\left(\cos\frac{\theta_\alpha}{2}\right)^{2\frac{(\alpha,\rho_0)}{(\alpha,\alpha)}-1}\sin\frac{\theta_\alpha}{2}d\theta_\alpha d\varphi_\alpha \prod_{1\leq j\leq r}d\Psi_j,$$

(5.18)

$$\text{where } \rho_0 \equiv \sum_{\alpha\in R^+}\alpha.$$

For the subgroups K of normal real forms of complex simple Lie groups generated by $X_\alpha - X_{-\alpha}$ the elements are parametrized as follows:

$$n = \prod_{\alpha>0}^{\Sigma^+}\exp(X_\alpha - X_{-\alpha})\theta_\alpha$$

(5.19)

and the invariant measure is

$$d\mu(n) = \prod_{\alpha>0}(\sin\theta_\alpha)^{2\frac{(\alpha,\rho_0)}{(\alpha,\alpha)}-2}d\theta_\alpha.$$

(5.20)

Notice that a factorization of type (5.16) and (5.19) is a natural generalization of known parametrizations of unitary and orthogonal groups, respectively. For the compact forms of arbitrary complex simple Lie groups the parameters θ_α, φ_α, Ψ_j play the role of generalized Euler angles. This factorization of the invariant measure is one of many remarkable properties of the universal parametrization.

§ 1.6 The highest vectors of irreducible representations of semisimple Lie groups *)

In this section we will consider finite-dimensional semisimple Lie groups though some results hold in infinite-dimensional case as well.

1.6.1 Generalities. The irreducible representations of semisimple Lie algebras considered in 1.4 were given by their highest (lowest) vector from which the other elements were constructed with the help of operators of negative (positive) weight. In the above notation the matrix elements of G are expressed in the form $\langle a | T^{\{l\}}(g) | b \rangle$ or just $\langle a | g | b \rangle$. In what follows we will need matrix elements of the form

$$\xi^{\{l\}}(g) = \langle L | g | L \rangle, \text{ where } |L\rangle \text{ is the highest vector.} \tag{6.1}$$

The matrix elements between arbitrary basic states are derived from the matrix element (6.1), called here the *highest (weight) matrix element*, as follows. In accordance with (6.1) and (4.16) we have

$$X_\alpha^R \xi^{\{L\}}(g) = \langle L | g X_\alpha | L \rangle = 0, \quad X_{-\alpha}^L \xi^{\{L\}}(g) = \langle L | X_{-\alpha} g | \alpha \rangle = 0,$$

$$h_j^R \xi^{\{L\}}(g) = \langle L | g h_j | L \rangle = L_j \xi^{\{L\}}(g) = -h_j^L \xi^{\{L\}}(g) = -\langle L | h_j g | L \rangle, \tag{6.2}$$

where $X_{\pm\alpha}^{R(L)}$ and $h_j^{R(L)}$ are infinitesimal operators of right (left) shifts on G corresponding to root vectors $X_{\pm\alpha}$ and to generators h_j of the Cartan subalgebra of $\mathfrak{g}$ respectively. Notice that (6.2) is another definition of the highest matrix element $\xi^{\{L\}}(g)$ given by (6.1). It follows from (6.2) that

$$\xi^{\{L\}}(g) = \prod_{1 \leq j \leq r} [\xi^{\{j\}}(g)]^{L_j}, \tag{6.3}$$

where $\xi^{\{j\}}(g)$ is the matrix element of the fundamental representation $\{1\}_j$, i.e., $\{1\}_j(h_i) \equiv \delta_{ij}$. Therefore to construct matrix elements of irreducible representations of G it suffices to know the matrix elements of the fundamental representations.

1.6.2 Expression of the highest matrix elements in terms of the adjoint representation. The highest matrix elements of irreducible representations of G can be represented as algebraic functions in certain matrix elements of the adjoint representation $a_{\alpha\beta} \equiv tr(X_{-\alpha} g X_\beta g^{-1})$, $X_\alpha \in \mathfrak{g}$, $g \in G$.

It is easy to see directly that the principal minors $D_i(a)$ of the matrix $(a_{s_i s_j})$, where $1 \leq i, j \leq r$ and s_i are Σ^+-ordered fundamental highest roots, satisfy

$$X_{+\alpha}^R D_i(a) = 0, \quad X_{-\alpha}^L D_i(a) = 0, \quad h_j^R D_i(a) = -h_j^L D_i(a) = \lambda_j^{(i)} D_i(a), \tag{6.4}$$

---

*) [35, 60, 61, 67, 68, 72, 73]

where

$$\lambda_j^{(i)} = \sum_{s_{r-i+1} \leq \alpha \leq s_r} \alpha(h_j). \tag{6.5}$$

Indeed, it follows from the definition of $a_{s_i s_j}$ and linearity of X_α^R that

$$X_\alpha^R a_{s_i s_j} = tr(X_{-s_i} g[X_\alpha, X_{s_j}] g^{-1}) = \begin{cases} N_{\alpha s_j} a_{s_i \alpha + s_j} & \text{for } \alpha + s_j \in \Sigma^+, \\ 0 & \text{otherwise} \end{cases} \tag{6.6}$$

and since this operation is nilpotent ($\alpha + s_j > s_j$ for $\alpha + s_j \in \Sigma^+$) the first subsystem in (6.4) clearly follows. Formula (6.5) is a direct corollary of the formula $h_k^R a_{s_i s_j} = s_j(h_k) a_{s_i s_j}$. The corresponding equations associated with the action of $X_{-\alpha}^L$ and h_j^L are proved similarly.

Therefore the definition of the highest matrix element $\xi^{\{L\}}(g)$ as a solution of (6.2) implies

$$\xi^{\{L\}}(g) = \prod_{1 \leq j \leq r} [D_j(a)]^{\mu_j}. \tag{6.7}$$

The linear independence of the system of fundamental highest roots allows one to select the powers μ_j of the principal minors $D_j(a)$ so as to get the highest matrix element of an arbitrary irreducible representation of G with highest weight $\{L\}$ from the condition $\sum_i \mu_i \lambda_j^{(i)} = L_j$.

1.6.3 A formal expression for the highest matrix elements of the fundamental representations. There is another more direct way to write an explicit formula for the highest matrix elements of the fundamental representations, which is applicable also to infinite-dimensional Lie algebras. To describe it introduce the "generators of highest vectors" $R_j^\pm$ of the fundamental representations of a simple Lie algebra. Let them satisfy the following conditions:

$$[X_{+i}, R_j^+] = 0, \quad [X_{-i}, R_j^-] = 0, \quad [h_j, R_j^\pm] = \pm \delta_{ij} R_j^\pm \tag{6.8}$$

and extend the Killing form so that

$$(R_i^-, R_j^+) = \delta_{ij}. \tag{6.9}$$

Then, as follows directly from (6.8) with the help of (2.8) for h_j and $X_{\pm j}$, $1 \leq j \leq r$, the bilinear form $(R_j^-, g R_j^+ g^{-1})$ with $g \in G$ satisfies (6.2). With the help of (4.19) this defines the matrix element

$$\xi^{\{j\}}(g) = tr(R_j^- g R_j^+ g^{-1}). \tag{6.10}$$

The latter expression is most convenient in applications if g is represented in Gauss decomposition (5.6) since then we have two equivalent expressions

$$\xi^{\{j\}}(g) = tr(z_+^{-1} R_j^- z_+ e^H z_- R_j^+ z_-^{-1} e^{-H})$$

$$= tr(z_-'^{-1} R_j^- z_-' e^{H'} z_+' R_j^+ z_+'^{-1} e^{-H'}) \tag{6.11}$$

$$= tr(R_j^- e^{H'} R_j^+ e^{-H'}) = e^{r_j},$$

where $e^H = a$, $e^{H'} = a'$, $H = \sum_j \tau_j h_j$, $H' = \sum_j \tau'_j h_j$. This expression can be rewritten as a series in the scalar products of vectors

$$\langle [X_{+j_k}[X_{+j_{k-1}}[\dots[X_{+j_1} R^-_{j_1}]\dots]]] \| [X_{-i_k}[X_{-i_{k-1}}[\dots[X_{-i_1} R^+_{i_1}]\dots]]]\rangle \quad (6.12)$$

with $i_1 = j_1 = j$ and expressed with the help of the Jacobi identity and relations (2.8) and (6.9). This series turns into a polynomial for finite-dimensional simple Lie algebras. The problem of its absolute convergence for infinite-dimensional Lie algebras will be considered in what follows for the cases associated with integrable systems.

1.6.4 Recurrence relations for the highest matrix elements of the fundamental representations.

The highest matrix elements of fundamental representations are connected by recursions which enable us to express them all in terms of the highest matrix element of the first fundamental representation corresponding to π_1. To do this consider the determinant

$$\Delta_i = \begin{vmatrix} \xi^{\{i\}} & X_i^L \xi^{\{i\}} \\ X_{-i}^R \xi^{\{i\}} & X_i^L X_{-i}^R \xi^{\{i\}} \end{vmatrix} \equiv \begin{vmatrix} \langle i|g|i \rangle & \langle i|X_i g|i \rangle \\ \langle i|g X_{-i}|i \rangle & \langle i|X_i g X_{-i}|i \rangle \end{vmatrix}. \quad (6.13)$$

It follows from (6.13), (2.8) and (4.19) and the fact that infinitesimal operators of left and right shifts commute that

$$X_{-j}^L \Delta_i = X_j^R \Delta_i = 0, \ h_j^R \Delta_i = -h_j^L \Delta_i = (2\delta_{ij} - k_{ij})\Delta_i, \ 1 \le i, j \le r, \quad (6.14)$$

which implies

$$\Delta_i = \hat{c} \prod_{j \ne i} [\xi^{\{j\}}(g)]^{-k_{ij}}. \quad (6.15)$$

Since, obviously, $\xi^{\{i\}}(e) \equiv \langle i|e|i \rangle \equiv 1$, where $e \in G$ is the unit element, it is easy to deduce that $\hat{c} = 1$. Thus we get

$$\Delta_i \equiv \det \begin{vmatrix} \xi^{\{i\}}(g) & X_i^L \xi^{\{i\}}(g) \\ X_{-i}^R \xi^{\{i\}}(g) & X_i^L X_{-i}^R \xi^{\{i\}}(g) \end{vmatrix} = \prod_{\substack{1 \le j \le r \\ j \ne i}} [\xi^{\{j\}}(g)]^{-k_{ij}}, \quad (6.16)$$

which is the desired recursion.

1.6.5 The highest matrix elements of irreducible representations expressed via generalized Euler angles.

The expressions for highest matrix elements given in the above subsections hold for any decomposition of the group. However, for the universal parametrization we can get the formulas factored with respect to parameters.

Consider a compact Lie group $\mathcal{K}$ and parametrize its elements as in (5.16). Then in the Gauss decomposition $k = z_- \cdot a \cdot z_+$ of the complex span of $\mathcal{K}$

an element a from its maximal abelian subgroup factors with respect to the parameters φ_α, θ_α, Ψ_j and takes the form

$$a = \prod_{\alpha>0} \exp\left\{i\frac{h_\alpha}{(\alpha\alpha)}\left[\Phi_\alpha - 2i\log\cos\frac{\theta_\alpha}{2}\right]\right\} \cdot \prod_{1\leq j\leq r} \exp\left(i\Psi_j\frac{h_j}{2}\right). \quad (6.17)$$

Proof of this statement follows directly from the Σ^+-ordering property mentioned in 2.5 and formula (5.17). Parametrizing the matrix $a_{\alpha\beta} = tr(X_{-\alpha}kX_\beta k^{-1})$ for the adjoint representation with the help of such a decomposition for k we get, since the $z_\pm$ are nilpotent,

$$D_i(a) = \prod_{\alpha>0} \exp\left[i\Phi_\alpha\frac{(\alpha\lambda^{(i)})}{(\alpha\alpha)}\right] \cdot \left(\cos\frac{\theta_\alpha}{2}\right)^{2\frac{(\alpha\lambda^{(i)})}{(\alpha\alpha)}} \cdot \prod_{1\leq j\leq r} \exp\left(i\Psi_j\frac{\lambda_j^{(i)}}{2}\right). \quad (6.18)$$

The final expression for the highest matrix element of the irreducible representation of K with highest weight L obtained by inserting (6.18) into (6.7) takes the form

$$\xi^{\{L\}}(k) = \prod_{\alpha>0} \exp\left[i\frac{(\alpha L)}{(\alpha\alpha)}\Phi_\alpha\right]\left(\cos\frac{\theta_\alpha}{2}\right)^{2\frac{(\alpha L)}{(\alpha\alpha)}} \cdot \prod_{1\leq j\leq r} \exp\left(i\Psi_j\frac{L_j}{2}\right). \quad (6.19)$$

Similarly, we derive the formula for the normal real form of a complex simple Lie group in the parametrization (5.19):

$$\xi^{\{L\}}(n) = \prod_{\alpha>0} (\sin\theta_\alpha)^{2\frac{(\alpha L)}{(\alpha\alpha)}}. \quad (6.20)$$

Therefore the universal parametrization based on $\sum^+$-ordering of roots of simple Lie algebras leads to expressions for the highest matrix elements of irreducible representations which are factored with respect to generalied Euler angles. These expressions enable us to simplify considerably a number of important deductions in representation theory. For example, the formula for the dimension N_L of an irreducible representation of an arbitrary compact group with highest weight L is inversely proportional to the norming constant of its highest matrix element

$$N_L = \left[\int d\mu(k)|\xi^{\{L\}}(k)|^2\right]^{-1}. \quad (6.21)$$

Therefore by normalizing the invariant measure (5.18) so that $\int d\mu(k) = 1$ and with (6.20) we immediately get by integration the known Weyl formula:

$$N_L = \prod_{\alpha>0} \frac{(\alpha, 2L+\rho_0)}{(\alpha, \rho_0)}. \quad (6.22)$$

§ 1.7 Superalgebras and superspaces *)

Since the concepts of a supermanifold, a Lie subgroup and a differential equation on a supermanifold are rather new in Mathematics and for completeness sake let us recall some basics, following mainly Ref. [80], [81]; there is no text book yet.

1.7.1 Superspaces. A space V (over a field k) endowed with a $\mathbb{Z}/2$-grading, i.e., a decomposition $V = V_{\bar 0} \oplus V_{\bar 1}$, is called a *superspace* and the $\mathbb{Z}/2$-grading is called *parity* and is denoted by p. (We bar the elements of $\mathbb{Z}/2$ to distinguish them from the elements of $\mathbb{Z}$.) The non-zero elements of $V_{\bar 0}$ and $V_{\bar 1}$ are called *homogeneous* (*even* and *odd*, respectively) elements of V and we write $p(v) = i$, $i \in \{\bar 0, \bar 1\}$, if and only if $v \neq 0$ and $v \in V_i$. A *subsuperspace* is a $\mathbb{Z}/2$-graded subspace W of the superspace V such that $W_i = W \cap V_i$.

Let V and W be superspaces. The superspace structure in the spaces $V \oplus W$, $V \otimes W$ and $\mathrm{Hom}(V,W)$ is naturally introduced, e.g. $(V \otimes W)_i = \oplus_{p+q=i} V_p \otimes W_q$, etc. The even homomorphisms of superspaces are called *morphisms*. Put $\pi(V)$ for the superspace defined by the formula $(\pi(V))_i = V_{i+\bar 1}$; its elements will be denoted by $\pi(v)$, where $v \in V$.

A *superalgebra* is a superspace A with a morphism $m\colon A \otimes A \to A$. An algebra homomorphism $\varphi\colon A \to B$, where A and B are superalgebras, is called a *superalgebra* homomorphism if $p(\varphi) = \bar 0$.

Conventions.

1) Sign Rule: We put $(-1)^{\bar 0} = 1$, $(-1)^{\bar 1} = -1$; the formulas which at first glance are defined only on homogeneous elements are actually defined everywhere by linearity; if something of parity p moves past something of parity q the sign $(-1)^{pq}$ accrues.

2) In what follows, classifying bilinear forms, etc., we will confine ourselves to homogeneous objects since the study of nonhomogeneous objects takes us beyond the limits of the investigated part of science.

The first convention makes it possible to superize without hesitation notions like commutator, Leibniz rule, Lie algebra, (co)homology, etc., e.g. the supercommutator is the map

$$[\cdot,\cdot]\colon \ a, b \mapsto ab - (-1)^{p(a)p(b)}ba.$$

Usually the term *superalgebra* is used for an associative superalgebra with unit. Let us give several examples of associative superalgebras. The superalgebra A is called *supercommutative* if $[a,b] = 0$ for any $a, b \in A$. An example

---

*) [8, 48, 49, 80, 81, 85–87, 163, 164]

of a commutative superalgebra is the *Grassmann superalgebra* $\wedge_C(n)$ in indeterminates $\xi_1, \ldots, \xi_n$, where $p(\xi_i) = \bar{1}$ for $1 \leq i \leq n$, over a commutative algebra C (we assume $p(c) = 0$ for $c \in C$).

Another important example: setting $p(\sqrt{-1}) = \bar{1}$ we make $\mathbb{C}$ into a *non-supercommutative* superalgebra over $\mathbb{R}$ denoted by $\mathbb{C}^s$.

The *tensor algebra* $T(V)$ of the superspace V is naturally defined, $T(V) = \oplus_{n \geq 0} T^n(V)$, where $T^0(V) = k$ and $T^n(V) = V \otimes \ldots \otimes V$ (n factors) for $n > 0$. The *symmetric algebra* of the superspace V is $S(V) = T(V)/I$, where I is the two-sided ideal generated by $v_1 \otimes v_2 - (-1)^{p(v_1)p(v_2)} v_2 \otimes v_1$ for $v_1, v_2 \in V$. The *exterior algebra* of the superspace V is $E(V) = S(\pi(V))$. Evidently, both the exterior and symmetric algebras of the superspace V are commutative superalgebras. It is worthwhile to mention that if $V_{\bar{0}} \neq 0$, $V_{\bar{1}} \neq 0$ then both $E(V)$ and $S(V)$ are infinite-dimensional.

A *Lie superalgebra* is a superalgebra $\mathfrak{g}$ (defined over a field or, more generally, a supercommutative superalgebra k) with a multiplication called *bracket* and usually denoted by $[.\,,.]$ or $\{.,.\}$ which satisfies the following conditions: $[X, X] = 0$ and $[Y, [Y, Y]] = 0$ for any $X \in (C \otimes \mathfrak{g})_{\bar{0}}$ and $Y \in (C \otimes \mathfrak{g})_{\bar{1}}$ and any supercommutative superalgebra C (we assume that the bracket in $C \otimes \mathfrak{g}$ is defined via the Sign Rule).

The condition on the bracket may be rewritten in an equivalent and more familiar form via the Sign Rule (as superskewcommutativity and the super Jacobi identity).

From an associative (super)algebra A construct a new (super)algebra A_L with the same (super)space and the multiplication $(a, b) \mapsto [a, b]$. Another method of getting a Lie superalgebra from an associative superalgebra A is to consider $\mathfrak{der}A$, the derivation algebra of A, defined via the Sign Rule.

From a Lie superalgebra $\mathfrak{g}$ we construct the associative superalgebra $U(\mathfrak{g}) = T(\mathfrak{g})/I$, where I is the two-sided ideal generated by the elements $x \otimes y - (-1)^{p(x)p(y)} y \otimes x - [x, y]$ for $x, y \in \mathfrak{g}$, called the *universal enveloping algebra* of the Lie superalgebra $\mathfrak{g}$. The *Poincaré-Birkhoff-Witt theorem* extends to Lie superalgebras with the same proof (beware the Sign Rule) and reads as follows:

if $\{X_i\}$ is a basis in $\mathfrak{g}_{\bar{0}}$ and $\{Y_j\}$ is a basis in $\mathfrak{g}_{\bar{1}}$ then the monomials $X_{i_1}^{n_1} \ldots X_{i_r}^{n_r} Y_{j_1}^{\epsilon_1} \ldots Y_{j_s}^{\epsilon_s}$, where $n_i \in \mathbb{Z}^+$ and $\epsilon_j = 0$ or 1, constitute a basis in $U(\mathfrak{g})$.

A superspace M is called a *left module* over a superalgebra A (or a *left A-module*) if there is given an even map $act: A \otimes M \to M$ such that $(ab)m = a(bm)$ and $1m = m$ if A is an associative superalgebra with unit (or $[a, b]m = a(bm) - (-1)^{p(a)p(b)}b(am)$ if A is a Lie superalgebra), where $a, b \in A$ and $m \in M$. The definition of a *right A-module* is similar. A module M over a supercommutative superalgebra C is supposed to be two-sided and the left module structure is obtained from the right one and vice versa according to the formula $cm = (-1)^{p(m)p(c)}mc$, where $m \in M$, $c \in C$: such modules will be called *C-modules*. There are two ways to apply the functor π to C-modules: $\pi(M)$ and $(M)\pi$, so to speak. The two-sided module structures on $\pi(m)$ and $(M)\pi$ are given via the Sign Rule. (Actually, there are *two* canonical ways to do this, see [80], [81]; the meaning of such an abundance is obscure.)

Sometimes instead of the map *act* a morphism ρ: $A \rightarrow \text{End}M$ is defined if A is an associative superalgebra (or ρ: $A \rightarrow (\text{End}M)_L$ if A is a Lie superalgebra); ρ is called a *representation* of A in M.

The simplest modules (in a sense) are those which are *irreducible* of general type (or *irreducible of G-type*); these do not contain invariant subspaces other than 0 and the whole module; and their "odd" counterparts, *irreducible modules of Q-type*, which do contain an invariant subspace that, however, is not a subsuperspace. Consequently, *Schur's lemma* states that over $\mathbb{C}$ the centralizer of a set of irreducible operators is either $\mathbb{C}$ or $\mathbb{C} \otimes \mathbb{C}^s = Q(1; \mathbb{C})$ (see the definition of the superalgebras Q below).

The next in terms of complexity are *indecomposable* modules, which cannot be represented as direct sum of invariant submodules.

A *C*-module is called *free* if it is isomorphic to a module of the form $C \oplus \ldots \oplus C \oplus \pi(C) \oplus \ldots \oplus \pi(C)$ (C occurs r times, $\pi(C)$ occurs s times). The *rank* of a free *C*-module M is the element $rkM = r + s\varepsilon$ from the ring $\mathbb{Z}[\varepsilon]/(\varepsilon^2 - 1)$ (over a field we usually write just dim $M = (r, s)$ or $r|s$ and call this number the *dimension* of M).

The module $M^* = Hom_C(M, C)$ is called *dual* to a *C*-module M. If $(. , .)$ is the pairing of modules M^* and M then to each operator $F \in Hom_C(M, N)$, where M and N are *C*-modules, there corresponds the dual operator $F^* \in Hom_C(N^*, M^*)$ defined by the formula

$$(F(m), n^*) = (-1)^{p(F)p(m)}(M, F^*(n^*)) \text{ for any } m \in M, n^* \in N^*.$$

A *supermatrix* is a matrix with entries from a *superspace* and a parity assigned to each row and column. Usually, the even rows and columns are written first followed by the odd ones giving rise to a block expression of matrices in the form $\begin{pmatrix} A & B \\ C & D \end{pmatrix}$; the elements of the matrices usually belong to a supercommutative superalgebra. Such an expression for matrices is called the *standard format*.

Write $\text{Mat}(n|m; C)$ for the set of all $(n, m) \times (n, m)$ matrices in the standard format with entries from a supercommutative superalgebra C.

Fix a standard (the even vectors first) basis $\{m_i\}$ of a free module M of rank $n|m$ over a supercommutative superalgebra C. To each operator $F \in \text{End}_C M$ assign the matrix $^mF = (F_{ji})$, where $Fm_i = \Sigma m_j F_{ji}$. Thus we obtain a one-to-one correspondence between $\text{End}_C(M)$ and $\text{Mat}(n|m; C)$.

Now it is evident that the space $\text{Mat}(n|m; C)$ is endowed with the natural superspace structure.

(The parity of $\begin{pmatrix} A & B \\ C & D \end{pmatrix}$ equals $\bar{0}$ (resp. $\bar{1}$) if and only if $p(A_{ij}) = p(D_{rs}) = \bar{0}$ (resp. $\bar{1}$), $p(B_{is}) = p(C_{rj}) = \bar{1}$ (resp. $\bar{0}$) and with an associative superalgebra structure.)

Put

$$Q(n; C) = \{X \in \text{Mat}(n|n; C)] : \left[X, \begin{pmatrix} 0 & 1_n \\ -1_n & 0 \end{pmatrix} \right] = 0\}.$$

The Lie superalgebras of series Q_L, or rather their queer traceless projectivizations, discovered by Gell-Mann, Mitchell and Radicatti, are examples of one of what V. Kac dubbed "strange" series, Pe and Q, which stand, perhaps, for *queer* and *peculiar*. This one is a "queer" analogue of the matrix algebra $\text{Mat}(n;k)$. The elements of $Q(n; C)$ preserve the complex structure given by an odd operator, cf. the definition of C^S and Schur's lemma.

Analogues of the trace tr : $\mathfrak{gl}(n) \to \mathfrak{gl}(1)$ are the *supertrace* str : $\mathfrak{gl}(n|m) \to \mathfrak{gl}(1|0) = \mathfrak{gl}(1)$ and the *queertrace* qtr : $\mathfrak{q}(n) \to \mathfrak{q}(1)$, where $\mathfrak{gl} = Mat_L$ and $\mathfrak{q} = Q_L$, defined by the formulas, cf. [80], [81]:

$$str \begin{pmatrix} A & B \\ D & E \end{pmatrix} = tr\ A - (-1)^{p(X)} tr\ E \text{ for } X = \begin{pmatrix} A & B \\ D & E \end{pmatrix};$$

$$qtr \begin{pmatrix} A & B \\ B & A \end{pmatrix} = (tr\ B) \begin{pmatrix} 0 & 1 \\ 1 & 0 \end{pmatrix}.$$

Both *str* and *qtr* are Lie superalgebra morphisms.

The even invertible elements from $\text{Mat}(n|m; C)$ constitute the *general linear group* $GL(n|m; C)$. Put $GQ(n; C) = Q(n; C) \cap GL(n|n; C)$.

On the group $GL(n|m; C)$ an analogue of the determinant is defined; it is called the *Berezinian* (in honour of F.A. Berezin who discovered it):

$$\text{Ber} \begin{pmatrix} A & B \\ D & E \end{pmatrix} = det(A - BE^{-1}D)det E^{-1}.$$

For the matrices from $GL(n|m; C)$ the identity $Ber\ XY = Ber\ X\ Ber\ Y$ holds, i.e., Ber: $GL(n|m; C) \to GL(1|0; C)$ is a group homomorphism.

As is well known, the determinant is connected with the trace by the formula $det X = \exp tr \log X$ that holds when both parts of the formula are defined. We also have $Ber\ X = \exp str \log X$ whenever the right hand side is defined. (This formula extends the domain of Ber onto nonhomogeneous matrices).

On the group $GQ(n; C)$ the Berezinian is identically equal to 1. However, on this "queer" analogue of GL the "queer" determinant is defined by the formula

$$\text{qet} \begin{pmatrix} A & B \\ B & A \end{pmatrix} = \exp \text{qtr} \log \begin{pmatrix} 1_n & A^{-1}B \\ A^{-1}B & 1_n \end{pmatrix}$$

satisfying $\text{qet} XY = \text{qet} X \cdot \text{qet} Y$ for $X, Y \in GQ(n; C)$. Put $SL(n|m; C) = \{X \in GL(n|m; C): \text{Ber} X = 1\}$, $SQ(n; C) = \{X \in GQ(n; C): \text{qet} X = 1\}$. These are *special linear groups*, and, as we will see, the functors $C \mapsto GL(n|m; C)$, $C \mapsto GQ(n; C)$, etc. are represented by supergroups.

A *bilinear form* is a map $B : M \times N \to C$, additive in each variable, such that $B(mc, n) = B(m, cn)$, $B(m, nc) = B(m, n)c$, where $m \in M$, $n \in N$, $c \in C$ and $p(B) = p(B(m, n) - p(m) - p(n))$. The superspace of bilinear forms is denoted by $Bil_C(M, N)$ or $Bil_C(M)$ if $M = N$. The *upsetting of forms* $uf: Bil_C(M, N) \to Bil_C(N, M)$ is defined by the formula

$$B^{uf}(n, m) = (-1)^{p(n)\,(p(B)+p(m))+p(B)p(m)} B(m, n).$$

(The adequate way to express bilinear forms in supercase would be $\langle m|B|n\rangle$.) A form $B \in Bil_C(M)$ is called *(skew) symmetric* if $B^{uf} = (-)B$.

Given bases $\{m_i\}$ and $\{n_j\}$ of C-modules M and N and a bilinear form $B : M \otimes N \to C$, we assign to B the matrix

$$(^{mf}B)_{ij} = (-1)^{p(m_i)p(B)} B(m_i, n_j).$$

If $X \in GL_C(M)$, $Y \in GL_C(N)$ then, with respect to the bases $\{Xm_i\}$ and $\{Yn_j\}$, an operator $F : M \to N$ and a bilinear form from B are given by the matrices

$$(^{m}F)' = (^{m}X)^{-1}(^{m}F)(^{m}Y), \quad (^{mf}B)' = (^{m}X)^{st}(^{mf}B)(^{m}Y),$$

where ^{m}X and ^{m}Y are the matrices of the operators X and Y with respect to the bases $\{m_i\}$ and $\{n_j\}$ (similarly for ^{m}F and ^{n}B) and where st denotes the *supertransposition* defined by the formula

$$X^{st} = \begin{pmatrix} A & B \\ C & D \end{pmatrix}^{st} = \begin{cases} \begin{pmatrix} A^t & C^t \\ -B^t & D^t \end{pmatrix} & \text{if } p(X) = \bar{0}; \\ \begin{pmatrix} A^t & -C^t \\ B^t & D^t \end{pmatrix} & \text{if } p(X) = \bar{1}. \end{cases}$$

Note that the order of the supertranspositon is equal to 4.

A *non-degenerate homogeneous symmetric bilinear form B over C can be reduced to the canonical form with the matrix* $diag(1_n, J_{2m})$ *or, if we wish to consider a split form (for which the maximal torus in the Lie superalgebra that preserves B is situated on the main diagonal), to the form*

$$diag(ant(1_n), J_{2m}) \text{ if } p(B) = \bar{0}$$

or to the form
$$J_{2n} \text{ if } p(B) = \bar{1}.$$

Similarly, *the skewsymmetric bilinear form can be reduced to canonical form with the matrix*
$$diag(J_{2m}, 1_n) \text{ or } diag(J_{2m}, \sigma_n) \text{ if } p(B) = \bar{0}$$

or to the form
$$P_{2n} \text{ if } p(B) = \bar{1}.$$

The *orthosymplectic* group is the group $Osp(n|2m; C)$ of automorphisms of the bilinear form with the even canonical matrix; the *peculiar* or, as A. Weil suggested, *periplectic* group $Pe(n; C)$ is the group of automorphisms of the odd canonical form. The *special peculiar (periplectic)* group is $SPe(n; C) = Pe(n; C) \cap SL(n|n; C)$. (The square root of the Berezinian, $Ser(X) = \sqrt{Ber(X)}$ is multiplicative on $SPe(n)$, see [81].)

A *real (quaternionic) structure* on a superalgebra C over $\mathbb{C}$ is an antilinear automorphism $cj \in Aut_{\mathbb{R}}(C)$ ("conjugation") such that $cj^2 = id$ (resp. $cj^4 = id$, $cj^2|C_{\bar{0}} = id$). An extension of a real (quaternionic) structure (from C) onto a C-module M, real and imaginary parts of an algebra, module and of their elements are naturally defined.

Given a real superspace, there are two ways to *complexify* it: both of them are obtained by tensoring with $\mathbb{C}$; however, $\mathbb{C}$ may be considered as a purely even superalgebra or as a *nonsupercommutative* superalgebra $\mathbb{C}^s$.

All notions of linear algebra have a straightforward generalization to the sesquilinear case, e.g., a bilinear form B on C-modules M and N is called *hermitian* if for $m \in M$, $n \in N$ and $c \in C$ we have
$$\langle cm|B|n \rangle = cj(c)\langle m|B|n \rangle, \quad \langle m|B|nc \rangle = \langle m|B|n \rangle c$$

or $\langle J_m m|B|J_N n \rangle = \langle m|B|n \rangle$ for operators J_M and J_N (of the same parity). This form defines complex structures in the real superspaces M, N. Denote by $Herm_C(M, N)$ or $Herm_C(M)$ for $M = N$ the superspace of hermitian forms.

Under a change of basis in the modules M and N with matrices (with respect to the old bases) X and Y respectively, hermitian forms are transformed according to the formula $({}^{mf}h)' = cj(X^{st})^{mf} \cdot h \cdot Y$.

A nondegenerate super(skew)symmetric hermitian form h over $\mathbb{C}$ can be reduced to the form with the matrix $diag(I_n^r, iI_m^s)$ (resp. $diag(iI_n^r, I_m^s)$) for $p(h) = \bar{0}$ or iP_{2n} (resp. P_{2n}) for $p(h) = \bar{1}$, where $I_n^r = diag(1_{n-r}, -1_r)$.

The *pseudosuperunitary* group is the group $U(n, r|m, s; C)$ of automorphisms of the (pseudo) hermitian form with the matrix $diag(I_n^r, iI_m^s)$. The group $U(n, 0|m, 0; C)$ is denoted by $U(n|m; C)$ and is called the *superunitary* group. (Do not confuse the superunitary group with the pseudounitary group $U(n, m)$ of automorphisms of the form with the matrix I_{n+m}^n on the $(n + m)$-dimensional *space*, not superspace!)

Important (pseudo)superunitary subgroups are the *special* and the *queer* ones:

$$SU(n, r|m, s) = \{X \in U(n, r|m, s) : BerX = 1\},$$

$$QU(n, r) = U(n, r|n, r) \cap Q(n),$$

$$SQU(n, r) = U(n, r|n, r) \cap SQ(n).$$

The *odd* version of the unitary group is the *peculiar (periplectic) unitary group* $PeU(n; C)$ of automorphisms of the hermitian form with either of the matrices iP_{2n} or P_{2n}, and $SPeU(n) = PeU(n) \cap SL(n|n)$.

If $G(C) \subset GL(n|m; C)$ is a subgroup containing the subgroup $Scal(C)$ of scalar matrices, then write $PG(C)$ for the group $G(C)/Scal(C)$. It is called the *projective group of type G* (unitary, special, etc.), e.g. $PGL(n|n)$, $PSQ(n)$.

1.7.2. Classical Lie superalgebras. The purpose of this section is to sketch the list of the known simple $\mathbb{Z}$-graded Lie superalgebras of finite growth over the field $\mathbb{C}$ of complex numbers and compare the list with the similar one we gave in 1.2.3 for selfsymmetric Lie algebras. In order to better describe some of these algebras we will also describe some of the nonsimple but close to simple Lie superalgebras and also a class of nonselfsymmetric Lie superalgebras: those of vector fields with polynomial coefficients.[1]) The most nontrivial phenomena here are: the possibility to realize a vectory Lie superalgebra as a Lie superalgebra of vector fields on *different* supermanifolds and the abundance of deformations that intertwine all classes of Lie superalgebras. For details, see [80], [81].

When dealing with superalgebras it sometimes becomes useful to know their definition. The naive one, *insufficient* to apply the method of this book, is obtained from the definition of a Lie algebra via the Sign Rule. In this section we will stick to it.

$\mathbb{C}$*-points of simple finite-dimensional Lie superalgebras* are classified by Kac [48]. *Vectory superalgebras* are not completely described. However, a conjecturially complete list is given in [81]. *Stringy superalgebras* constitute an important particular case of Lie algebras of vector fields and generalize $\mathfrak{vect}^L(1)$. Over $\mathbb{C}$, these algebras were classified modulo deformations, see [48] and the references therein. *Loop superalgebras* (corresponding to outer automorphisms of simple finite-dimensional Lie superalgebras) are listed in [163], [164]. It is interesting that in order to get a selfsymmetric loop superalgebra one must sometimes start with a nonselfsymmetric superalgebra.

---

[1]) We will call Lie (super)algebras of vector fields "vectory" for short and, moreover, unless otherwise stated, assume that their coefficients are polynomials.

Over $\mathbb{R}$ a correct description of the real structures of all known simple $\mathbb{Z}$-graded Lie superalgebras of finite growth both finite- and infinite-dimensional is obtained in [163].[2]).

First, let us give examples of matrix Lie superalgebras. Recall that $\mathfrak{gl}(n|m) = Mat(n|m)_L$ and $\mathfrak{sl}(n|m) = \{X \in \mathfrak{gl}(n|m): str X = 0\}$ are called the *general linear* and *special linear* Lie superalgebras: $\mathfrak{q}(n) = Q(n)_L$ and $\mathfrak{sq}(n) = \{X \in \mathfrak{q}(n);\ qtr X = 0\}$ are called the *queer* and *special queer* Lie superalgebras.

If $\mathfrak{s}$ is Lie algebra of scalar matrices and $\mathfrak{g} \subset \mathfrak{gl}(n|m)$ is a Lie subsuperalgebra containing $\mathfrak{s}$, then the *projective* Lie superalgebra of type $\mathfrak{g}$ is $\mathfrak{pg} = \mathfrak{g}/\mathfrak{s}$. Projectivization sometimes leads to new Lie superalgebras: $\mathfrak{pgl}(n|n)$, $\mathfrak{pq}(n)$, $\mathfrak{psl}(n)$, $\mathfrak{psq}(n)$. The minimal dimension of the superspace of an irreducible linear realization of these superalgebras is attained for the adjoint representation.

Let $B_{n,2m} = diag(1_n, J_{2m})$ (or $diag(ant(1_n), J_{2m})$). The Lie superalgebra

$$\mathfrak{osp}(n|2m) = \{X \in \mathfrak{gl}(n|2m) :\ X^{st}B_{n,2m} + B_{n,2m}X = 0\}$$

is called the *orthosymplectic* Lie superalgebra. The Lie superalgebra

$$\mathfrak{pe}(n) = \{X \in \mathfrak{gl}(n|n) :\ X^{st}J_{2n} + (-1)^{p(x)}J_{2n}X = 0\}$$

is called the *peculiar* or *periplectic* Lie superalgebra and $\mathfrak{spe}(n) = \mathfrak{pe}(n) \cap \mathfrak{sl}(n|n)$ is its special subsuperalgebra.

Now let us give examples of classical Lie superalgebras of *formal vector fields* (*vectory* superalgebras). First, note that in Lie algebras of (formal) vector fields there is a natural filtration of the form

$$\mathcal{L} = \mathcal{L}_{-d} \supset \mathcal{L}_{-1} \supset \mathcal{L}_0 \supset \mathcal{L}_1 \supset \cdots,$$

where $\mathcal{L}_0$ is the unique maximal subalgebra of finite codimension. The Lie algebra $\mathcal{L}$ itself is realized by vector fields on $(\mathcal{L}/\mathcal{L}_0)^*$ preserving certain structure. It is a wonderful and rather unexpected fact that simple vectory superalgebras possess *several* maximal subalgebras of finite codimension.

Standard realizations. The Lie superalgebra $\mathfrak{vect}(n|m) = \mathfrak{der}\mathbb{C}[[x]]$, where $x = (u_1, \ldots, u_n, \xi_1, \ldots, \xi_m)$, is called the *general vectory* superalgebra, while

$$\mathfrak{sve}(n|m) = \{\mathcal{D} \in \mathfrak{vect}(n|m) :\ L_{\mathcal{D}}v_x = 0\} = \{\mathcal{D} \in \mathfrak{vect}(n|m) :\ div\mathcal{D} = 0\},$$

where the *divergence* of a field $\mathcal{D} = \Sigma f_i \frac{\partial}{\partial u_i} + \Sigma g_j \frac{\partial}{\partial \xi_j}$ is the series

$$div\mathcal{D} = \Sigma \frac{\partial f_i}{\partial u_i} + \Sigma (-1)^{p(g_j)} \frac{\partial g_j}{\partial \xi_j},$$

---

[2]) Notice that for superalgebras the notion of a real structure is more sophisticated than in nonsuper case: there are real and quaternionic forms and several notions of unitarity.

is called the *special* or *divergence-free vectory* superalgebra.

Let

$$\alpha_1 = dt + \sum_{1 \le i \le n} (p_i dq_i - q_i dp_i) + \sum_{1 \le j \le s} \xi_j d\xi_j, \quad \omega_0 = d\alpha_1.$$

Sometimes instead of ω_0 and α_1 it is more convenient to take $\tilde{\omega}_0 = d\tilde{\alpha}_1$, where

$$\tilde{\alpha}_1 = dt + \sum_{1 \le i \le n} (p_i dq_i - q_i dp_i) + \begin{cases} \sum_{1 \le j \le r} (\xi_j d\eta_j + \eta_j d\xi_j) & \text{for } s = 2r; \\ \sum_{1 \le j \le r} (\xi_j d\eta_j + \eta_j d\xi_j) + \theta d\theta & \text{for } s = 2r + 1. \end{cases}$$

Further, let

$$\alpha_0 = d\tau + \sum_{1 \le i \le n} (q_i d\xi_i + \xi_i dq_i), \quad \omega_1 = d\alpha_0.$$

The Lie superalgebra

$$\mathfrak{k}(2n + 1|m) = \{ \mathcal{D} \in \mathfrak{vect}(2n + 1|m) : L_{\mathcal{D}}\alpha_1 = f_{\mathcal{D}} \cdot \alpha_1 \}$$

is called the *contact* Lie superalgebra. The Lie superalgebra

$$\mathfrak{m}(n) = \{ \mathcal{D} \in \mathfrak{vect}(n|n + 1) : L_{\mathcal{D}}\alpha_0 = f_{\mathcal{D}} \cdot \alpha_0 \}$$

is called the *odd contact* Lie superalgebra. The Lie superalgebra

$$\mathfrak{po}(2n|m) = \{ \mathcal{D} \in \mathfrak{k}(2n + 1|m); \ L_{\mathcal{D}}\alpha_1 = 0 \}$$

is called the *Poisson* superalgebra.

$$\mathfrak{b}(n) = \{ \mathcal{D} \in \mathfrak{m}(n) : L_{\mathcal{D}}\alpha_0 = 0 \}$$

is the *Buttin* superalgebra. The Lie superalgebras

$$\mathfrak{sm}(n) = \mathfrak{m}(n) \cap \mathfrak{sve}(n|n + 1), \quad \mathfrak{sb}(n) = \mathfrak{b}(n) \cap \mathfrak{sve}(n|n + 1)$$

are called *special odd contact* and *special Buttin* superalgebras.

It is more convenient to define contact, Poisson and Buttin superalgebras in terms of generating functions. For the series $\mathfrak{k}(2n + 1|m)$ in the realization with the form $\tilde{\alpha}_1$ set

$$K_f = 2\Delta(f)\frac{\partial}{\partial t} + H_f + \frac{\partial}{\partial t}E,$$

where

$$H_f = \sum \left(\frac{\partial f}{\partial q_i} \frac{\partial}{\partial p_i} - \frac{\partial f}{\partial p_i} \frac{\partial}{\partial q_i} \right) - (-1)^{p(f)} \sum \frac{\partial f}{\partial \xi_j} \frac{\partial}{\partial \xi_j},$$

$\Delta(f) = 2f - E(f)$ and $E = \Sigma y_i \frac{\partial}{\partial y_i}$, and where the y are all the coordinates except t for the realization of the series $\mathfrak{k}$ with the form α_1 (for the form $\tilde{\alpha}_1$ the required modifications are evident).

Similarly, set

$$M_f = \Delta(f)\frac{\partial}{\partial \tau} + Le_f + (-1)^{p(f)}\frac{\partial f}{\partial \tau}E,$$

where

$$Le_f = \sum \left(\frac{\partial f}{\partial q_i}\frac{\partial}{\partial \xi_i} + (-1)^{p(f)}\frac{\partial f}{\partial \xi_i}\frac{\partial}{\partial q_i} \right),$$

where the y are all the coordinates except τ for the series $\mathfrak{m}$. As is not difficult to verify, the contact brackets of the generating functions correspond to the supercommutator of vector fields:

$$\{f,g\}_{k.b} = \Delta(f)\frac{\partial g}{\partial t} - \frac{\partial f}{\partial t}\Delta(g) - \{f,g\}_{P.b.}$$

and

$$\{f,g\}_{m.b.} = \Delta(f)\frac{\partial g}{\partial \tau} - (-1)^{p(f)}\frac{\partial f}{\partial \tau}\Delta(g) - \{f,g\}_{B.b.}$$

where the Poisson and Buttin (a.k.a. Schouten) brackets are defined by the formulas

$$\{f,g\}_{P.b.} = \sum \left(\frac{\partial f}{\partial q_i}\frac{\partial g}{\partial p_i} - \frac{\partial f}{\partial p_i}\frac{\partial g}{\partial q_i} \right) - (-1)^{p(f)}\sum \frac{\partial f}{\partial \xi_j}\frac{\partial g}{\partial \xi_j}$$

and

$$\{f,g\}_{B.b.} = \sum \left(\frac{\partial f}{\partial q_i}\frac{\partial g}{\partial \xi_i} + (-1)^{p(f)}\frac{\partial f}{\partial \xi_i}\frac{\partial g}{\partial q_i} \right)$$

respectively.

Denote by

$$\mathfrak{h}(2n|m) = \{D \in \mathfrak{vect}(2n|m) : \ L_D(\omega_0) = 0\}$$

the Lie superalgebra of *Hamiltonian* fields. Put

$$\mathfrak{le}(n) = \{D \in \mathfrak{vect}(n|n) : \ L_D(\omega_1) = 0\} = \langle Le_f : \ f \in \mathbb{C}[[q,\xi]]\rangle,$$

$$\mathfrak{sle}(n) = \mathfrak{le}(n) \cap \mathfrak{sve}(n|n),$$

$$\mathfrak{sle}^0(n) = \mathfrak{sle}(n)/\langle M_{\xi_1 \ldots \xi_n}\rangle,$$

$$\mathfrak{s}^0(n) = \mathfrak{sve}(1|n)/\langle \xi_1 \ldots \xi_n \frac{\partial}{\partial t}\rangle.$$

Clearly,

$$\mathfrak{k}(2n+1|m) = \langle K_f : f \in \mathbb{C}[[t,p,q,\xi]]\rangle,$$

$$\mathfrak{m}(n) = \langle M_f : f \in \mathbb{C}[[\tau,q,\xi]]\rangle,$$

$$\mathfrak{po}(2n|m) = \mathbb{C}[[p,q,\xi]],$$

$$\mathfrak{b}(n) = \Pi(\mathbb{C}[[q,\xi]]).$$

Clearly, the Lie superalgebras from the series $\mathfrak{vect}$, $\mathfrak{sve}$, $\mathfrak{s}^0$, $\mathfrak{h}$ and $\mathfrak{po}$ are finite-dimensional for $m = 0$.

A Lie superalgebra from series $\mathfrak{h}$ ($\mathfrak{le}$ and $\mathfrak{sle}$) is the quotient of the corresponding $\mathfrak{po}$ ($\mathfrak{b}$ and $\mathfrak{sb}$) modulo the one-dimensional center $\mathfrak{z}$, which consists of constants in the realization by generating functions. Put

$$\mathfrak{spo}(m) = \{K_f \in \mathfrak{po}(o|m) : \int f v_\xi = 0\}, \quad \mathfrak{sh}(m) = \mathfrak{spo}(m)/\mathfrak{z}.$$

Nonstandard realizations. Let us list $\mathbb{Z}$-gradings in the above Lie superalgebras of vector fields $\mathcal{L}$ associated with filtrations constructed with respect to a maximal subsuperalgebra of finite codimension. The standard realization is denoted by $(*)$; note that it corresponds to the case of the minimal codimension of $\mathcal{L}_0$. Put deg $x = {}^0x$ and note that gradings in the series $\mathfrak{vect}$ ($\mathfrak{m}$ and $\mathfrak{k}$) induce gradings in series $\mathfrak{sve}$, $\mathfrak{s}^0$ ($\mathfrak{sm}$, $\mathfrak{le}$, $\mathfrak{sle}$, $\mathfrak{sle}^0$, $\mathfrak{b}$, $\mathfrak{sb}$ and $\mathfrak{po}$, $\mathfrak{h}$ respectively). Except for several exceptional gradings the following are all nonstandard gradings (see [80], [81]):

Lie superalgebra	$\mathbb{Z}$-grading
$\mathfrak{vect}(n\|m;r)$, $0 \le r \le m$	${}^0u_i = {}^0\xi_j = 1$ $(*)$
	${}^0\xi_j = 0$ for $1 \le j \le r$; ${}^0u_i = {}^0\xi_{r+j} = 1$ for $j \ge 1$
$\mathfrak{k}(1\|2n;n)$	${}^0t = {}^0\xi_i = 1$, ${}^0\xi_{n+1} = 0$ for $1 \le i \le n$
$\mathfrak{m}(n;r), 0 \le r \le n$	${}^0\tau = 2$, ${}^0q_i = {}^0\xi_i = 1$ $(*)$
	${}^0\tau = {}^0q_i = 1$, ${}^0\xi_i = 0$
	${}^0\tau = {}^0q_i = 2$, ${}^0\xi_i = 0$ for $1 \le i \le r < n$; ${}^0u_{r+j} = {}^0\xi_{r+j} = 1$
$\mathfrak{k}(2n+1\|m;r)$, $0 \le r \le \left[\dfrac{m}{2}\right]$	${}^0t = 2$, ${}^0p_i = {}^0q_i = {}^0\xi_j = 1$ $(*)$
	${}^0t = {}^0\xi_i = 2$, ${}^0\xi_{r+1} = 0$ for $1 \le i \le r \le \left[\dfrac{m}{2}\right]$
	${}^0p_i = {}^0q_i = {}^0\xi_{2r+j} = 1$ for $j \ge 1$

Remark. Experts in string theories have now rediscovered some of the exceptional gradings that are responsible for the fact that $\mathfrak{vect}(1|1)$ and $\mathfrak{k}(1|2)$ are isomorphic as abstract subalgebras. These superalgebras are also isomorphic to $\mathfrak{m}(1)$, cf. [80], [81].

The following table lists the distribution of the known classical Lie superalgebras into types with respect to their selfsymmetricity, or the presence of a Cartan matrix or a bilinear symmetric form.

Table. Selfsymmetricity of Lie superalgebras
A) Selfsymmetric Lie superalgebras

without any bilinear form or Cartan matrix			
all stringy algebras except $k^L(1/6)$			
with Cartan matrix		without Cartan matrix	
A symmetrizable with even form	A non symmetrizable with odd form	with even bilinear form	with odd bilinear form
$\mathfrak{sl}(n/m), m \leq n$; $\mathfrak{psl}(n), n > 1$; $\mathfrak{osp}(m/2n)$ $\mathfrak{ag}_2; \mathfrak{ab}_3; \mathfrak{d}(\alpha)$ $\mathfrak{g}^{(1)}$ with above $\mathfrak{g}$; $(\mathfrak{p})\mathfrak{sl}(m/n)^{(2\epsilon)}_{-st}$, where $\epsilon \cong mn(2)$ $\mathfrak{osp}(2m/2n)^{(2)}$	$\mathfrak{psq}(n)^{(2)}$	$\mathfrak{k}^L(1/6)$	$\mathfrak{psq}(n), n > 2$ $\mathfrak{psq}(n)^{(1)}, n > 2$

B) Nonselfsymmetric Lie superalgebras (no Cartan matrix)

without any form	
$\mathfrak{spe}(n)$ and vectory; twisted loops thereof except those mentioned below	
$(\mathfrak{p})\mathfrak{sl}^{(2)}_{\pi}$; $(\mathfrak{p})\mathfrak{sl}^{(2)}_{-st\Pi}$; $\mathfrak{d}((-1 \pm i\sqrt{3})/2)^{(3)}$; $\mathfrak{psq}(n)^{(4)}, n > 2$	
with even bilinear form	with odd bilinear form
$\mathfrak{sh}(2n); \mathfrak{sh}(2n)^{(1)}; \mathfrak{sh}(2n)^{(2)}$	$\mathfrak{sh}(2n + 1), \mathfrak{sh}(2n + 1)^{(1)}$

Chapter 2
Representations of complex semisimple Lie groups and their real forms

All the Lie algebras and Lie groups considered in this chapter are finite-dimensional; sometimes without mentioning this specifically we confine ourselves to a reductive Lie group, i.e., to a direct product of a simple group by a 1-dimensional center.

§ 2.1 Infinitesimal shift operators on semisimple Lie groups*)

2.1.1 General expressions of infinitesimal operators. Under left and right shifts on G the set of its parameters transforms thanks to (1.5.11) as follows:

$$g^L = g_0^{-1} g, \ g^R = g g_0, \tag{1.1}$$

or, infinitesimally,

$$\delta_L g \cdot g^{-1} = -\sum_a \omega_a^L \cdot \mathcal{F}_a \cdot \delta\varepsilon, \ g^{-1} \cdot \delta_R g = \sum_a \omega_a^R \cdot \mathcal{F}_a \cdot \delta\varepsilon, \tag{1.2}$$

which is expressed symbolically as $\dot{g} g^{-1} = -\mathcal{F}$, $g^{-1}\dot{g} = \mathcal{F}$, cf. (1.5.18). Here $g \in G$ and $\mathcal{F}_a$ is a generator of an infinitesimal transformation with parameter $\omega_a^{L(R)} \cdot \delta\varepsilon$. The parametrization of g by the set $\{\alpha_a : 1 \leq a \leq N\}$ determined from (1.2) enables us to calculate the values of $\dot{\alpha}_a \equiv \frac{d\alpha_a}{d\varepsilon}$, i.e. to calculate the derivative along the infinitesimal shift $\delta\varepsilon$ in direction $\omega_a^{L(R)}$ and represent infinitesimal shift operators on G (in what follows they are simply called generators) in the form (cf. (1.5.14))

$$F^L = \sum_a \frac{d\alpha_a(\omega^L)}{d\varepsilon} \frac{\partial}{\partial \alpha_a(\omega^L)}, \ F^R = \sum_a \frac{d\alpha_a(\omega^R)}{d\varepsilon} \frac{\partial}{\partial \alpha_a(\omega^R)}. \tag{1.3}$$

---

*) [61]

The generators corresponding to the tangent matrix $\mathcal{F}_a$ in (1.2) are obtained from the general expression (1.3) for $w_b^{R,L} = \delta_{ab}$, $1 \leq b \leq N$, i.e.,

$$F_a^L = \sum_b {}_L\dot{\alpha}_b^{(a)} \frac{\partial}{\partial \alpha_b}, \quad F_a^R = \sum_b {}_R\dot{\alpha}_b^{(a)} \frac{\partial}{\partial \alpha_b}, \tag{1.4}$$

where ${}_{L,R}\dot{\alpha}_b^{(a)}$ are solutions of (1.2) for $w_b^{L,R} = \delta_{ab}$.

Thanks to definition (1.3) and equations (1.2) the generators of left and right shifts commute and are related (see 1.5.3) with the help of the matrix of the adjoint representation, i.e. $\delta_L g \cdot g^{-1} \equiv g \cdot (g^{-1} \cdot \delta_R g) \cdot g^{-1}$. Hence

$$F_a^L = -\sum_b (\mathcal{F}_a, g\mathcal{F}^b g^{-1}) \cdot F_b^R, \tag{1.5}$$

where $\mathcal{F}^b \in \mathfrak{g}^*$ satisfies

$$(\mathcal{F}_a, \mathcal{F}^b) = \delta_{ab}. \tag{1.6}$$

All the above relations hold for general Lie groups. For the semisimple Lie groups in Gauss, Iwasawa or Cartan decomposition we can get explicit expressions for the shift generators in the corresponding parametrization. The calculation procedure and the structure of the final answer are recursive: shift generators are expressed via the shift generators and the matrix of the adjoint representation on the subgroups in the corresponding factorization. The technique of calculation is absolutely similar for all these parametrizations and therefore we will illustrate it for the Iwasawa decomposition alone; for the other decompositions we give only the final answer. For definiteness sake let us consider generators of left shifts; those for the right ones are obtained absolutely similarly.

In the Iwasawa decomposition ($g = k \exp(h(\tau)) \cdot z_+$, where $k \in K$, $z_+ \in Z^+$, $\exp h(\tau) \in H$, and τ_i are real parameters of H) the left-hand side of (1.2) takes the form

$$-\mathcal{F} = \dot{g}g^{-1} = \dot{k}k^{-1} + k \cdot \dot{h}(\tau) \cdot k^{-1} + k \exp(h(\tau))\dot{z}_+ z^{-1} \exp(-h(\tau)) \cdot k^{-1} \tag{1.7$_1$}$$

or, equivalently,

$$-k^{-1}\mathcal{F}k = k^{-1} \cdot \dot{k} + h(\dot{\tau}) + \exp(h(\tau))\dot{z}_+ z_+^{-1} \exp(-h(\tau)). \tag{1.7$_2$}$$

The generator of the left shift on G corresponding to a tangent transformation $\mathcal{F}$ is given by the formula

$$F_I^L = \sum_a (k^{-1}\dot{k}, \mathcal{K}^a)K_a^R + \sum_j (h(\dot{\tau}), h^j)\partial/\partial\tau_j + \sum_a (\dot{z}_+ z_+^{-1}, Z^a)Z_a^L, \tag{1.8}$$

where K_a^R and Z_a^L are generators of right (left) shifts on $\mathcal{K}$ and Z^-, respectively; $\mathcal{K}^a$ and Z^a are the corresponding infinitesimal operators in the dual space. Therefore the problem is to find the coefficients of the decomposition of

$k^{-1}\dot{k}$, $h(\dot{\tau})$ and $\dot{z}_+ z_+^{-1}$ with respect to the basis elements of the corresponding algebras. Clearly, the generators of Z^L (resp. K^R) act as differential operators on the parameters of z_+ (resp. parameters of k).

Consider the complex semisimple Lie groups and their normal real forms. Then $k^{-1}\dot{k}$, $h(\dot{\tau})$ and $\dot{z}_+ z_+^{-1}$ are factored thanks to (1.2) with respect to the root subspaces of the Lie algebras $\mathfrak{k}$, $\mathfrak{h}$ and $\mathfrak{z}^+$ respectively:

$$k^{-1}\dot{k} = \sum_{\alpha>0}(\omega_\alpha \cdot \mathcal{X}_\alpha + \omega_{-\alpha} \cdot \mathcal{X}_{-\alpha}) + \sum_j \omega_j h_j, \quad h(\dot{\tau}) = \sum_j \dot{\tau}_j h_j,$$

$$\dot{z}_+ z_+^{-1} = -\sum_{\alpha>0} \theta_\alpha \cdot \mathcal{X}_\alpha. \tag{1.9}$$

The maximal compact subalgebra of a complex semisimple Lie algebra is generated by $\mathcal{X}_\alpha - \mathcal{X}_{-\alpha}$, $i(\mathcal{X}_\alpha + \mathcal{X}_{-\alpha})$ and ih_j, while the maximal compact subalgebra of its normal form is generated by $\mathcal{X}_\alpha - \mathcal{X}_{-\alpha}$. In these cases the parameters in (1.9) satisfy the following conditions, respectively:

$$\omega_{-\alpha}^* = -\omega_\alpha, \quad \omega_j \equiv i\varphi_j, \quad \varphi_j^* = \varphi_j, \tag{1.10$_1$}$$

$$\omega_\alpha^* = \omega_\alpha = -\omega_{-\alpha}, \quad \omega_j = 0. \tag{1.10$_2$}$$

Inserting (1.9) into (1.7) and using the relation

$$\exp h(\tau) \cdot \mathcal{X}_\alpha \exp(-h\tau) = \exp \alpha(\tau)\mathcal{X}_\alpha$$

we get

$$-k^{-1}\mathcal{F}k = \sum_{\alpha>0}(\omega_\alpha \mathcal{X}_\alpha + \omega_{-\alpha} \cdot \mathcal{X}_{-\alpha}) + \sum_j(\dot{\tau}_j + i\dot{\varphi}_j)h_j - \sum_{\alpha>0} \exp(\alpha(\tau))\theta_\alpha \cdot \mathcal{X}_\alpha. \tag{1.11}$$

The dual space to the semisimple Lie algebra is spanned by $\mathcal{X}^{\pm\alpha}$ and h^j normed as follows

$$(\mathcal{X}^\alpha, \mathcal{X}_\beta) = \delta_{\alpha+\beta,0}, \quad (\mathcal{X}^\alpha \cdot h_j) = (h^j, \mathcal{X}_\beta) = 0, \quad (h^i, h_j) = \delta_{ij}, \tag{1.12}$$

and related with the root vectors $\mathcal{X}_{\pm\alpha}$ and the basis $h_j \in \mathfrak{h}$ by the relations

$$\mathcal{X}^{\pm\alpha} = v_\alpha^{-1}\mathcal{X}_{\pm\alpha}, \quad h^i = \sum_j (vk)_{ij}^{-1} h_j \tag{1.13}$$

(see (1.2.10), (1.2.11)). The numbers v_α are determined up to a common factor and can be chosen, say, as $v_\alpha = (\alpha, \alpha)^{-1}$ in which case they symmetrize the Cartan matrix in accordance with (1.2.17).

The system (1.11) with relations (1.12) allows one to find the coefficients of the decomposition (1.9) via the elements of the matrix of the adjoint representation of the maximal compact subgroup K of G:

$$\omega_{-\alpha} = -(\mathcal{X}^\alpha, k^{-1}\mathcal{F}k), \quad \omega_\alpha - \exp(\alpha(\tau))\theta_\alpha = -(\mathcal{X}^{-\alpha}, k^{-1}\mathcal{F}k),$$

$$\dot{\tau}_j + i\dot{\varphi}_j = -(h^j, k^{-1}\mathcal{F}k). \tag{1.14}$$

Making use of (1.10), inserting the solutions of (1.14) into (1.9) and thanks to (1.8) we get the following final expressions for the generators of left shifts corresponding to the tangent transformation $\mathcal{F}$ on a complex semisimple group:

$$
\begin{aligned}
_C F_I^L = &\sum_{\alpha>0} (\mathcal{X}^{-\alpha}, k^{-1} \mathcal{F}^\dagger k) \cdot X_\alpha^R - \sum_{\alpha>0} (\mathcal{X}^\alpha, k^{-1} \mathcal{F} k) X_{-\alpha}^R \\
&- \sum_{\alpha>0} \exp(-\alpha(\tau)) (\mathcal{X}^{-\alpha}, k^{-1}(\mathcal{F} + \mathcal{F}^\dagger) k) Z_\alpha^L \\
&- (1/2) \sum_j (h^j, k^{-1}(\mathcal{F} + \mathcal{F}^\dagger) k) \partial/\partial \tau_j \\
&+ (i/2) \sum_j (h^j, k^{-1}(\mathcal{F} - \mathcal{F}^\dagger) k) h_j^R,
\end{aligned}
\tag{1.15}
$$

or for the generators of left shifts on its normal real form:

$$
\begin{aligned}
_N F_I^L = &\sum_{\alpha>0} (\mathcal{X}^\alpha, k^{-1} \mathcal{F} k)(X_\alpha^R - X_{-\alpha}^R) - \sum_j (h^j, k^{-1} \mathcal{F} k) \partial/\partial \tau_j \\
&- \sum_{\alpha>0} (\mathcal{X}^\alpha + \mathcal{X}^{-\alpha}, k^{-1} \mathcal{F} k) \exp(-\alpha(\tau)) Z_\alpha^L,
\end{aligned}
\tag{1.16}
$$

where $Z_\alpha^L \equiv X_\alpha^L$.

In Gauss decomposition ($g = z_- \exp(h(t)) z_+$, where $z_\pm \in Z^\pm, \exp h(t) \in H$ and t_j are parameters of H), by a complete analogy with the above scheme we get the expression for the generators of left shifts on the complex semisimple Lie group and its normal real form

$$
\begin{aligned}
_{C,N} F_G^L = &-\sum_{\alpha>0} (\mathcal{X}^\alpha, z_-^{-1} \mathcal{F} z_-) X_{-\alpha}^R - \sum_{\alpha>0} \exp(-\alpha(t)) (\mathcal{X}^{-\alpha}, z_-^{-1} \mathcal{F} z_-) X_\alpha^L \\
&- \sum_j (h^j, z_-^{-1} \mathcal{F} z_-) \partial/\partial \tau_j,
\end{aligned}
\tag{1.17_1}
$$

where $X_{-\alpha}^R (X_\alpha^L)$ are operators of right (left) shifts on the nilpotent subgroup Z^- (resp. Z^+) corresponding to negative (positive) roots of $\mathfrak{g}$. The parameters for the real forms of these subgroups are real and $t_j = t_j^* \equiv \tau_j$.

As has been mentioned in 1.6 the Iwasawa and Gauss decompositions are applicable to complex semisimple Lie groups and their normal real forms. Indeed, for the compact form the Iwasawa decomposition is an identity and for the Gauss decomposition the elements $z_\pm$ belong to the complex envelope of the corresponding compact group and satisfy $z_+^\dagger = z_-^{-1}$. The Cartan decomposition ($g = k_1 \exp(h(\tau)) k_2$, where k_1, k_2 belong to the maximal compact subgroup $\mathcal{K}$ and $\exp(h(\tau))$ to the maximal abelian noncompact subgroup of G, see (1.6.11)) is more universal.

In the Cartan parametrization a relation of type (1.7) takes the form

$$-k_1^{-1}\mathcal{F}k_1 = k_1^{-1}\dot{k}_1 + h(\dot{\tau}) + \exp(h(\tau))\dot{k}_2 k_2^{-1}\exp(-h(\tau)). \qquad (1.18)$$

For complex semisimple Lie groups the calculations are the simplest and the maximal compact subgroup coincides with the corresponding compact form whose centralizer is abelian: $GL(1) \times \ldots \times GL(1)$. The corresponding parameters $\varphi_j^{(1)}$, $\varphi_j^{(2)}$ of k_1 and k_2 respectively can be considered as take into account for $h(t)$, where $t_j \equiv \tau_j + i\varphi_j$, $\varphi_j \equiv \varphi_j^{(1)} + \varphi_j^{(2)}$. Then with conditions (1.10_1) we get

$$k_1^{-1}\dot{k}_1 = \sum_{\alpha>0}(\omega_\alpha \mathcal{X}_\alpha - \omega_\alpha^* \mathcal{X}_{-\alpha}) + i\sum_j \varphi_j^{(1)} h_j, \ h(\tau) = \sum_j \dot{\tau}_j h_j,$$

$$\exp h(\tau)\dot{k}_2 k_2^{-1}\exp(-h(\tau)) \qquad (1.19)$$

$$= -\sum_{\alpha>0}(\theta_\alpha \exp(\alpha(\tau))\mathcal{X}_\alpha - \theta_\alpha^* \exp(-\alpha(\tau))\mathcal{X}_{-\alpha} + i\sum_j \varphi_j^{(2)} h_j,$$

and therefore

$$-k^{-1}\mathcal{F}k_1 = \sum_{\alpha>0}(\omega_\alpha \mathcal{X}_\alpha - \omega_\alpha^* \mathcal{X}_{-\alpha}) + \sum_j (\dot{\tau}_j + i\dot{\varphi}_j)h_j$$

$$- \sum_{\alpha>0}(\theta_\alpha \exp(\alpha(\tau))\mathcal{X}_\alpha - \theta_\alpha^* \exp(-\alpha(\tau))\mathcal{X}_{-\alpha}),$$

implying

$$\omega_\alpha = \frac{1}{2\operatorname{sh}\alpha(\tau)}(\mathcal{X}^{-\alpha}, k_1^{-1}(\exp(-\alpha(\tau))\mathcal{F} + \exp(\alpha(\tau))\mathcal{F}^\dagger)k_1),$$

$$\theta_\alpha = \frac{1}{2\operatorname{sh}\alpha(\tau)}(\mathcal{X}^{-\alpha}, k_1^{-1}(\mathcal{F} + \mathcal{F}^\dagger)k_1), \ \dot{\tau}_j + i\dot{\varphi}_j = -(h^j, k_1^{-1}\mathcal{F}k_1). \qquad (1.20)$$

Substituting solutions of (1.20) into (1.19) we get the following final expression for the generators of left shifts on a complex semisimple group:

$$_C F_G^L = \sum_{\alpha \gtrless 0} \frac{1}{2\operatorname{sh}\alpha(\tau)}(\mathcal{X}^{-\alpha}, k_1^{-1}(\exp(-\alpha(\tau))\mathcal{F} + \exp(\alpha(\tau))\mathcal{F}^\dagger)k_1)X_\alpha^R$$

$$- \sum_{\alpha \gtrless 0} \frac{1}{2\operatorname{sh}\alpha(\tau)}(\mathcal{X}^{-\alpha}, k_1^{-1}(\mathcal{F} + \mathcal{F}^\dagger)k_1)X_{-\alpha}^L$$

$$- \frac{1}{2}\sum_j (h^j, k_1^{-1}(\mathcal{F} + \mathcal{F}^\dagger)k_1)\frac{\partial}{\partial\tau_j} + \frac{i}{2}\sum_j (h^j, k_1^{-1}(\mathcal{F} - \mathcal{F}^\dagger)k_1)\frac{\partial}{\partial\varphi_j},$$

$$(1.21)$$

where X_α^R (X_α^L) are combinations of generators of right (left) shifts on K corresponding to $\mathcal{X}_\alpha$. Notice that all $\alpha(\tau)$ are nonzero since otherwise there would have existed a root vector $\mathcal{X}_\alpha$ commuting with $\mathfrak{h}$, which is impossible. In accordance with the partition of the elements of $\mathfrak{g}$ into a compact F and a noncompact J (see 1.2.6) we have similarly decomposed the operators $\mathcal{F}$ which are related with root vectors $\mathcal{X}_{\pm\alpha}$ and $h_j \in \mathfrak{h}$ via the relations

$$\mathcal{F}_\alpha = i^{\varepsilon-1}\mathcal{X}_\alpha - (-i)^{\varepsilon-1}\mathcal{X}_{-\alpha}, \quad \mathcal{F}_j = ih_j,$$

$$\mathcal{J}_\alpha = i^\varepsilon \mathcal{X}_\alpha + (-i)^\varepsilon \mathcal{X}_{-\alpha}, \quad \mathcal{J}_j = h_j; \; \varepsilon = 1, 2. \tag{1.22}$$

Therefore the generators of left shifts on G associated with the compact tangent operators $F_{\alpha,j}$ coincide identically with the generators of left shifts on $\mathcal{K}$ and the formula (1.21) for them expresses a relation between the generators of left and right shifts on $\mathcal{K}$ of type (1.5):

$$-K^L = \sum_\mu (Y^\mu, k^{-1}\mathcal{F}k)K_\mu^R, \quad k \equiv k_1,$$

where the elements $Y^\mu \equiv \{\mathcal{X}^\alpha, \mathcal{X}^{-\alpha}; h^j\}$, $\mu = \pm\alpha, j$, are considered as a complete system of generators, i.e.

$$-K^L = \sum_j (h^j, k^{-1}\mathcal{F}k)h_j^R + \sum_{\alpha>0}[(\mathcal{X}^\alpha, k^{-1}Fk)\mathcal{X}_{-\alpha}^R + (X^{-\alpha}, k^{-1}\mathcal{F}k)X_\alpha^R].$$

Notice that if we consider the action of this relation on the highest vector of weight $\{L\}$ in the representation space of the group $\mathcal{K}$, we see that

$$-K^L = \sum_j (h^j, k^{-1}\mathcal{F}k)L_j + \sum_{\alpha>C}(\mathcal{X}^\alpha, k^{-1}\mathcal{F}k)\mathcal{X}_{-\alpha}^R. \tag{1.23}$$

For the normal real forms the tangent space of a compact subgroup is spanned by the root vectors $\mathcal{X}_\alpha - \mathcal{X}_{-\alpha}$; the parameters $k_1^{-1}k_1$ and $k_2 k_2^{-1}$ satisfy (1.10_2) and therefore

$$_N F_C^L = \sum_{\alpha \gtrless 0} \frac{1}{2\,\mathrm{sh}\,\alpha(\tau)}(\mathcal{X}^{-\alpha}, k_1^{-1}(\exp(-\alpha(\tau))\mathcal{F} + \exp(\alpha(\tau))\mathcal{F}^\dagger)k_1)X_\alpha^R$$

$$- \sum_{\alpha \gtrless 0} \frac{1}{2\,\mathrm{sh}\,\alpha(\tau)}(\mathcal{X}^{-\alpha}, k_1^{-1}(\mathcal{F} + \mathcal{F}^\dagger)k_1)X_\alpha^L - \sum_j (h^j, k_1^{-1}\mathcal{F}k_1)\frac{\partial}{\partial\tau j}. \tag{1.24}$$

In general to solve (1.18) one should take into account the fact that for the compact tangent transformations we have $\mathcal{F} = \sigma(\mathcal{F})$, and for noncompact we have $\mathcal{J} = -\sigma(\mathcal{J})$ for involutive anti-automorphism σ which distinguishes the real form under consideration. Under the action of σ the Cartan subalgebra of the corresponding compact algebra splits into two subspaces, one generated by $\{ih_j\}$, corresponding to the compact part, and one by $\{h_s\}$, corresponding to the noncompact part.

The complete system of noncompact elements is obtained by commuting the h_s with the elements of the maximal compact subalgebra of the real form under consideration. An infinitesimal transformation from the maximal compact subgroup of G factors in turn with respect to the root system of the compact group from which G is obtained. These facts being taken into account and since there are three types of roots, see 1.2.6, we get the following representation for a tangent transformation of the maximal compact subgroup $\mathcal{K}$ of an arbitrary real form G:

$$
k_1^{-1}\dot{k}_1 = i \sum_j \dot{\varphi}_j h_j + \sum_\alpha [\omega_\alpha(\mathcal{X}_\alpha + \sigma(\mathcal{X}_\alpha)) - \omega_\alpha^*(\mathcal{X}_{-\alpha} + \sigma(\mathcal{X}_{-\alpha}))]
$$
$$
+ \sum_\nu [\omega_\nu^0 \mathcal{X}_\nu - \omega_{\nu}^{0*} \mathcal{X}_{-\nu}],
$$

(1.25)

and similarly for $\dot{k}_2 k_2^{-1}$ with ω_α, ω_ν^0, $\dot{\varphi}_j \equiv \dot{\varphi}_j^{(1)}$ being replaced by θ_α, θ_ν^0, $\dot{\varphi}_j^{(2)}$. Here the first of the right-hand sums runs over the basis of the Cartan subalgebra of the maximal compact subgroup of G, the second one runs over all the different roots of multiplicity 1 and 2, where the latter ones are counted only once, and the third sum runs over the roots invariant with respect to the invariance subgroup of G. The parameters θ_ν^0 and $\varphi_j^{(2)}$ are connected with the centralizer and can be included into the set of parameters ω_ν^0 and $\varphi_j^{(1)}$. An element $h(\dot{\tau})$ decomposes with respect to the elements h_s.

A solution of (1.18) takes the form (taking (1.25) into account):

$$
\omega_\nu^0 = -(\mathcal{X}^{-\nu}, k_1^{-1}\mathcal{F}k_1), \quad \dot{\varphi}_j = i(h^j, k_1^{-1}\mathcal{F}k_1),
$$
$$
\dot{\tau}_s = -(h^s, k_1^{-1}\mathcal{F}k_1), \quad \theta_\alpha = \frac{1}{2\,\mathrm{sh}\,\alpha(\tau)}(\mathcal{X}^{-\alpha}, k_1^{-1}(\mathcal{F} - \sigma(\mathcal{F}))k_1),
$$
$$
\omega_\alpha = -\frac{1}{2\,\mathrm{sh}\,\alpha(\tau)}[\exp(\alpha(\tau))(\mathcal{X}^{-\alpha}, k_1^{-1}\sigma(\mathcal{F})k_1)
$$
$$
- \exp(-\alpha(\tau))(\mathcal{X}^{-\alpha}, k_1^{-1}\mathcal{F}k_1)],
$$

(1.26)

implying the following formula for the generators of left shifts:

$$
{}_R F_C^L = -\sum_s (h^s, k_1^{-1}\mathcal{F}k_1)\frac{\partial}{\partial \tau_s} + \sum_j (h^j, k_1^{-1}\mathcal{F}k_1)h_j^R - \sum_{\nu \geqq 0}(\mathcal{X}^{-\nu}, k_1^{-1}\mathcal{F}k_1)X_\nu^R
$$
$$
- \sum_{\alpha \geqq 0}\frac{\delta_\alpha}{2\,\mathrm{sh}\,\alpha(\tau)}(\mathcal{X}^{-\alpha}, k_1^{-1}(\mathcal{F} - \sigma(\mathcal{F}))k_1)X_\alpha^L
$$
$$
- \sum_{\alpha \geqq 0}\frac{\delta_\alpha}{2\,\mathrm{sh}\,\alpha(\tau)}[\exp(\alpha(\tau))\,\mathcal{X}^{-\alpha}, k_1^{-1}\sigma(\mathcal{F})k_1)
$$
$$
- \exp(-\alpha(\tau))(\mathcal{X}^{-\alpha}, k_1^{-1}\mathcal{F}k_1)]X_\alpha^R,
$$

(1.27)

where α runs over all the different roots of multiplicity 1 for $\delta_\alpha = 1/2$ and those of multiplicity 2 for $\delta_\alpha = 1$ and ν runs over all the roots of the invariance subalgebra. The operators X_α^R generate right shifts on the compact subgroup which appear after restriction of the algebra of the real form onto the algebra corresponding to the maximal compact subgroup; the X_α^L generate left shifts corresponding to the parameters of k_2.

The formulas for the generators of right shifts on G are similar to the above ones with the obvious replacement of k_1 by k_2 and X^R by X^L.

In conclusion let us give an expression for infinitesimal generators of left shifts on a noncompact real form G^R of a complex semisimple or reductive Lie group G with Lie algebra $\mathfrak{g} = \bigoplus_{i \in \mathbf{Z}} \mathfrak{g}_i$. First denote by $\bar\alpha$, α_0, α_0^q, $\underline\alpha$ and α the positive roots corresponding to root subspaces of $\mathfrak{g}_0$, $\mathfrak{g}_0^0$, $\mathfrak{g}_0^q$, $\bigoplus_{a \geq 1} \mathfrak{g}_a$ and $\mathfrak{g}_1$ respectively and let R_0, R_0^0, R_+ and $\mathbf{R}_+^1$ be the sets of roots $\bar\alpha$, α_0, $\underline\alpha$ and α. The elements from $\mathfrak{g}_0$, i.e. the set of elements h_j, $1 \leq j \leq r$, of the Cartan subalgebra and the root vectors $\mathcal{X}_{\pm\bar\alpha}$ will be denoted by $\mathcal{X}_{\mu_0}$. As in (1.12) denote by $\mathcal{X}^{\pm\underline\alpha}$ and $\mathcal{X}^{\mu_0}$ the dual basis, normed so that $(X_{\pm\alpha}, X^{\pm\beta}) = \delta_{\alpha,\beta}$, etc. In accordance with (1.6.2) the Gauss decomposition for an arbitrary regular element $g \in G^R$ is $g = \hat{z}_- g_0 \hat{z}_+$, where $g_0 \in G_0$ and $\hat{z}_\pm$ are generated by $X_{\pm\underline\alpha}$; G_0 is a subgroup of G^R generated by $\mathcal{X}_{\mu_0}$. Then the generators of left shifts on G^R corresponding to the tangent transformation $\mathcal{F} \in \mathfrak{g}$ are given by the formula

$$
\begin{aligned}
{}_R F_{MG}^L = & \sum_{\underline\alpha} (\mathcal{X}^{\underline\alpha}, \hat{z}_-^{-1} \mathcal{F} \hat{z}_-) Z_{-\underline\alpha}^R + \sum_{\mu_0} (\mathcal{X}^{\mu_0}, \hat{z}_-^{-1} \mathcal{F} \hat{z}_-) F_{\mu_0}^R \\
& + \sum_{\underline\alpha\underline\beta} (\mathcal{X}^{\underline\alpha}, g_0 \mathcal{X}_{-\underline\beta} g_0^{-1})(\mathcal{X}^{-\underline\beta}, \hat{z}_-^{-1} \mathcal{F} \hat{z}_-) Z_{\underline\alpha}^L,
\end{aligned}
$$

(1.17₂)

where $Z_{-\underline\alpha}^R$ $(Z_{\underline\alpha}^L)$ are generators of right (left) shifts corresponding to root vectors $\mathcal{X}_{-\underline\alpha}$ $(\mathcal{X}_{\underline\alpha})$ and $F_{\mu_0}^R$ are generators of right shifts on G_0 corresponding to $\mathcal{X}_{\mu_0}$.

2.1.2 The asymptotic domain. The infinitesimal operators of left (right) shifts on complex semisimple Lie groups and their real forms enable us, in principle, to find matrix elements of irreducible representations of these groups, distinguish the unitary components and calculate their main characteristics. To this end one should construct from the generators the algebra of Casimir operators (see 1.5.3) and find their common eigenfunctions selecting therefore operator-irreducible representations given by the eigenvalues of Casimir operators.

To get the desired result pass to the limit of the noncompact parameters as $\alpha(\tau) \to \infty$ in the positive Weyl chamber, i.e. $\alpha(\tau) > 0$:

$$-c\overset{\infty L}{F_C} = -\sum_{\alpha>0}(\mathcal{X}^{-\alpha}, k^{-1}\mathcal{F}^\dagger k)X_\alpha^R + \sum_{\alpha>0}(\mathcal{X}^\alpha, k^{-1}\mathcal{F}k)X_{-\alpha}^R$$

$$+ (1/2)\sum_j (h^j, k^{-1}(\mathcal{F} + \mathcal{F}^\dagger)k)\partial/\partial\tau_j \tag{1.28}$$

$$+ (1/2)\sum_j (h^j, k^{-1}(\mathcal{F} - \mathcal{F}^\dagger)k)h_j^R,$$

$$-_R\overset{\infty L}{F}_C = \sum_{\alpha>0}\delta_\alpha(\mathcal{X}^{-\alpha}, k^{-1}\sigma(\mathcal{F})k)X_\alpha^R + \sum_{\alpha>0}\delta_\alpha(\mathcal{X}^\alpha, k^{-1}\mathcal{F}k)X_{-\alpha}^R$$

$$+ \sum_s (h^s, k^{-1}\mathcal{F}k)\partial/\partial\tau_s + \sum_j (h^j, k^{-1}\mathcal{F}k)h_j^R \tag{1.29}$$

$$+ \sum_{\nu \gtrless 0}(\mathcal{X}^{-\nu}, k^{-1}\mathcal{F}k)X_\nu^R;\ k \equiv k_1,$$

$$h_j^R \equiv -i\partial/\partial\varphi_j.$$

Comparing (1.21), (1.27) and (1.28), (1.29) we see that the rather involved analytic dependence of the first two formulas on parameters of k_1, k_2 and $\{\tau\}$ simplifies considerably in the asymptotic domain. Namely, the generators depend only on noncompact parameters in the form of derivatives along $\partial/\partial\tau_s$ and the dependence on parameters of k_2 (or k_1 for right shifts) vanishes altogether. This circumstance was the base for the asymptotic method of constructing representations of semisimple Lie groups.

It is interesting to notice an analogy between this method and potential scattering theory. It is well known (see, e.g. [3]) that all the necessary dynamical information of the potential theory is contained in the $\hat{S}$-matrix which is a ratio of Jost functions — preexponential factors in the asymptotic expression for the Schrödinger wave function. The Regge behaviour of the amplitude of the potential scattering is a corollary of the polynomial (exponential) asymptotics of Legendre functions (matrix elements of $SU(1,1)$) with respect to energy. In the general representation theory of noncompact semisimple Lie groups there is a similar situation, where the role of Jost functions is played by the coefficients of the leading terms of the asymptotic expansion of matrix elements of the corresponding representations. In the vicinity of infinite values of noncompact parameters these elements are of exponential nature (for details see 2.3 and 2.4).

The operators $\partial/\partial\tau_s$, $1 \le s \le r_A$, and right shifts X_j^R, $X_{\pm\nu}^R$ generating the invariance subgroup S of rank r_S (the centralizer of A in K) of the group G, commute, clearly, with all the generators of left shifts (1.28) or (1.29) of the limit representation. Therefore for parameters determining the constructed representation of the algebra with elements (1.28) and (1.29) it is convenient to take the eigenvalues ρ_s of the operators $\partial/\partial\tau_s$ and the weights $\{l\}$ of the irreducible representations of S whose number is equal exactly to rkG, i.e. $r \equiv r_A + r_S$. The eigenvalues of Casimir operators of the considered groups are expressed via these parameters $\{\rho, l\}$. Thus, the algebras generated by (1.28) or (1.29) can be represented in the space of functions

$$\Phi(\tau, k) \equiv \exp\left(\sum_s \tau_s \rho_s\right) \varphi^{\{l\}}(k), \tag{1.30}$$

where $\{l\}$ fixes an irreducible representation of S.

Notice that in the language of the functional algebra (see 1.1.2) the passage to the limit (resulting in the most convenient form of shift generators for constructing irreducible representations of the Lie group) is connected with a canonical transformation in the phase space. In new variables we associate with irreducible representations motions in the phase space along the special surfaces with fixed values of canonical momenta which are conjugated to the cyclic variables, (e.g. for complex groups such cyclic variables are the parameters φ_j and τ_j and the corresponding momenta are $\hat{l}_j \equiv i\partial/\partial\varphi_j$, $\hat{\rho}_j \equiv \partial/\partial\tau_j$ in (1.28)). For the functions $\varphi^{\{l\}}(k)$ in (1.30) it is convenient to take the matrix elements satisfying

$$X_\nu^R \varphi^{\{l\}}(k) = 0, \quad h_\nu^R \varphi^{\{l\}}(k) = \nu(l)\varphi^{\{l\}}(k), \tag{1.31}$$

i.e. the highest matrix elements with respect to the right shifts from S. Indeed, the action of the generators (1.29) on the basis functions (1.31) can be presented in the form

$$-_R\overset{\infty}{\hat{F}} = \left(\sum_{1\leq j\leq r_A} \rho_j h^j + \sum_{1\leq j\leq r_S} l_j h^j, \ k^{-1}\mathcal{F}k\right) + \sum_{\nu>0}(\mathcal{X}^\nu, \ k^{-1}\mathcal{F}k)X_{-\nu}^R$$

$$+ \sum_{\alpha>0}\delta_\alpha(\mathcal{X}^{-\alpha}, k^{-1}\sigma(\mathcal{F})k)X_\alpha^R + \sum_{\alpha>0}\delta_\alpha(\mathcal{X}^\alpha, k^{-1}\mathcal{F}k)X_{-\alpha}^R. \tag{1.32}$$

The latter expression can be rewritten as

$$-\overset{\infty}{\hat{F}} = (\rho(h) + l(h), k^{-1}\mathcal{F}k) + \sum_{\alpha>0}\delta_\alpha(\mathcal{X}^\alpha, k^{-1}\mathcal{F}k)X_{-\alpha}^R, \tag{1.33}$$

where $\delta_\alpha = 1$ for roots of multiplicity 0 and 2, while $\delta_\alpha = 1/2$ for those of multiplicity 1. In the passage from (1.32) to (1.33) we have also taken into account the fact that $\sigma(\mathcal{X}_{-\alpha})$ is a root vector corresponding to a positive root of the considered Lie algebra $\mathfrak{g}$ (see 1.2.6). This expression also holds for the complex semisimple Lie groups whose roots are all of multiplicity 2 and whose invariance subgroup is $S = GL(1) \times \ldots \times GL(1)$ r_S times, where $r_A \equiv r_S$.

The generators of the limit representation (1.33) are in very close connection with generators of left shifts on the compact real form. Indeed, comparing (1.23) with (1.33) for a complex semisimple Lie group ($\delta_\alpha \equiv 1$, $\rho(h) + l(h) = \sum\limits_{1 \leq j \leq r_A} h_j(\rho_j + l_j)$) we see that the generators of the limit representation are obtained from (1.23) by replacing the highest weights $L(h)$ of the irreducible representations of the compact group, where $L_j \in \mathbb{Z}$, by the weights $\rho(h) + l(h)$ whose components are arbitrary complex numbers. The generators of real forms are obtained by the corresponding restriction procedure.

The previously mentioned relation of the generators of the limit representation with the relation (1.23) for the shift generators on the compact subgroup will prove to be essential in the integration of the representation of a semisimple algebra with elements (1.33) to operator-irreducible representations of the corresponding groups (see 2.4).

In conclusion, notice that until we do not require our representation to be (pseudo) unitarity, etc. the parameters $\{\rho\}$ are arbitrary complex numbers. Imposing constraints (1.4.3), we come to the restrictions onto these parameters. Requiring square integrability of functions given on the maximal compact subgroup $\mathcal{K} \subset G$ which belong to the space of the regular representation of G uniquely determines this representation provided $l(h) \in \mathbb{Z}$. As we will show in 2.4 the representations constructed above are operator-irreducible and the separation of their topologically irreducible components is based there on the study of analytic properties of the kernels of intertwining operators.

§ 2.2 ·Casimir operators and
the spectrum of their eigenvalues*)

2.2.1 General formulation of the problem. As we have mentioned in 1.4.3, Casimir operators of an arbitrary semisimple Lie group are expressed as polynomials in traces of powers of the matrix of generators of the corresponding representation

$$\hat{\mathcal{F}}^{L(R)} = \sum_\mu F_\mu{}^{L(R)} Y_\mu, \tag{2.1}$$

where $Y_\mu \equiv \{\mathcal{X}_\alpha, \mathcal{X}_{-\alpha}; h_j\}$ is the complete set of generators normalized by relations (1.12), (1.13), $F_\mu^{L(R)} = (Y^\mu, \hat{\mathcal{F}}^{L(R)})$. For the operator-irreducible representations the Casimir operators reduce to the numerical functions of

_____

*) [7, 35, 51, 61, 62, 108]

their highest weights which are polynomials in the components of the weights. An important property of Casimir operators is that for a semisimple Lie algebra of rank r there are exactly r independent Casimir operators and their eigenvalues uniquely determine an irreducible representation.

A constructive solution of the eigenvalue problem for the whole complete set of Casimir operators is needed in a number of physical applications of representation theory, in particular, in the study of the quantization of non-linear dynamical systems associated with Lie algebras (see Chapter 7). The choice of a group factorization leads in general to physically nonequivalent quantum systems whose Hamiltonians are identified with quadratic Casimir operators and whose wave functions are identfied with operators eigenvalues.

To construct a system of equations for the eigenfunctions of Casimir operators considered as dynamical variables in involution requires the realization of Casimir operators as differential operators in group parameters with the subsequent passage to the variables of the phase space and the description of the spectrum of eigenvalues of these operators. For a quadratic Casimir operator of an arbitrary semisimple Lie group the eigenvalues are given by the following remarkably simple and neat Racah formula

$$C_2(l) = \sum_{1 \leq j \leq r} [(l_j + \rho_0^j)^2 - (\rho_0^j)^2], \qquad (2.2)$$

where l_j and ρ_0^j are components of the highest weight of the irreducible representation and the half sum of positive roots respectively. For a Casimir operator of an arbitrary order the description of the spectrum of its eigenvalues can be performed in principle via Harish-Chandra's approach or Berezin's formula which are ill suited for concrete physical applications.

To calculate eigenvalues as functions of weights of irreducible representations $\{\rho, l\}$ it is possible in a number of cases to refer to some calculation simplifying considerations like the traditional method of applying Casimir operators to the highest vector. However, this method works only when the irreducible representations of the considered groups contain at least one non-trivial finite-dimensional representation with simple spectrum.

Obviously, the most direct way to find explicitly the spectrum of eigenvalues of an arbitrary Casimir operator for an arbitrary irreducible representation of a complex or real semisimple Lie group is to calculate explicitly the traces of powers of shift generators whose dependence on the weights of the representations is most simple. The generators in the asymptotic domain (obtained in the preceding subsection) of an arbitrary irreducible representation with highest weight allow one to realize this programme completely.

The passage to the limit as noncompact parameters tend to infinity realizes a canonical transformation in the phase space which provides a coordinate system most convenient for calculations.

The results obtained hold for any representation; thanks to 2.1.2 they hold also for finite-dimensional unitary representations (degenerate as well) of compact groups not only with simple but with an arbitrary spectrum.

2.2.2 Quadratic Casimir operators. In the following study of the quantization problem of nonlinear dynamical systems we will need explicit expressions for second order Casimir operators of normal real forms of complex semisimple Lie groups, which we will calculate here for several parametrizations. In the Gauss, Iwasawa and Cartan decompositions the matrix of the generators of left shifts (2.1) for the normal form is given, thanks to (1.17), (1.16) and (1.24), in the form

$$-\hat{\mathcal{F}}_G = \sum_{\alpha>0}(z_-\mathcal{X}^\alpha z_-^{-1})X_{-\alpha}^R + \sum_{\alpha>0}\exp(-\alpha(\tau))(z_-\mathcal{X}^{-\alpha}z_-^{-1})X_\alpha^L$$

$$+ \sum_j (z_- h^j z_-^{-1})\frac{\partial}{\partial\tau_j},$$

$$-\hat{\mathcal{F}}_I = -\sum_{\alpha>0}(k\mathcal{X}^\alpha k^{-1})(X_\alpha^R - X_{-\alpha}^R)$$

$$+ \sum_{\alpha>0}\exp(-\alpha(\tau))(k(\mathcal{X}^\alpha + \mathcal{X}^{-\alpha})k^{-1})X_\alpha^L + \sum_j (kh^j k^{-1})\frac{\partial}{\partial\tau_j},$$

$$-\hat{\mathcal{F}}_c = -\sum_{\alpha>0}\frac{1}{2\mathrm{sh}\alpha(\tau)}[\exp(-\alpha(\tau))(k\mathcal{X}^{-\alpha}k^{-1})$$

$$+ \exp(\alpha(\tau))(k\mathcal{X}^\alpha k^{-1})](X_\alpha^R - X_{-\alpha}^R)$$

$$+ \sum_{\alpha>0}\frac{1}{2\mathrm{sh}\alpha(\tau)}(k(\mathcal{X}^\alpha + \mathcal{X}^{-\alpha})k^{-1})(X_\alpha^L - X_{-\alpha}^L) + \sum_j (kh^j k^{-1})\frac{\partial}{\partial\tau_j},$$

where $X_\alpha^{R(L)} - X_{-\alpha}^{R(L)}$ are generators of right (left) shifts on $\mathcal{K}$, i.e., $K_\alpha^{R(L)} \equiv X_\alpha^{R(L)} - X_{-\alpha}^{R(L)}$. Since in these decompositions the action of shift generators on the subgroups Z^- and $\mathcal{K}$ is given by relations

(G): $X_\alpha^L z_- = 0,\quad X_{-\alpha}^R z_- = z_-\mathcal{X}_{-\alpha}(X_{-\alpha}^R z_-^{-1} = -\mathcal{X}_{-\alpha}z_-^{-1});$

(I): $X_\alpha^L k = 0,\quad X_\alpha^R k = k\mathcal{X}_\alpha;$ (2.3)

(C): $X_\alpha^L k = 0,\quad X_\alpha^R k = k\mathcal{X}_\alpha,$

the matrices of the generators can be rewritten as follows

$$-\hat{\mathcal{F}}_G = z_-\left\{\sum_{\alpha>0}[\mathcal{X}^\alpha X_{-\alpha}^R + \mathcal{X}^\alpha\mathcal{X}_{-\alpha} + \exp(-\alpha(\tau))\mathcal{X}^{-\alpha}X_\alpha^L]\right.$$

$$\left. + \sum_j h^j\frac{\partial}{\partial\tau_j}\right\}z_-^{-1};$$ (2.4₁)

$$-\hat{\mathcal{F}}_I = k\left\{\sum_{\alpha>0}\left[-\mathcal{X}^\alpha(X_\alpha^R - X_{-\alpha}^R) - \mathcal{X}^\alpha(\mathcal{X}_\alpha - \mathcal{X}_{-\alpha})\right.\right.$$

$$\left.\left. + \exp(-\alpha(\tau))\ (\mathcal{X}^\alpha + \mathcal{X}^{-\alpha})X_\alpha^L\right] + \sum_j h^j \frac{\partial}{\partial \tau_j}\right\} k^{-1};\qquad (2.4_2)$$

$$-\hat{\mathcal{F}}_C = k\left\{\sum_{\alpha>0}\frac{1}{2\,\mathrm{sh}\,\alpha(\tau)}\left[-(\exp(-\alpha(\tau))\mathcal{X}^{-\alpha}\right.\right.$$

$$+ \exp(\alpha(\tau))\mathcal{X}^\alpha)(X_\alpha^R - X_{-\alpha}^R) + (\mathcal{X}^\alpha + \mathcal{X}^{-\alpha})(X_\alpha^L - X_{-\alpha}^L)$$

$$\left. - (\exp(-\alpha(\tau))\mathcal{X}^{-\alpha} + \exp(\alpha(\tau))\mathcal{X}^\alpha)(\mathcal{X}_\alpha - \mathcal{X}_{-\alpha})\right.\qquad (2.4_3)$$

$$\left.+ \sum_j h^j \frac{\partial}{\partial \tau_j}\right\} k^{-1}.$$

This implies with the help of (2.3) and (1.12), (1.13) the final expression for quadratic Casimir operators $C_2 \equiv \frac{1}{2}\,\mathrm{tr}\,\hat{\mathcal{F}}^2$:

$$C_2^G = \frac{1}{2}\sum_{i,j}(vk)_{ij}^{-1}\left(\frac{\partial}{\partial \tau_i} + \delta_i\right)\frac{\partial}{\partial \tau_j}$$

$$+ \sum_{\alpha>0}v_\alpha^{-1}\exp(-\alpha(\tau))X_{-\alpha}^R X_\alpha^L,\qquad (2.5_1)$$

$$C_2^I = \frac{1}{2}\sum_{i,j}(vk)_{ij}^{-1}\left(\frac{\partial}{\partial \tau_i} + \delta_i\right)\frac{\partial}{\partial \tau_j}$$

$$+ \sum_{\alpha>0}v_\alpha^{-1}\left[-\exp(-\alpha(\tau))K_\alpha^R X_\alpha^L + \exp(-2\alpha(\tau))X_\alpha^L X_\alpha^L\right],\qquad (2.5_2)$$

$$C_2^C = \frac{1}{2}\sum_{i,j}(vk)_{ij}^{-1}\left(\frac{\partial}{\partial \tau_i} + \delta_i(\tau)\right)\frac{\partial}{\partial \tau_j}$$

$$+ \sum_{\alpha>0}v_\alpha^{-1}\frac{1}{4\,\mathrm{sh}^2\,\alpha(\tau)}\left[K_\alpha^R K_\alpha^R - 2\,\mathrm{ch}\,\alpha(\tau)K_\alpha^R K_\alpha^L + K_\alpha^L K_\alpha^L\right],\qquad (2.5_3)$$

where

$$\delta_i(\tau) \equiv -\sum_{\alpha>0}v_\alpha\cdot\mathrm{ctg}\,\alpha(\tau)\cdot\mathrm{tr}(h^\alpha, h_i),$$

$$\delta_i \equiv \sum_{\alpha>0}v_\alpha\,\mathrm{tr}(h^\alpha, h_i) = \lim_{\alpha(\tau)\to-\infty}\delta_i(\tau).$$

For the modified Gauss decomposition (1.5.2) it follows from (1.17$_2$), after an appropriate canonical transformation which kills terms of type δ_i that

$$C_2^{MG} = C_0 + \sum_{\underline{\alpha},\underline{\beta}}(\mathcal{X}^{\underline{\alpha}}, g_0\mathcal{X}^{-\underline{\beta}}g_0^{-1})Z_{-\underline{\beta}}^R Z_{\underline{\alpha}}^L,\qquad (2.5_4)$$

where C_0 is a quadratic Casimir operator of G_0.

2.2.3 Construction of Casimir operators for semisimple Lie groups. First let us describe in detail how to calculate eigenvalues of Casimir operators for complex semisimple Lie groups. For generality we use the Cartan decomposition. To this end let us separate the operators (1.28) into two sets $F_{\pm}$, corresponding to commuting systems of tangent matrices

$$\mathcal{F}_{\pm} \equiv -(\mathcal{J} \pm i\mathcal{F})/2, \quad [\mathcal{F}_+, \mathcal{F}_-] = 0, \quad \text{i.e.,}$$

$$F_{\pm} = \sum_j (h^j k^{-1} \mathcal{J} k) d_{\pm j} \pm \sum_{\alpha>0} (\mathcal{X}^{\pm\alpha}, k^{-1} \mathcal{J} k) X^R_{\mp\alpha}, \tag{2.6}$$

where $d_{\pm j} \equiv (\rho_j \pm l_j)/2$. The algebra of generators $F_{\pm}$ is the direct sum of the corresponding algebras. Therefore, Casimir operators of a complex semisimple Lie group are divided into two subsystems of Casimir operators $C_j^{\pm}$ of the corresponding compact groups constructed from F_+ and F_-, respectively. Consider, say, the calculation of C_j^+. (The operators C_j^- are calculated similarly.)

Let us express the matrix $\hat{\mathcal{F}}_+$ of $F_+ = (Y^\mu, \hat{\mathcal{F}}_+)$ in the form

$$\hat{\mathcal{F}}_+ = k \left[\sum_j h^j d_{+j} + \sum_{\alpha>0} (\mathcal{X}^\alpha X_{-\alpha} + \mathcal{X}^\alpha X^R_{-\alpha}) \right] k^{-1}. \tag{2.7}$$

Here we have taken into account the definition (2.3) of right shift generator. Then

$$C_j^+ = \operatorname{tr} \mathcal{F}_+^j = \operatorname{tr} k \Psi^j k^{-1},$$

$$\Psi = \sum_j h^j d_j + \sum_{\alpha>0} (\mathcal{X}^\alpha X_{-\alpha} + \mathcal{X}^\alpha X^R_{-\alpha}). \tag{2.8}$$

Further simplification of the right-hand side of (2.8) is performed when we move k^{-1} from right to the left, i.e., consecutively commute k^{-1} with Ψ:

$$\Psi_{ab}(k^{-1})^e_c = (k^{-1})^e_c \Psi_{ab} + \sum_m (k^{-1})^e_m \theta_{ab;cm},$$

$$\theta_{ab;cm} \equiv -\sum_{\alpha>0} (\mathcal{X}^\alpha)_{ab} (X_{-\alpha})_{cm}. \tag{2.9}$$

Since the trace of a product containing any (nonzero) number of root vectors with generator of $\mathfrak{h}$ is 0, then (2.8) and (2.9) imply

$$C_j^+ = \sum \overset{\circ}{\Psi}_{ab} \cdot (D^j)^{ec}_{ab} \cdot \left(\sum_j h^j d_j \right)_{ec}, \tag{2.10}$$

where

$$D^{ec}_{ab} = \delta_{cb} \overset{\circ}{\Psi}_{ae} + \theta_{ae;cb}, \quad \overset{\circ}{\Psi}_{ab} \equiv \sum_j h^j_{ab} d_j + \sum_{\alpha>0} (\mathcal{X}^\alpha X_{-\alpha})_{ab}. \tag{2.11}$$

In all the above formulas the representation of h^j, $\mathcal{X}^{\pm\alpha}$ was arbitrary and we did not use any of their properties. Realizing the action of the matrix D on the "column-vector" U with two indices (i.e., U is actually a matrix) we get

$$DU = \Psi U - \sum_{\alpha>0} \mathcal{X}^{\alpha} U \mathcal{X}_{-\alpha}.$$

Hence DU commutes with $h \in \mathfrak{h}$, if $[U, h] = 0$.

Therefore, let us introduce a complete orthonormal system of matrices S_i (i.e. $\operatorname{tr} S_i S^j = \delta_i^j$) commuting with $\mathfrak{h}$. We have

$$C_j^+ = \sum_{p,q} \left[\sum_j h^j d_j + \sum_{\alpha>0} \mathcal{X}^{\alpha} \mathcal{X}_{-\alpha} \right]_{pp} \cdot A_{pq}^{j-2} \left[\sum_j h^j d_j \right]_{qq}, \qquad (2.12)$$

where A is a numerical matrix with elements

$$A_{pq} = \sum_j \operatorname{tr}(S_p S^q h^j) d_j + \sum_{\alpha>0} \operatorname{tr}(S_p S^q \mathcal{X}^{\alpha} \mathcal{X}_{-\alpha})$$
$$+ \sum_{\alpha>0} \operatorname{tr}(\mathcal{X}^{\alpha} S_p \mathcal{X}_{-\alpha} S^q). \qquad (2.13)$$

For the classical complex semisimple Lie groups the dimension of A coincides with the dimension of the representation of h^j and $\mathcal{X}^{\pm\alpha}$ (with simple spectrum) and (2.13) takes the form

$$C_j^+ = \sum_{p,q} (A^j) pq, \quad \text{where} \quad A_{pq} = \mathring{\Psi}_{pq} + \theta_{pq;qp}. \qquad (2.14)$$

This expression holds also for finite-dimensional unitary degenerate representations (with nonsimple spectrum) of compact groups. Therefore, the Casimir operators of infinite-dimensional representations with weights $d_{\pm j}$ of complex semisimple Lie groups are obtained by formal replacement of integer components of weights l_j by $d_{\pm j}$ in the expressions for Casimir operators of the corresponding compact forms.

The main steps of the calculations leading to formulas (2.10) and (2.11) do not essentially vary under the passage to real forms. Indeed, as is clear from (1.33), the difference is the replacement of the maximal compact subgroup of a complex group by the maximal compact subgroup of its real form. Therefore, the trace of a power of a matrix of the generators for real groups is also represented by a sum of elements of a numerical matrix.

For convenience of applications Table 2.2 contains an explicit form of the matrix A for all the classical complex semisimple Lie groups and their real forms.

G	A	m
$L(n, \mathbb{C})$		$m_{\pm a} = (\rho_a \pm x_a)/2$
$L(n, \mathbb{R})$		$m_a = \rho_a$
$U^*(2N)$ $n = 2N$	$A_{ab} = \delta_{ab} A_{aa} - \theta(a, b),$ $A_{aa} = m_a + n - a,$ where $1 \le a \le n$	$m_{2k} = (\rho_k - x_k)/2,\ m_{2k-1} = (\rho_k + x_k)/2,$ $1 \le k \le N$
$U(p,q)$ $p \ge q,\ p + q = n$		$m_a = (\rho_a + x_a)/2,\ 1 \le a \le q,$ $m_a = l_{p-a+1},\ q+1 \le a \le p,$ $m_a = (\rho_{n-a+1} - x_{n-a+1})/2$
$O(N, \mathbb{C})$ $N = 2n$	$A_{ab} = \delta_{ab} A_{aa} - \theta(a,b)(1 - \delta_{a,2n-a+1})$ $A_{ii} = m_i + n(\mathrm{sgn}(i) - 1) - (i+1)$ where $i = \begin{cases} a, & a \le n \\ a - 2n - 1, & a \ge n+1 \end{cases}$	$m_i^\pm = \rho_i \pm x_i,\ m_{-i} \equiv -m_i,\ 1 \le i \le n$
$O^*(2n)$ $n = 2k+1$		$m_{2i} = (\rho_i - x_i)/2,\ m_{2i-1} = (\rho_i + x_i)/2,$ $1 \le i \le k$ $m_{-i} \equiv -m_i,\ m_{k+1} = x_{k+1},\ n = 2k+1$
$O(p,q)$ $p - q = 2k,\ p + q = n$		$m_i = \rho_i,\ 1 \le i \le q;\ m_i = l_{n-i+1},$ $q+1 \le i \le n;\ m_{-i} \equiv m_i$
$O(N, \mathbb{C})$ $N = 2n+1$	$A_{ab} = \delta_{ab} A_{aa} - \theta(a,b)(1 - \delta_{a,2n-a+2})$ $A_{ii} = m_i + (n + {}^1\!/{}_2)(\mathrm{sgn}(i) - 1) - (i+1)$ where $i = \begin{cases} a, & a \le n \\ a - 2n - 2, & a \ge n+1 \end{cases}$	$m_i^\pm = (\rho_i \pm x_i)/2,\ m_{-i} \equiv m_i,$ $1 \le i \le n;\quad m_0 = 0$
$O(p,q)$ $p - q = 2k+1,\ p + q = n$		$m_i = \rho_i,\ 1 \le i \le q;\ m_i = l_{i-q},$ $q+1 \le i \le (n-1)/2$
$Sp(2n, \mathbb{C})$	$A_{ab} = \delta_{ab} A_{aa} - \theta(a,b)(1 + \delta_{a,2n-a+1})$ $A_{ii} = m_i + (n+1)(\mathrm{sgn}(i) + 1) - i,$ where $i = \begin{cases} a, & a \le n \\ a - 2n - 2, & a \ge n+1 \end{cases}$	$m_i^\pm = \rho_i \pm x_i,\ m_{-i} \equiv m_i,\ 1 \le i \le n$
$Sp(2n, \mathbb{R})$		$m_{-i} \equiv m_i,$ $m_i = \rho_i,\ 1 \le i \le n$
$Sp(2p, 2q)$ $p \ge q,\ p + q = n$		$m_i = \begin{cases} m_{2k} = (\rho_k - x_k)/2 \\ m_{2k-1} = (\rho_k + x_k)/2,\ 1 \le k \le q \\ m_{n-1+j} = l_j,\ 1 \le j \le p - q \end{cases}$

§ 2.3   Representations of semisimple Lie groups*)

2.3.1 Integral form of realization of operator-irreducible representations. The
problem of constructing representations of a Lie group from a representation of its Lie algebra is sometimes equivalent to the solution of a system of
Lie equations of type (1.1) connecting initial parameters with those obtained
under an arbitrary group transformation. An explicit form of infinitesimal
operators (1.33) enables one to perform the integration of these equations
and get explicit expressions for the integral operator $T^{\{\rho,l\}}$ realizing a representation of a semisimple Lie group with weight $\{\rho, l\}$. The knowledge of
the transformation law for group parameters is sufficient to find important
characteristics of a group representation such as the matrix elements (1.4.4),
the characters of irreducible representations and their Plancherel measure
introduced in 2.5.

As we noted in 2.1 the constructed representations of a semisimple Lie
group G in the Cartan decomposition are realized on the space $D^{\{\rho\}}$ of differentiable functions determined on the maximal compact subgroup $\mathcal{K}$ of G:

$$T^{\{\rho\}}(g)f(k) = \prod_{1 \leq j \leq r_A} [R_j(g,k)]^{\rho_j} f(\tilde{k}), \tag{3.1}$$

where $R_j \equiv \exp(\tilde{\tau}_j - \tau_j)$ and the parameters $\tilde{k}$ and $\tilde{\tau}$ are obtained from the
initial ones, k and τ, under the action of $g \in G$. The formulas connecting
initial and transformed parameters are found by solving the Lie equations
with generators (1.33).

The representation (3.1) is reducible since its generators commute with
the transformations from the invariant subgroup S, the centralizer of $\mathcal{A}$ in
$\mathcal{K}$. To select operator-irreducible components, expand f functions from $D^{\{\rho\}}$
with respect to matrix elements of S.

Let us represent $k \in \mathcal{K}$ in the form $k = k_{-s} \cdot s$, where $s \in S$ and $k_{-s} \in \mathcal{K}/S$.
The invariance of $T^{\{\rho\}}$ with respect to S means that under an arbitrary
transformation from G an element s is transformed into $\tilde{s} = s \cdot N(k_{-s}g)$,
where $N \in S$ does not depend on s. Making use of this we get

$$T^{\{\rho,l\}}(g)f_{\{m\}}(k_{-s}) = \prod_{1 \leq j \leq r_A} [R_j(k_{-s},g)]^{\rho_j} \sum_{\{m'\}} D^{\{l\}}_{\{m\},\{m'\}}(N)f_{\{m'\}}(\tilde{k}_{-s}),$$

$$\tag{3.2}$$

where $D^{\{l\}}_{\{m\},\{m'\}}(N) \equiv \langle m|N|m' \rangle$ is the matrix element of an irreducible rep-
resentation of S with the highest weight $\{l\}$ between the states $|m'\rangle$ and $\langle m|$.
The expression (3.2) is an integral form of realization of the representation
$\{\rho, l\}$ of G corresponding to the algebra constructed in 2.1 with elements

---

*) [61]

(1.33). Proof of the operator irreducibility of the representation $\{\rho, l\}$ determined by (3.2) will be given in 2.4.

In accordance with (3.2) the operator $T^{\{\rho, l\}}(g)$ can be considered as an integral operator with singular kernel

$$
\begin{aligned}
&T^{\{\rho,l\}}_{\{m\},\{m'\}}(k_{-s}, k'_{-s}) \\
&= \prod_{1 \leq j \leq r_A} [R_j(g, k_{-s})]^{\rho_j} D^{\{l\}}_{\{m\},\{m'\}}(N) \cdot \delta(k'_{-s} \cdot \tilde{k}^{-1}_{-s}),
\end{aligned}
\tag{3.3}
$$

where $\delta(k'_{-s} \cdot \tilde{k}^{-1}_{-s})$ is the δ-function on $\mathcal{K}/S$ given by the formula

$$
\int f(k'_{-s}) \delta(k'_{-s} \cdot k^{-1}_{-s} d\mu(k'_{-s}) = f(k_{-s}),
$$

where $d\mu(k_{-s})$ is an invariant measure on $\mathcal{K}/S$.

2.3.2 The matrix elements of finite transformations. As a basis in $D^{\{\rho,l\}}$ it is convenient to take the matrix elements $D^{\{l\}}_{\{L\};\{M\},\{N\}}(k)$ of irreducible unitary representations of $\mathcal{K}$ between the states with quantum numbers $\{M\}$ and $\{N\}$ which are eigenvectors of S. Then formula (3.1) implies the following integral representation for the matrix elements $D^{\{\rho,l\}}(g)$ of the representation $\{\rho, l\}$ of G:

$$
D^{\{\rho,l\}}_{\{{}^{L}_{MN}\};\{{}^{L'}_{M'N'}\}}(g) = N_L^{-1} \int d\mu(k) \cdot \prod_{1 \leq j \leq r_A} [R_j(k,g)]^{\rho_j} D^{*\{l\}}_{\{L'\}}{}_{\{M'\},\{N\}}(\tilde{k}) \cdot D^{\{l\}}_{\{L\}}{}_{\{M\},\{N\}}(k),
\tag{3.4}
$$

where N_L is the dimension of the representation $\{L\}$ of $\mathcal{K}$, and $d\mu(k) = d\mu(k_{-s}) \cdot d\mu(s)$ the invariant measure normalized by unit on $\mathcal{K}$. The number of continuous parameters of the discrete basis introduced exactly equals the number of quantum numbers, including the components of the weights of the representation.

For further reduction and concretization of the expression obtained for the matrix elements of operator-irreducible representations of G we have to know explicitly how the transformed parameters $\tilde{k}$ and $\tilde{\tau}$ depend on g, k and τ. Let us first consider complex semisimple Lie groups.

Comparing (1.23) and (2.6) we see that

$$
E^{\{l\}}_{\{m\}}(k,\tau) \equiv \exp(\sum_j \tau_j l_j) \cdot D^{\{l\}}_{\{m\},\{l\}}(k),
\tag{3.5}
$$

which is the highest vector with respect to right shifts on K, i.e.,

$$
X^R_\alpha E^{\{l\}}_{\{m\}} = 0, \ \alpha > 0; \ h^R_j E^{\{l\}}_{\{m\}} = l_j E^{\{l\}}_{\{m\}},
$$

transforms thanks to (3.1) in accordance with a finite-dimensional nonunitary representation:

$$E^{\{l\}}_{\{m\}}(\tilde{k}, \tilde{\tau}) = \sum_{\{m'\}} D^{\{l\}}_{\{m\},\{m'\}}(g) \cdot E^{\{l\}}_{\{m'\}}(k, \tau). \qquad (3.6)$$

Making use of the orthogonality of matrix elements of irreducible representations of $\mathcal{K}$ we get the following equations, corollaries to (3.5) and (3.6):

$$\prod_{j=1}^{r_A} [R_j(k, g)]^{l_j} = \left[\sum_{\{m\}} \left| D^{\{l\}}_{\{m\},\{l\}}(gk) \right|^2 \right]^{1/2} ,$$

$$D^{\{l\}}_{\{l\},\{l\}}(k^{-1}\tilde{k}) = D^{\{l\}}_{\{l\},\{l\}}(k^{-1}gk) \left/ \left[\sum_{\{m\}} \left| D^{\{l\}}_{\{m\},\{l\}}(gk) \right|^2 \right]^{1/2} \right. . \qquad (3.7)$$

Taking for $\{l\}$ the system of fundamental representations of the corresponding group it is not difficult to derive from this the relations between the transformed $\tilde{k}$, $\tilde{\tau}$ and initial k, τ parameters. In the asymptotic domain for the noncompact parameters tending to infinity the right-hand sides of both the equations (3.7) are considerably simplified and in particular for $g = \exp \sum_j \varepsilon_j h_j \in \mathcal{A}$ are expressed in terms of the highest matrix elements $\xi^{\{l\}}(k) \equiv D^{\{l\}}_{\{l\},\{l\}}(k)$ of the corresponding representations of $\mathcal{K}$. Therefore for the universal parametrization of (1.5.16) the problem gets an explicit solution thanks to (1.6.19).

For real forms of complex groups relations (3.5) and (3.7) retain their form with $E^{\{l\}}_{\{m\}}(k, \tau)$ transformed via finite-dimensional nonunitary representations of the corresponding real form, where not all the parameters τ_j iñ (3.5) are linearly independent and nonzero and k belongs to the maximal compact subgroup of the corresponding real form.

Relations thus obtained completely solve the integration problem for a system of Lie equations for an arbitrary semisimple Lie group. For any classical series which has a matrix realization these relations can be written in a tensor basis.

In what follows we will need relations of type (3.7) for normal forms when $g = \exp \sum_j h_j \varepsilon_j \in \mathcal{A}(H)$. They are of the form

$$R_j^2(k, \varepsilon) = \xi^{\{j\}}(k^{-1} \exp(2h(\varepsilon)) \cdot k),$$

$$\xi^{\{j\}}(\tilde{k}) = \exp \varepsilon_j \cdot \xi^{\{j\}}(k) \cdot [\xi^{\{j\}}(k^{-1} \exp(2h(\varepsilon)) \cdot k]^{-1/2}, \qquad (3.8)$$

where $\xi^{\{j\}}$ is the restriction onto $\mathcal{K}$ of the highest vector of the j-th fundamental representation of G.

§ 2.4 Intertwining operators and the invariant bilinear form*)

2.4.1 Intertwining operators and problems of reducibility, equivalence and unitarity of representations.
One of the cardinal problems of representation theory of semisimple Lie groups is the description of all the irreducible unitary representations. For the real forms this problem is especially difficult, in particular because there are several types of principal series of unitary representations caused by the existence of several nonconjugate Cartan subgroups and since there are realizations in the spaces of analytic functions of several complex variables with complicated structures of irreducible subrepresentations.

Among the different approaches used in construction of unitary representations of semisimple Lie groups the most constructive are, it seems, the investigations of asymptotic properties of matrix elements and study of kernels of Hermitian forms. These properties can be related in turn with the study of the analytic properties of intertwining operators (1.4.2) in the weight space. The study of reducibility and equivalence of representations (see 1.4) also leads to intertwining operators. Search for explicit expressions for Plancherel's measure for the principal series (see (2.5)) is also most easily performed with the help of intertwining operators.

Notice also that these operators play an important role in the study of quantum dynamical systems both in the approach discussed in this book and via the ISP-method.

The starting point for the explicit construction and study of intertwining operators is their main property

$$\hat{B}^{\{\omega,\Lambda\}}T^{\{\Lambda\}}(g) = T^{\{\Lambda'\}}(g)\hat{B}^{\{\omega,\Lambda\}}, \tag{4.1}$$

where $\{\Lambda'\} = \omega\{\Lambda\}$ for $\omega \in W_G$.

As has been mentioned, both operator and topological irreducibility can be connected as had been mentioned with analytic properties of intertwining operators in the weight space. To clarify this it is convenient to realize $\hat{B}$ as an integral operator on the space of functions (determined on $\mathcal{K}$) from the space of the corresponding representation G in the Cartan decomposition:

$$\hat{B}^{\{\omega,\Lambda\}}f(k_1) = \int_K d\mu(k_2)B^{\{\omega,\Lambda\}}(k_1, k_2)f(k_2). \tag{4.2}$$

The kernel of this operator is, generally, a distribution. Then the operator irreducibility means that the kernel is proportional to $\delta(k_2^{-1}k_1)$. For integer

---

*) [19, 35, 36, 51, 61, 113]

components of $\{\Lambda\}$ the representation $T^{\{\Lambda\}}(g)$ may, nevertheless, be topologically reducible.

The possibility to distinguish completely irreducible representations in this approach is based on the study of analytic properties of intertwining operators in the space of weights of the representation $T^{\{\Lambda\}}(g)$,

$$B_{\{L\}}^{\{\omega,\Lambda\}} = \int d\mu(k) B^{\{\omega,\Lambda\}}(k) \overset{*\{L\}}{\hat{D}}(k), \qquad (4.3)$$

where $D^{\{L\}}(k)$ is the matrix element of the representation $T^{\{L\}}(k)$ of maximal compact subgroup K of G. As we will explain in what follows, $B^{\{\omega,\Lambda\}}(k_1, k_2)$ depends only on $k_2^{-1} k_1 \equiv k$.

To study the explicit analytic dependence of intertwining operators on the weights of the representation it is convenient to pass to the operator form of (4.3):

$$\hat{j}^{\{\omega,\Lambda\}} = \int_K B^{\{\omega,\Lambda\}}(k) \cdot k \, d\mu(k), \qquad (4.4)$$

where the compact generators $i(\mathcal{X}_\alpha + \mathcal{X}_{-\alpha})$ and ih_j entering the element k (see 1.5.16) have integer spectrum of eigenvalues.

The problem of equivalence of representations with highest weights $\{\Lambda\}$ and $\{\Lambda'\}$ reduces to establishing the isomorphism of the spaces of the representations $D^{\{\Lambda\}}$ and $D^{\{\Lambda\}}$ under the action of the operator $\hat{B}^{\{\omega,\Lambda\}}$, where $\{\Lambda'\} = \omega\{\Lambda\}$ and finally the problem is determined by the norming constant of this operator.

As has been already mentioned, the intertwining operators contribute also to the asymptotics of matrix elements, an important characterization of representations of noncompact groups. In operator form this can be expressed as follows:

$$T^{\{\Lambda\}}(g) \underset{\{\tau\} \to \infty}{\longrightarrow} \sum_{\omega \in W_G} (\hat{B}^{\{\omega,\Lambda\}} \cdot \exp \sum_j \tau_j \rho_j^\omega (\hat{B}^{\{\omega,\Lambda\}})^{-1}) \hat{B}^{\{\underline{\omega},\Lambda\}}, \qquad (4.5)$$

where $\underline{\omega}$ corresponds to the complete (cyclic) permutation of all the components of the weight $\{\Lambda\}$.

The asymptotic decomposition of the matrix element of $T^{\{\Lambda\}}(g)$ between the states with fixed values of the weights of $T^{\{L\}}(k)$ follows obviously from (4.5). The coefficients at the exponentials are then expressed via the functions (4.3) and carry complete information on the unitary components of $\{\Lambda\}$ whose matrix elements asymptotically decrease in a certain way.

In particular, to the principal series of unitary representations $\{\Lambda\} \equiv \{\rho, l\}$ there correspond matrix elements square integrable over the invariant measure (1.5.12), $d\mu(g) = d\mu(k) d\mu(k') D(\tau) \prod_{1 \le j \le r_A} d\tau_j$, in the Cartan decomposition for G with $D(\tau) \to \exp(2 \sum_j \rho_j^0 \tau_j)$.

Depending on the structure of the range of the components of ρ_j, $1 \leq j \leq r_{\mathcal{A}}$, of representations of the principal series the latter are called *continuous, discrete* or *semidiscrete*. The contribution to the asymptotics of the continuous principal series $(\rho_j = -\rho_j^0 + i\sigma_j$, $\sigma_j = \sigma_j^*$, $-\infty < \sigma_j < \infty)$ is given by all the terms in (4.5), whereas for (semi)discrete series (if any) the absence of terms which decrease slower than $\exp(-\sum_j \rho_j^0 \tau_j)$ is justified by the vanishing of the corresponding functions B.

The vanishing of the corresponding functions is directly determined by their analytic properties (distribution of zeros in the weight space). This is the most clear analogy of the asymptotic method of the representation theory of noncompact groups with potential scattering theory in which the functions B play the role of the Jost functions mentioned in 2.1.2.

The study of analytic properties in the complex space $\{\rho\}$ (to distinguish completely irreducible and unitary representations) is similar to the study of bound states, resonances, etc. with the help of analytic properties of Jost functions $f(\lambda, k)$ in the complex k-plane and their physical interpretation (cf. e.g. [3]).

The problem of constructing kernels of invariant Hermitian forms can be reduced to the study of kernels of the corresponding intertwining operators. This enables one to distinguish (pseudo) unitary representations if any.

Therefore to use intertwining operators effectively in the above problems of representation theory it is necessary, first, to know their explicit analytic expressions and, second, to diagonalize the functions $B_{\{L\}}^{\{\omega,\Lambda\}}$ which are finite-dimensional matrices in the weight space of the maximal compact subgroup.

The first problem is dealt with in what follows and we obtain uniform expressions for the kernels of intertwining operators of semisimple Lie groups and operator-valued functions (4.4).

The diagonalization of matrices of the form (4.3) is, generally, a much more difficult problem. However, in particular cases, e.g., for $U(m,1)$ it is explicitly solved with the help of intertwining operators.

2.4.2 Construction of intertwining operators. Let us pass to the direct construction of intertwining operators for the representations of semisimple Lie groups in the asymptotic approach discussed above. For this let us rewrite (4.1) in infinitesimal form using (4.2):

$$[_1 F^{\{\Lambda'\}} -_2 F^{T\{\Lambda\}}] B^{\{\omega,\Lambda\}}(k_1, k_2) = 0, \tag{4.6}$$

where the indices 1 and 2 of the shift generators F correspond to the indices of their variables. The subsystem corresponding to compact generators of G shows that the kernel only depends on $k \equiv k_2^{-1} k_1$. Then making use of the expression for noncompact generators of G in the asymptotic domain (see 2.1.2) we can reformulate the remaining equations explicitly.

For example, let us perform this for a complex group G. In accordance with (2.6) we have the following system of equations for $B(k)$:

$$
\left[\sum_i \mathrm{tr} \left(k^{-1} \left\{ \begin{matrix} h_j \\ \mathcal{X}_{-\beta} \\ \mathcal{X}_{\beta} \end{matrix} \right\} k h_i \right) d_i + \sum_{\alpha>0} \mathrm{tr} \left(k^{-1} \left\{ \begin{matrix} h_j \\ \mathcal{X}_{-\beta} \\ \mathcal{X}_{\beta} \end{matrix} \right\} k \mathcal{X}_\alpha \right) X^R_{-\alpha} \right.
$$

$$
\left. + \left\{ \begin{matrix} -d'_j - \rho^0_j \\ \mathcal{X}_{-\beta} \\ 0 \end{matrix} \right\} \right] B(k) = 0. \tag{4.7}
$$

(For real groups the system for $B(k)$ is obtained from (4.7) by replacing the maximal compact subgroup of G by the maximal compact subgroup of the corresponding real form G^R of G and $\{\Lambda; G\}$ by $\{\Lambda; G^R\}$).

In the parametrization (1.5.16) the kernel of the intertwining operator for G contains δ-functions in parameters $\{\theta_\beta, \varphi_\beta\}$, where $\beta \in \Sigma^- \cap \omega(\Sigma^-)$; up to factors corresponding to these parameters the kernel is of the form

$$
B^{\{\omega,\Lambda\}}(k) \sim \prod_{\alpha \in \Sigma^+ \cap \omega(\Sigma^-)} (\cos \theta_\alpha/2)^{2(\alpha\rho)/(\alpha\alpha)} \exp[i\varphi_\alpha(\alpha l)/(\alpha\alpha)]
$$

$$
\times \prod_j \exp[i\Psi_j(\pi_j l)/2]; \ l \equiv \{l_j\}, \ \rho \equiv -\{\rho_j + 2\rho^0_j\}. \tag{4.8}
$$

Inserting (1.5.16), (1.5.18) and (4.8) into (4.4) we get

$$
\hat{j}^{\{\omega,\Lambda\}} = \prod_{\alpha \in \Sigma^+ \cap \omega(\Sigma^-)} \delta[h_\alpha - (\alpha l)] J_\alpha[\mathcal{X}_{\pm\alpha}, \rho] \prod_j \delta[h_j - (\pi_j l)], \tag{4.9}
$$

where

$$
J_\alpha[\mathcal{X}_{\pm\alpha}, \rho] \equiv \int \sin \frac{\theta_\alpha}{2} \left(\cos \frac{\theta_\alpha}{2} \right)^{2\frac{(\alpha,\rho+\rho_0)}{(\alpha,\alpha)}-1} \exp \left[\frac{i\theta_\alpha(\mathcal{X}_\alpha + \mathcal{X}_{-\alpha})}{\sqrt{2(\alpha,\alpha)}} \right] d\theta_\alpha. \tag{4.10}
$$

Formula (4.9) may be rewritten as follows:

$$
\hat{j}^{\{\omega,\Lambda\}} = \hat{j}^1 \cdot \hat{j}^{\{\omega,\Lambda\}}_{(-2)}, \ \hat{j}^1 \equiv \prod_{\alpha \in \Sigma^+_1 \cap \omega(\Sigma^-)} \delta[h_\alpha - (\alpha l)] J_\alpha \prod_j \delta[h_j - (\pi_j l)], \tag{4.11}
$$

where $\hat{j}^{\{\omega,\Lambda\}}_{(-2)}$ is the corresponding operator for $G_{(-2)}$. The latter group is obtained from G by deleting π_1 and π_r from the Dynkin diagram of G and $\Sigma^+_{(-2)} \equiv \Sigma^+/\Sigma^+_1$. The intertwining operators of the normal real form of G are expressed by the formulas similar to (4.9) and (4.11) in the corresponding complex case and do not contain factors corresponding to the weights of the centralizer of G.

As an example of intertwining operators for an arbitrary real semisimple Lie group let us give their expression for $U(p,q)$, $p \geq q$. In this case the integral formula (4.4) takes the form

$$_{U(p,q)}\hat{J}^{\{\omega,\Lambda\}} = \hat{J}_L^{(1)} \cdot \hat{J}_U \cdot \hat{J}_L^{(2)} \cdot {}_{U(p-1,q-1)}\hat{J}^{\{\omega,\Lambda\}}, \qquad (4.12)$$

where $\hat{J}_L^{(\mu)}$ are expressions (4.11) similar to $_{GL(n,\mathbb{C})}\hat{J}^1$ with Σ_1^+ replaced by Σ_μ^+ and $\mathcal{X}_{\pm\alpha}$ by $\mathcal{X}_{\pm\alpha} + \mathcal{X}_{\pm\bar{\alpha}}$ (here $\bar{\pi}_j \equiv \pi_{p+q-j}$ for $1 \leq j \leq q-1$); $\hat{J}_U$ is the operator corresponding to $U(p-q+1,1)$ and $_{U(p-1,q-1)}\hat{J}^{\{\omega,\Lambda\}}$ is the intertwining operator for the subgroup $U(p-1,q-1) \subset U(p,q)$ which does not contain values with index 1. Therefore for $U(p,q)$ the problem reduces to the search for an operator $\hat{J}_U$ for $U(m,1)$ with $m = p-q+1$ of real rank 1. This group is distinguished among the pseudo-unitary groups since the restriction of its irreducible representation onto the maximal compact subgroup $U(m) \times U(1)$ contains the irreducible representations of the latter at most once. This allows us to diagonalize the kernel of the intertwining operator in the form (4.3) and therefore to solve completely the above-mentioned questions.

Let us consider $U(m,1)$ in detail and illustrate with it the general arguments of the preceding section. The kernel of the intertwining operator of the representation $\{\rho,\chi; p_1,\ldots,p_{m-1}\}$, where $\{p_1,\ldots,p_{m-1}\}$ and χ are weights of subgroups $U(m-1)$ and $U(1)$ of the centralizer $S = U(m-1) \otimes U(1)$, for a nontrivial element of Weyl group has the form

$$B^{\{d_\pm; p\}} = [\exp i\varphi - m_1^1]^{-(d_- + m)}[\exp(-i\varphi) - m_1^{*1}]^{-(d_+ + m)}$$
$$\times \prod_{1 \leq j \leq m-1} (\det_j \tilde{m})^{p_j - p_{j+1}}, \qquad (4.13)$$

where

$$\tilde{m}_\beta^\alpha = m_\beta^\alpha + m_1^\alpha m_\beta^1/(\exp(i\varphi) - m_1^1),$$

$$2 \leq \alpha, \beta \leq m, \ p_m = 0, \ d_\pm \equiv (\rho \pm \chi)/2,$$

$$m_\beta^\alpha \in U(m), \ \exp(i\varphi) \in U(1),$$

$\det_j m$ are principal minors of the matrix m, i.e. $\det_1 m = m_m^m$, $\det_2 m = m_m^m \cdot m_{m-1}^{m-1} - m_m^{m-1} m_{m-1}^m$, etc. Notice that the weights Λ_α of the representation are connected with parameters $d_\pm$ and p_i by relations

$$\Lambda_1 = d_+ + m; \ \Lambda_{i+1} = p_i + m - i \text{ for } 1 \leq i \leq m-1; \ \Lambda_{m+1} = -d_-.$$

In accordance with (4.13) a matrix element (4.3) diagonal with respect to weights of S

$$B_{\{l\}}^{\{\rho,\chi;p\}} = N_{\{l\}}^{-1} \int dm \int d\varphi B^{\{d_\pm,p\}}(m,\varphi) \overset{*\{l\}}{D}_{\{p\},\{p\}}(m) \exp(-ir\varphi),$$

$$\sum_1^m l_i + r = \chi + \sum_1^{m-1} p_i, \qquad (4.14)$$

is of the form

$$B_{\{l\}}^{\{\rho, \chi; p\}} = (-1)^{\sum_1^m l_\alpha} N_{\{l\}}^{-1} \Gamma(-d_+ - d_- - m) \Gamma^{-1}(-d_- - l_m)$$

$$\times \Gamma^{-1}(-d_+ + l_m - m + 1) \prod_1^{m-1} (-d_+ + p_\alpha - \alpha)^{-1}$$

$$\times B(-d_- - p_\alpha - m + \alpha, -d_+ + p_\alpha - \alpha + 1)$$

$$\times B^{-1}(-d_- - l_\alpha - m + \alpha, -d_+ + l_\alpha - \alpha + 1).$$

(The detailed deduction of formulas (4.13) and (4.14) is given in [61].) The analytic properties of (4.14) are directly determined for integers $d_\pm$ by the signs of the arguments of the Γ-functions which constitute (4.14). This enables us to list all the irreducible components, including the "strange" series. The matrix form of expression of the asymptotic decomposition (4.5) for a non-compact transformation with parameter τ from $U(m, 1)$ for the maximally occupied weights of S is of the form

$$D_{\{l\}, \{l\}}^{\{\rho, s\}}(\tau) \xrightarrow[\tau \to \infty]{} \exp(\tau \rho) B_{\{l\}}^{\{-\rho - 2m, s\}} \theta(\operatorname{Re} \rho + m)$$

$$+ \exp[-\tau(\rho + 2m)](-1)^{\sum_1^m l_\alpha} B_{\{l\}}^{\{\rho, s\}} \theta(-\operatorname{Re} \rho - m),$$

(4.15)

where the coefficients $B_{\{l\}}^{\{\rho, s\}}$ of the principal terms of the asymptotic expansion are given by formula (4.14). This enables us to distinguish all the unitary representations starting from the structure of the arguments of the Γ-functions.

The factorizability of expressions (4.11) and (4.12) that reduce to the product of the corresponding operators for groups of rank 1 is a concrete realization of Schiffmann's suggestion of the possibility of reducing the intertwining operators of a group of arbitrary rank to real groups of rank 1. However, unlike the abstract expression of intertwining operators in the form of a convolution of the corresponding operators for simple reflections we give here an explicit expression for an arbitrary transformation from the Weyl group. The operators (4.4) are represented as products of known functions of type (4.9) in generators of compact subgroups having known (integer) spectrum of eigenvalues.

In conclusion notice that the operator-irreducibility of the representations $\{\rho, s\}$ constructed in 2.3 follows directly from the above analysis. Indeed, for $\{\Lambda'\} = \{\Lambda\}$ we see that $B^{\{\Lambda\}}(k)$ are of δ-like form since $\Sigma^+ \cap \Sigma^- = \emptyset$.

2.4.3 The invariant Hermitian form. The problem of selecting unitary representations may be connected with the search for an invariant Hermitian form and subsequent study to determine whether it is positive definite (see 1.4).

Let us express a Hermitian form in the space of the representation $D^{\{\Lambda\}}$ as follows:

$$(f_1, f_2)^{\{\Lambda\}} = \iint f_1^*(k_1) K^{\{\Lambda\}}(k_1, k_2) f_2(k_2) d\mu(k_2) d\mu(k_1). \qquad (4.16)$$

The kernel is generally a distribution. The Hermitian property and invariance take the form

$$(f_1, f_2)^{\{\Lambda\}} = (f_2, f_1)^{\{\Lambda\}*} \qquad (4.17)$$

$$(f_1, \hat{F} f_2)^{\{\Lambda\}} = (\hat{F} f_1, f_2)^{\{\Lambda\}}. \qquad (4.18)$$

The representations $\{\Lambda\}$ satisfying these conditions are called *pseudo-unitary*; the *unitary* representations are those which also satisfy

$$(f, f) > 0 \text{ for any } f \neq 0. \qquad (4.19)$$

If functions from the space of representation $D^{\{\Lambda\}}$ are square integrable then the representation belongs to one of the principal series and enters in the decomposition of the regular ones; otherwise it belongs to the complementary series. The equalities (4.17) and (4.18) can be rewritten as the following conditions on the kernel K:

$$K^*(k_1, k_2) = K(k_2, k_1) \qquad (4.20)$$

$$[_1\hat{F}^\dagger -_2 \hat{F}^T] K(k_1, k_2) = 0, \qquad (4.21)$$

and the subsystem (4.21) connected with the generators of compact transformations from G shows that the kernel depends only on $k \equiv k_2^{-1} k_1$ and therefore

$$K^*(k) = K(k^\dagger). \qquad (4.22)$$

The invariance condition

$$T^{\{\Lambda\}\dagger}(g) \hat{K}^{\{\Lambda\}} T^{\{\Lambda\}}(g) = \hat{K}^{\{\Lambda\}} \qquad (4.23)$$

whose infinitesimal form is (4.21) for an invertible $\hat{K}$ takes the form

$$T^{\{\Lambda\}\dagger}(g^{-1}) = \hat{K}^{\{\Lambda\}} T^{\{\Lambda\}}(g) (K^{\{\Lambda\}})^{-1} \qquad (4.24)$$

which, in particular, implies that the characters of $T^{\{\Lambda\}\dagger}$ and $T^{\{\Lambda\}}$ coincide, i.e.,

$$\pi^{\{\Lambda\}*}(\lambda_\alpha^{-1}) = \pi^{\{\Lambda\}}(\lambda_\alpha) \qquad (4.25)$$

(see 2.5.2). This enables us to list restrictions imposed by the condition of invariance onto the highest weight of the representation, the characters of irreducible representations being given explicitly.

For a general semisimple Lie group G, from the invariance condition (4.21) and making use of the explicit form of the generators, we deduce a system of equations for the kernel $K^{\{\Lambda\}}(k)$ of type (4.7) whose compatibility condition is $\{\rho, s\}_\omega = \{-\overset{*}{\rho} - 2\rho^0, s\}$, where $\omega \in W_G$. A solution for $K^{\{\Lambda\}}(k)$ coincides with the kernel of the corresponding intertwining operator.

To see this, it suffices to compare (4.6) and (4.21). Therefore, the obtained expressions enable us to distinguish the pseudo-unitary components from all the irreducible representations.

Let us illustrate the above with $U(m, 1)$. It follows from solvability of the corresponding system that for pseudo-unitary representations either $\rho = -m + i\sigma$, $\sigma = \sigma^*$, or ρ is real. In the first case, corresponding to the principal continuous series, the kernel is just a δ-function and the positive definiteness of the form is obvious. In the second case, starting from the explicit expression (4.14) and imposing conditions of positive definiteness we get restrictions on the weights $(l_1, \ldots, l_m)$ of irreducible unitary representations of the compact subgroup $U(m)$ entering the decomposition of the representation $\{\Lambda_1, \ldots, \Lambda_{m+1}\}$ of $U(m, 1)$ with integer weights. Therefore we can completely describe the discrete series of unitary representations of this group.

The positive definiteness implies $f_\alpha \leq \min(\Lambda_1, \Lambda_{m+1}) - 1$ or $f_\alpha \geq \max(\Lambda_1, \Lambda_{m+1})$, $f_\alpha \equiv l_\alpha + m - \alpha$. Hence the residue of the corresponding Γ-function in the denominator of (4.14) deletes $(-1)^{l_\alpha}$ from the form. Of the whole 2^m possible sets of inequalities only $m + 1$ are compatible with the chain $f_1 > f_2 > f_3 > \ldots > f_{m-1} > f_m$, namely

$$\min(\Lambda_1, \Lambda_{m+1}) > f_1 > \ldots > f_m;$$

$$f_1 > \ldots > f_m \geq \max(\Lambda_1, \Lambda_{m+1}); \quad f_1 > \ldots > f_{k-1} \geq \max(\Lambda_1, \Lambda_{m+1})$$

$$\geq \min(\Lambda_1, \Lambda_{m+1}) > f_k > \ldots > f_m; \quad 2 \leq k \leq m.$$

Making use of the Gelfand-Tsetlin inequalities for the unitary group, i.e. $f_1 > \Lambda_2 \geq f_2 > \Lambda_3 \geq f_3 > \ldots > \Lambda_m \geq f_m$, we finally see that parameters of the discrete series of unitary representations of $U(m, 1)$ satisfy

$$\min(\Lambda_1, \Lambda_{m+1}) > f_1 > \Lambda_2 \geq f_2 > \ldots > \Lambda_m \geq f_m;$$

$$f_1 > \Lambda_2 \geq f_2 > \ldots > \Lambda_m \geq f_m \geq \max(\Lambda_1, \Lambda_{m+1});$$

$$f_1 > \Lambda_2 \geq f_2 > \ldots > \Lambda_{k-1} \geq f_{k-1} \geq \hat{T}(\Lambda_1, \Lambda_k, \Lambda_{m+1})$$

$$> f_k > \Lambda_{k+1} \geq \ldots > \Lambda_m \geq f_m,$$

$$(4.26)$$

where $\hat{T}$ is the ordering symbol.

Therefore, being restricted onto $U(m)$, the space of a representation of a discrete series of $U(m, 1)$ splits into the sum of nonequivalent irreducible invariant subspaces $D(l_1, \ldots, l_m)$ determined by (4.26). On each of these subspaces the invariant Hermitian form is uniquely determined up to a factor and is positive definite.

§ 2.5 Harmonic analysis on semisimple Lie groups*)

2.5.1 General method. One of the main aspects of application of the group-theoretical approach to the study of dynamical systems is harmonic analysis on the group (or on the homogeneous space with the given motion group). For a generalization of the classical Fourier analysis it is most useful and effective as applied to quantum systems where the main objects are expansions in wave functions. These functions are given on a group G or its homogeneous space corresponding to the physical problem. They are eigenfunctions for Casimir operators of the corresponding representation and are square integrable on G with the invariant Haar measure $d\mu(g)$, i.e., they belong to $L^2(G; \mu)$.

In accordance with this the main problem of harmonic analysis on G is the construction of generalized Fourier analysis for functions from a subspace dense in $L^2(G; \mu)$:

$$f(g) = \int_\Lambda \sum_{\{\mathcal{M}\}} f_{\{\mathcal{M}\}}(\Lambda) \cdot e_{\{\mathcal{M}\}}(\Lambda, g) d\omega(\Lambda)$$

$$\text{where } f_{\{\mathcal{M}\}}(\Lambda) = \int_G f(g) \cdot \overset{*}{e}_{\{\mathcal{M}\}}(\Lambda, g) d\mu(g) \tag{5.1}$$

satisfying Plancherel formula

$$\int_G \overset{*}{f}{}^1(g) f^2(g) d\mu(g) = \int_\Lambda \sum_{\{\mathcal{M}\}} \overset{*}{f}{}^1_{\{\mathcal{M}\}}(\Lambda) f^2_{\{M\}}(\Lambda) d\omega(\Lambda). \tag{5.2}$$

Here $\{\Lambda\}$ denotes the weights of irreducible representations entering the decomposition of the regular one in terms of which the eigenvalues of Casimir operators are expressed and $\{\mathcal{M}\}$ is the set of the remaining weights (basis indices) characterizing basis vectors $e_{\{\mathcal{M}\}}(\Lambda, g)$ of $\{\Lambda\}$. The measure $d\omega(\Lambda)$ on the spectrum $\{\Lambda\}$ which contains in general both continuous and discrete components is called *Plancherel measure*. The integral sign in (5.1) and (5.2) denotes integration over continuous and summation over discrete part of the spectrum (the same applies to $\{\mathcal{M}\}$). For basis vectors we have the following relations of completeness and orthogonality:

$$\int_\Lambda d\omega(\Lambda) \sum_{\{\mathcal{M}\}} \overset{*}{e}_{\{\mathcal{M}\}}(\Lambda, g) \cdot e_{\{\mathcal{M}\}}(\Lambda, g') = \delta(gg'^{-1}), \tag{5.3}$$

$$\int_G d\mu(g) \overset{*}{e}_{\{\mathcal{M}\}}(\Lambda, g) \cdot e_{\{\mathcal{M}'\}}(\Lambda', g) = \omega^{-1}(\Lambda)\delta(\Lambda - \Lambda')\delta_{\{\mathcal{M}\},\{\mathcal{M}'\}}, \tag{5.4}$$

---

*) [7, 13, 19, 21, 35, 36, 51, 61, 95, 109, 112, 113]

where $\delta(\Lambda - \Lambda')$ is the δ-function for continuous and the Kronecker symbol for discrete components of the spectrum.

Notice that the mathematically rigorous formulation of the above expressions which is a generalization of Peter-Weyl's theorem for compact groups requires the spectral theory for semisimple Lie groups. Therefore we confine ourselves to stating that all the integrals which appear should be understood as regularized distributions and integrals in (5.3) and (5.4) as products of distributions on the corresponding class. In what follows we will consider basis vectors as regular distributions and Plancherel's measure as absolutely continuous with respect to Lebesgue measure, i.e., $d\omega(\Lambda) = \omega(\Lambda)d\Lambda$ with a continuous weight function $\omega(\Lambda)$. (The latter fact is used in particular in (5.4).) Formula (5.1) realizes the decomposition of the regular representation into irreducible unitary components of the principal series.

For basis $e_{\{\mathcal{M}\}}(\Lambda, g)$ we can take matrix elements of irreducible unitary representations of the principal series, i.e., in the Cartan decomposition for G we take expressions of the form (3.4) for the corresponding restrictions onto the weights $\{\Lambda\} \equiv \{\rho, l\}$, which for brevity we will denote by $D^{\{\Lambda\}}_{\{\mathcal{L}\},\{\mathcal{L}'\}}$. (Notice that in certain applications the corresponding decompositions are convenient to perform with respect to generating functions of matrix elements which often have more lucid analytic structure and simple properties.)

The matrix elements are normed by the condition $D^{\{\Lambda\}}_{\{\mathcal{L}\},\{\mathcal{L}'\}}(g_0) = \delta_{\{\mathcal{L}\},\{\mathcal{L}'\}}$ in accordance with (3.4) and satisfy the completeness and orthogonality conditions in the form (5.3) and (5.4) and also the conjugacy and summation relations:

$$D^{\{\Lambda\}}_{\{\mathcal{L}\},\{\mathcal{L}'\}}(g^{-1}) = \overset{*}{D}{}^{\{\Lambda\}}_{\{\mathcal{L}'\},\{\mathcal{L}\}}(g), \tag{5.5}$$

$$D^{\{\Lambda\}}_{\{\mathcal{L}\},\{\mathcal{L}'\}}(g_1 g_2) = \sum_{\{\mathcal{L}''\}} D^{\{\Lambda\}}_{\{\mathcal{L}\},\{\mathcal{L}''\}}(g_1) D^{\{\Lambda\}}_{\{\mathcal{L}''\},\{\mathcal{L}'\}}(g_2). \tag{5.6}$$

Thus, to use harmonic analysis effectively in application it is necessary to know the main ingredients of (5.1)–(5.4) explicitly, i.e., to know matrix elements of the principal series of unitary representations, the invariant Haar measure and the Plancherel measure. In a number of cases, to get the necessary information on certain properties of a physical system it is sufficient to consider the formulation in which the dependence on weights $\{\mathcal{M}\}$ is summarized. In other words, in some cases it is sufficient for us to consider "the decomposition of unity", i.e., to consider representation of (5.3) is in the form

$$\int_{\{\Lambda\}} \pi^{\{\Lambda\}}(g) d\omega(\Lambda) = \delta(g). \tag{5.7}$$

Here $\pi^{\{\Lambda\}}(g) = \sum\limits_{\{\mathcal{L}\}} D^{\{\Lambda\}}_{\{\mathcal{L}\},\{\mathcal{L}\}}(g)$ is the character of the irreducible representation considered as a distribution on G. Notice that properties of characters are essential in the formulation of some results for nonlinear dynamical systems.

2.5.2 Characters of operator-irreducible representations. To every irreducible representation of a semisimple Lie group G we can assign a function (a distribution for infinite-dimensional representations) on G:

$$\pi^{\{\Lambda\}}(g) = \text{tr } T^{\{\Lambda\}}(g). \qquad (5.8)$$

This functional completely determines a representation up to equivalence and is called its (*central*) *character*. Obviously, the character is a distribution invariant with respect to inner automorphisms of G, i.e.

$$\pi^{\{\Lambda\}}(g) = \pi^{\{\Lambda\}}(g_0^{-1}gg_0) \quad \text{for} \quad g_0 \in G$$

or, infinitesimally,

$$F_a^R \pi^{\{\Lambda\}}(g) = -F_a^L \pi^{\{\Lambda\}}(g). \qquad (5.9)$$

The character is also an eigenfunction for the Casimir operators

$$C_j \pi^{\{\Lambda\}}(g) = c_j(\Lambda)\pi^{\{\Lambda\}}(g), \qquad (5.10)$$

where $c_j(\Lambda)$ is an eigenvalue. Not every solution of (5.9) and (5.10) determines the character of a representation; to distinguish characters additional selection rules are needed (of boundary condition type).

Starting from definition (5.8) and formula (3.3) we get the following integrable representation for the character of an operator-irreducible representation of G:

$$\pi^{\{\rho,l\}}(g) = \int d\mu(k_{-s}) \cdot \delta(k_{-s} \cdot \tilde{k}_{-s}^{-1}) \prod_{1 \leq j \leq r_A} [R_j(g, k_{-s})]^{\rho_j} \pi^{\{l\}}(N), \qquad (5.11)$$

where $\pi^{\{l\}}(N)$ is a character of an irreducible representation of the centralizer S expressed via eigenvalues $\exp\varphi_j$, $1 \leq j \leq r_S$, of N via the known Weyl formula for compact groups:

$$\pi^{\{l\}}(\varphi) = \frac{\sum\limits_{\omega \in W_S} \det\omega \cdot \exp[i\sum\limits_j \varphi_j(l_j + (1/2)\rho_j^0)^\omega]}{\sum\limits_{\omega \in W_S} \det\omega \cdot \exp[\frac{i}{2}\sum\limits_j \varphi_j\rho_j^{0\omega}]}. \qquad (5.12)$$

(On introducing smoothing functions to correctly define the character as a generalized function in eigenvalues of g, see e.g. [19].)

Notice that the formal calculation of the trace in (3.4), i.e. $\sum\limits_{\{\mathcal{L}\}} D^{\{\Lambda\}}_{\{\mathcal{L}\},\{\mathcal{L}\}}(g)$ leads also to (5.11) although the justification of the change in order of summation and integration requires additional consideration. Let us give the final formula for the character of an operator-irreducible representation $\{\rho, l\}$ of G:

$$\pi^{\{\rho,l\}}(g) = \sum_{\omega \in W/W_S} \frac{\exp \sum_{1 \leq j \leq r_A} \tau_j \rho_j^\omega}{\prod_{\alpha > 0} {}'|1 - \exp(-\alpha(t))|^{\delta_\alpha}} \pi^{\{l^\omega\}}(\varphi), \qquad (5.13)$$

where τ_j, $1 \leq j \leq r_A$, are noncompact parameters of $\mathcal{A}$ in the Cartan decomposition. The eigenvalues $\lambda_a \equiv \exp t_a$ of g are parametrized via τ and φ according to the standard rules. The product $\prod'_{\alpha > 0}$ runs over all positive roots α of multiplicity δ_α except the roots of S; all $\alpha(t)$ are different. For a complex group ($W_S = 1$, $\delta_\alpha = 2$) formula (5.13) takes the form

$$\pi^{\{\rho,l\}}(g) = \sum_\omega \exp\left[\sum_j (\tau_j \rho_j^\omega + i\varphi_j l_j^\omega)\right] \bigg/ \prod_{\alpha > 0} |1 - \exp(-\alpha(t))|^2. \quad (5.14)$$

Notice that for topologically reducible representations $\{\Lambda\}$ formula (5.13) gives the sum of characters of irreducible components. Besides, it follows from (4.25) that the characters of the dual representations $T^{\{\Lambda\}^\dagger}$ and $T^{\{\Lambda\}}$ coincide, i.e.,

$$\overset{*}{\pi}{}^{\{\Lambda\}}(\lambda^{-1}) = \pi^{\{\Lambda\}}(\lambda_a). \qquad (5.15)$$

This allows us to list restrictions imposed by (pseudo) unitarity on the highest weight of the representation if we know the characters of irreducible representations explicitly.

2.5.3 Plancherel measure of the principal continuous series of unitary representations.

With known expressions for matrix elements for semisimple Lie groups we can calculate the weight function of Plancherel's measure of the principal series of unitary representations via (5.4). In this case ($\rho_j = -\rho_j^0 + i\sigma_j$, $1 \leq j \leq r_A$) the orthogonality relation (5.4) takes the form

$$\int_G d\mu(g) \overset{*}{D}{}^{\{\rho,l\}}_{\{\mathcal{L}_1\},\{\mathcal{L}_2\}}(g) \cdot D^{\{\rho',l'\}}_{\{\mathcal{L}_3\},\{\mathcal{L}_4\}}(g)$$

$$= \delta_{\{\mathcal{L}_1\},\{\mathcal{L}_3\}} \delta_{\{\mathcal{L}_2\},\{\mathcal{L}_4\}} \delta_{\{l\},\{l'\}} \cdot \prod_{j=1}^{r_A} \delta(\sigma_j - \sigma_j') \cdot \omega^{-1}(\sigma, l), \qquad (5.16)$$

and the corresponding calculations can be performed till the end by using a method applied by Fock to calculate the norming constant of the continuous spectrum of the hydrogen atom. Its essence is to pass to the asymptotic limit of integrands whose properties are supposed to be known. After that for the Cartan decomposition it remains only to integrate over parameters of the maximal compact subgroup which considerably simplifies all the calculations and leads to final expressions.

For the Cartan decomposition (1.5.11) and the corresponding invariant measure (1.5.12)

$$\int f(g)d\mu(g) = \int_A \prod_{1 \le j \le r_A} d\tau_j D(\tau) \int \int f(k \cdot a(\tau) \cdot k')d\mu(k')d\mu(k) \quad (5.17)$$

in the asymptotic domain and with a given ordering over τ_j, e.g. $\tau_j - \tau_{j+1} \to \infty$, $1 \le j \le r_A$, $\tau_{r_A+1} \equiv 0$, we get $D(\tau) \to \exp 2 \sum_j \rho_j^0 \tau_j$.

Let us normalize the invariant measures on $\mathcal{K}$ to have total measure one. To find the norming constant $\omega^{-1}(\sigma, l)$ in (5.16), which does not depend on the weights $\{\mathcal{L}_1\}$ and $\{\mathcal{L}_2\}$ of basis vectors of the representation, it is convenient to take a certain set $\{\mathcal{L}_1\} = \{\mathcal{L}_2\}$ compatible with conditions on matrix elements of $\mathcal{K}$ with the centralizer S. Namely, for the matrix element $D_{\{L\};\{M\},\{N\}}^{\{l\}}(k)$ of the maximal compact subgroup entering the integral representation (3.4) of $D_{\{\mathcal{L}_1\},\{\mathcal{L}_2\}}^{\{\rho,l\}}(g)$, take the matrix element between the highest vectors of the corresponding representations, i.e., set

$$\{\mathcal{L}\} \equiv \left\{ \begin{matrix} L \\ MN \end{matrix} \right\} \equiv \left\{ \begin{matrix} l \\ ll \end{matrix} \right\}.$$

We can then make use of the explicit expressions (1.6.19) for the highest matrix elements of irreducible representations of compact groups and invariant measure (1.5.18), ((1.6.20) and (1.5.20) for normal real forms, respectively), in order to calculate integrals of type (5.16). The asymptotic expression for $D_{\{l\},\{l\}}^{\{\rho,l\}}(g)$ is derived from (3.4) by the usual methods and is of the form

$$D_{\{l\},\{l\}}^{\{\rho,l\}}(g) \to \exp\left[-\sum_j \tau_j \rho_j^0 \right] \cdot \sum_{\omega \in W_G} \exp[i\tau_j \cdot \sigma_j^\omega]$$

$$\times \int d\mu(k) \cdot \prod_{j=1}^{r_A} [R_j^{(a)}]^{\rho_j} \overset{*}{D}_{\{l\},\{l\}}^{\{l\}}(k) \cdot D_{\{l\},\{l\}}^{\{l\}}(\tilde{k}^{(a)}), \quad (5.18)$$

where the index a indicates that the corresponding variables are taken in the asymptotic domain. The explicit expressions for $R^{(a)}$ and $\tilde{k}^{(a)}$ follow directly from (3.7). Inserting into (5.16) the asymptotic values of the matrix elements (5.18) and integrating over τ_j using (5.17) (which leads to δ-functions in the right-hand side of (5.16)) we get

$$\omega^{-1}(\sigma, l) = (2\pi)^{r_A} \int d\mu(k) \int d\mu(k') \cdot \prod_{j=1}^{r_A} [R_j^{(a)}(k)]^{\rho_j}$$

$$\times [R_j^{(a)}(k')]^{\rho_j} D_{\{l\},\{l\}}^{*\{l\}}(k'^{-1}k) \cdot D_{\{l\},\{l\}}^{\{l\}}((\tilde{k}'^{(a)})^{-1}\tilde{k}^{(a)}). \quad (5.19)$$

Further simplification of the formula for $\omega(\sigma, l)$ uses properties of the matrix elements $D_{\{l\},\{l\}}^{\{l\}}$. Expanding $D_{\{l\},\{l\}}^{\{l\}}(k'^{-1}k)$ with respect to the complete

Group		ρ_s
$GL(n,\mathbf{C})$	$\displaystyle\prod_{i>j}[(\sigma_i - \sigma_j)^2 + (x_i - x_j)^2]$	$\rho_s = -(n - 2s + 1) + i\sigma_s$
$GL(n,\mathbf{R})$	$\displaystyle\prod_{i>j}(\sigma_i - \sigma_j)\,\mathrm{th}\left[\frac{\pi}{2}(\tilde\sigma_i - \tilde\sigma_j)\right];\ \tilde\sigma_s \equiv \sigma_s + i\xi_s,\ \xi_s = 0,1;$	$\rho_s = -(n+1)/2 + s + i\sigma_s$
$*U(2n)$	$\displaystyle\prod_{i>j}[(\sigma_i - \sigma_j)^2 + (x_i - x_j)^2]\,[(\sigma_i - \sigma_j)^2 + (x_i + x_j + 1)^2];$	$\rho_s = -2n + 4s - 2 + i2\sigma_s$
$U(p,q)$ $p \le q$	$\displaystyle\prod_{1\le i\le p}\sigma_i\,\mathrm{th}\left(\frac{\pi}{2}\tilde\sigma_i\right)\cdot\prod_{i>j}[(\sigma_i - \sigma_j)^2 + (x_i - x_j)^2]\,[(\sigma_i + \sigma_j)^2]\,[(\sigma_i + \sigma_j)^2$ $+(x_i - x_j)^2]\displaystyle\prod_{1\le s\le p}\ \prod_{1\le a\le q-p}[\sigma_s^2 + (x_s - l_a)^2];$	$\rho_s = -(p + q - 2s + 1) + i2\sigma_s$
$O(n,\mathbf{C})$	$n = 2k$ $\displaystyle\prod_{i>j}[(\sigma_i + \sigma_j)^2 + (x_i + x_j)^2]\,[(\sigma_i - \sigma_j)^2 + (x_i - x_j)^2];$ $n = 2k+1$ $\displaystyle\prod_s[\sigma_s^2 + x_s^2]\prod_{i>j}[(\sigma_i + \sigma_j)^2 + (x_i + x_j)^2]\,[(\sigma_i - \sigma_j)^2 + (x_i - x_j)^2];$	$\rho_s = -2n + 2s + i\sigma_s$ $\rho_s = -n + 2s + i\sigma_s$
$*\tilde{O}(2n)$	$n = 2k$ $\displaystyle\prod_{1\le s\le k}\sigma_s\,\mathrm{th}\left(\frac{\pi}{2}\tilde\sigma_s\right)\cdot\prod_{i>j}[(\sigma_i - \sigma_j)^2 + (x_i - x_j)^2]\,[(\sigma_i - \sigma_j)^2 +$ $+(x_i + x_j + 1)^2]\,[(\sigma_i + \sigma_j)^2 + (x_i + x_j + 1)^2];$ $n = 2k+1$ $\displaystyle\prod_s\sigma_s\,\mathrm{th}\left(\frac{\pi}{2}\tilde\sigma_s\right)\cdot[\sigma_s^2 + (x_s + 1/2)^2]\,[\sigma_s^2 + (x + x_s + 1/2)^2]\times$ $\times\displaystyle\prod_{i>j}[(\sigma_i - \sigma_j)^2 + (x_i - x_j)^2]\,[(\sigma_i - \sigma_j)^2 + (x_i + x_j + 1)^2]\,[(\sigma_i \mid \sigma_j)^2 +$ $+(x_i - x_j)^2]\,[\,][(\sigma_i + \sigma_j)^2 + (x_i + x_j + 1)^2];$	$\rho_s = -2n - 1 + 4s + i2\sigma_s$ $\rho_s = -2n - 1 + 4s + i2\sigma_s$

Group	Weight function	ρ_s
$O(p,q)$ $p\le q$	$q-p=2k$: $$\prod_{p>i>j}(\sigma_i+\sigma_j)\,\text{th}\left[\frac{\pi}{2}(\tilde\sigma_i+\tilde\sigma_j)\right](\sigma_i-\sigma_j)\,\text{th}\left[\frac{\pi}{2}(\tilde\sigma_i-\tilde\sigma_j)\right]\times$$ $$\times\prod_{s,a}\left[\sigma_s^2+\left(l_a+\frac{q-p}{2}-a\right)^2\right];$$ $q-p=2k+1$: $$\prod_{1\le s\le p}\sigma_s\,\text{th}\left(\frac{\pi}{2}\tilde\sigma_s\right)\cdot\prod_{i>j}(\sigma_i+\sigma_j)\,\text{th}\left[\frac{\pi}{2}(\tilde\sigma_i+\tilde\sigma_j)\right]\times$$ $$\times(\sigma_i-\sigma_j)\,\text{th}\left[\frac{\pi}{2}(\tilde\sigma_i-\tilde\sigma_j)\right]\cdot\prod_{s,a}\left[\sigma_s^2+\left(l_a+\frac{q-p}{2}-a\right)^2\right];$$	$\rho_s=(p+q)/2+s+i\sigma_s$ $\rho_s=-(p+q)/2+s+i\sigma_s$
$\text{Sp}(2n,\mathbf{R})$	$$\prod_{1\le i\le n}\sigma_i\,\text{th}\left(\frac{\pi}{2}\tilde\sigma_i\right)\prod_{s>k}(\sigma_s+\sigma_k)\,\text{th}\left[\frac{\pi}{2}(\tilde\sigma_s+\tilde\sigma_k)\right](\sigma_s-\sigma_k)\times$$ $$\times\,\text{th}\left[\frac{\pi}{2}(\tilde\sigma_s-\tilde\sigma_k)\right];$$	$\rho_s=-(n-s+1)+i\sigma_s$
$\text{Sp}(2n,\mathbf{C})$	$$\prod_{1\le s\le n}(\sigma_s^2+x_s^2)\prod_{i>j}\left[(\sigma_i+\sigma_j)^2+(x_i+x_j)^2\right]\left[(\sigma_i-\sigma_j)^2+(x_i-x_j)^2\right];$$	$\rho_s=-2(n-s+1)+i\sigma_s$
$\text{Sp}(2n,2m)$ $n\le m$	$$\prod_{1\le s\le n}\sigma_s\,\text{th}\left(\frac{\pi}{2}\tilde\sigma_s\right)\cdot\left[\sigma_s^2+(x_s+\tfrac{1}{2})^2\right]\prod_{i>j}\left[(\sigma_i-\sigma_j)^2+(x_i-x_j)^2\right]\left[(\sigma_i+\sigma_j)^2+(x_i+x_j+1)^2\right]\times$$ $$\times\left[(\sigma_i-\sigma_j)^2+(x_i-x_j)^2\right]\left[(\sigma_i+\sigma_j)^2+(x_i+x_j+1)^2\right]\times$$ $$\times\prod_{k,\alpha}\left[\sigma_k^2+(x_k+\tfrac{1}{2}-l_\alpha)^2\right]\left[\sigma_k^2+(x_k+\tfrac{1}{2}+l_\alpha)^2\right]$$	

Table 2.6 Weight functions of the Plancherel measure of the principal continuous series of unitary representations of the classical Lie groups

system of states via (5.6) and integrating over the parameters of S we derive from (5.19)

$$\omega^{-1}(\sigma, l) = \frac{(2\pi)^{r_A}}{N_l} \left| \int_{K/S} d\mu(k) \prod_{1 \leq j \leq r_A} [R_j^{(a)}(k)]^{\rho_j} \overset{*\{l\}}{D}_{\{l\},\{l\}} (\tilde{k}^{(a)^{-1}} k) \right|^2, \quad (5.20)$$

where N_l is the dimension of the representation of S given by (1.6.21) in terms of the highest weight $\{l\}$.

Therefore, the problem of calculating Plancherel's measure of the principal continuous series of unitary representations of semisimple Lie groups reduces to integrals of the form

$$J(\sigma, l) = \int_{K/S} d\mu(k) \prod_{1 \leq j \leq r_A} [R_j^{(a)}(k)]^{\rho_j} \overset{*\{l\}}{D}_{\{l\},\{l\}} (\tilde{k}^{(a)^{-1}} k), \quad (5.21)$$

for $\rho_j = -\rho_j^0 + i\sigma_j$. Strictly speaking, these should be understood as distributions since the integration is performed at the limit of convergence.

Let us illustrate how to calculate (5.21) for complex groups. In this case it takes the form

$$J(\sigma, l) = \int \prod_{1 \leq j \leq r_A} |\det_j k|^{i(\sigma_j - \sigma_{j+1}) + (l_j - l_{j+1})} d\mu(k), \quad (5.22)$$

where the integrand is nothing but an analytic continuation of the squared modulus of the highest matrix element of the representation $\{L\}$ of the maximal compact subgroup to the domain of complex values $(l_j + i\sigma_j)/2$ of the highest weight of the corresponding representation. Therefore the weight function of the Plancherel measure is of the form

$$\omega(\sigma, l) = (2\pi)^{-r_A} |N_L|^2 \Big|_{\{L\} \equiv \{(l+i\sigma)/2\}},$$

where N_L is the dimension of the irreducible representation of K given by (1.6.21).

For an arbitrary semisimple Lie group the weight function for the principal continuous series is given by the formula

$$\omega(\sigma, l) = (2\pi)^{-r_A} N_{\{l\}} |J(\sigma, l)|^2,$$

$$J(\sigma, l) = \prod_{\alpha > 0} {}' B[\xi_\alpha/2, (\alpha, (\rho + \rho_0/2))]/B[\xi_\alpha/2, (\alpha, \rho_0/2)], \quad (5.23)$$

where the product runs over positive roots (except those of the centralizer) for which all the $|(\alpha, \rho + \rho_0)|$ are different and ξ_α is the multiplicity of the corresponding root (the explicit expressions of $\omega(\sigma, l)$ for the classical series are given in Table 2.6).

Notice that for representations of type I (the case of Riemannian symmetric spaces of nonpositive curvature) the expression for the weight function (5.23) can be obtained by a considerably more simple procedure [21]. Formula (5.23) holds also for an arbitrary (not necessarily classical) complex semismple Lie group.

The expressions found for weight functions of Plancherel's measure of the principal continuous series of unitary representations of semisimple Lie groups are analytic functions in parameters of the representation ρ which have poles of order no greater than 1 (thanks to hyperbolic tangents).

The positions of poles and their number are in one-to-one correspondence with the set of principal series of unitary representations of the groups considered.

§ 2.6 Whittaker vectors*)

In the following study of the quantization of the generalized Toda lattice we will need explicit expressions for the *Whittaker vectors* introduced by Kostant. Conventionally they are defined as the eigenvectors of the shift operator of the principal series of unitary representations of a semisimple or reductive Lie group G:

$$T^{\{\Lambda\}}(z)\varphi_+ = \varphi(z)\varphi_+, \tag{6.1}$$

where z belongs to a maximal nilpotent subgroup of G. Infinitesimally (6.1) takes the form (for left shifts)

$$\hat{Z}_\alpha^L \varphi_+ = i\delta_{j\alpha}\varphi_+, \tag{6.2_1}$$

or, thanks to the linearity of $\hat{Z}_\alpha^L$ in derivative of the parameters of G,

$$\hat{Z}_\alpha^L \log \varphi_+ = i\delta_{j\alpha}. \tag{6.2_2}$$

Here $\delta_{j\alpha}$ means that the operators of infinitesimal left shifts annihilate a Whittaker vector for any α which is not a simple root.

For our goals it suffices at the moment to confine ourselves to the normal real forms of complex semisimple Lie groups G in the Iwasawa decomposition. Then the φ_+ are given on the maximal compact subgroup $\mathcal{K}$ of G and the $\hat{Z}_\alpha^L$ are determined thanks to (2.1.16) by the expression

$$\hat{Z}_\alpha^L = \sum_{\beta>0}(\mathcal{X}^\beta, k^{-1}\mathcal{X}_\alpha k)(X_\alpha^R - X_{-\alpha}^R) - \sum_j (h^j, k^{-1}\mathcal{X}_\alpha k)\rho_j. \tag{6.3}$$

An operator $\hat{Z}_\alpha^L$ of the form (6.3) is, in accordance with (2.1.23), an analytic continuation (with respect to the weight) of an infinitesimal operator of the

---

*) [23, 24, 53, 108]

left regular representation of $\mathcal{K}$ corresponding to a positive root α. Therefore thanks to (1.6.2) and (1.6.3) a solution of the homogeneous part of (6.2$_2$) is given by $\ln \xi^{\{\rho\}}(k) = \ln \prod_{1 \leq j \leq r} [\xi^{\{j\}}(k)]^{\rho_j}$, where $\xi^{\{j\}}(k)$ is the restriction onto $\mathcal{K}$ of the highest matrix element of the j-th fundamental representation of G. It is also obvious that the δ-like nonhomogeneity in (6.2$_2$) is generated by the terms in $\log \varphi_+$ obtained from $\log \xi^{\{j\}}(k)$ under the action of operators corresponding to root vectors of negative simple roots. As is easy to see, the elements of the representation $\{\Lambda\}$ obtained under the action on $\xi^{\{j\}}$ of operators of greater height do not satisfy (6.2).

Substituting $\Psi_+ \equiv \sum_{i,k} c_{ik}\varphi_{ik}(k)$, where $\varphi_{ik} \equiv \hat{X}^L_{-i} \log \xi^{\{k\}}(k)$, into (6.2$_2$), where only the terms with $i = k$, i.e., $(c_{ik} \equiv \delta_{ik}c_i)$ should be taken into account since $\hat{X}^L_{-i}\xi^{\{k\}}(k) = 0$ for $i \neq k$ we get, thanks to (1.4.19), the chain of equalities

$$\hat{Z}^L_\alpha \Psi_+ \equiv \hat{X}^L_\alpha \Psi_+ = \sum_k c_k \hat{X}^L_\alpha \hat{X}^L_{-k} \ln \xi^{\{k\}}(k)$$

$$= \sum_k c_k \delta_{\alpha k} \hat{h}^L_k \ln \xi^{\{k\}}(k) = \sum_k \delta_{\alpha k} c_k.$$

Therefore, Ψ_+ satisfies the equations under consideration only when $c_k = \sqrt{-1}$ for all k.

Therefore, setting $\varphi_+(1) = 1$ we get the following final expression for Whittaker vector:

$$\varphi_+(k) = \xi^{\{\rho\}}(k) \exp i \sum_j \hat{X}^L_{-j} \log \xi^{\{j\}}$$

$$\equiv \xi^{\{\rho\}}(k) \exp \left\{ -i \sum_j \hat{K}^L_j \log \xi^{\{j\}} \right\}, \qquad (6.4)$$

$$\hat{K}^L_j \equiv \hat{X}^L_j - \hat{X}^L_{-j}.$$

As we will see in Chapter 7, where we quantize the system (3.2.8) in Schrödinger's representation containing a finite nonperiodic Toda lattice as the simplest particular case, the role of the basis functions in the representation space of the corresponding real group is played by the generalized Whittaker vectors. To construct them consider the Iwasawa decomposition (1.5.9) for an arbitrary regular element $g = kT(\tau)\hat{z}_+$, $T(\tau) = \exp(\sum_j h_j\tau_j)$, $1 \leq j \leq rk\mathcal{A}$, of a noncompact real form $G^{\mathbf{R}}$, where the parameters of the centralizer S are considered excluded from $k \in \mathcal{K}$ and included in $T(\tau)$, i.e., $g = k_{-s} \cdot \exp B \cdot \hat{z}_+$ (see 2.3.1). Here k_{-s} contains only parameters corresponding to roots of $\mathfrak{g}/\mathfrak{g}_0$, whereas B takes values in $\mathfrak{g}_0$ and in addition to $\{\tau_j\}$

has parameters χ_s, $1 \le s \le d_0 - r$. In such a parametrization the expression for the quadratic Casimir operator takes the form (see 2.2)

$$C = C_0 - \sum_{\underline{\alpha},\underline{\beta}} (\mathcal{X}^{\underline{\beta}}, \exp B \cdot \mathcal{X}_{-\underline{\alpha}} \cdot \exp(-B)) \mathcal{K}_{\underline{\alpha}}^R Z_{\underline{\beta}}^L$$

$$+ \sum_{\underline{\alpha},\underline{\beta}} (\exp(-B)\mathcal{X}^{\underline{\alpha}} \exp B, \ \exp B \mathcal{X}^{-\underline{\beta}} \exp(-B)) Z_{\underline{\alpha}}^L Z_{\underline{\beta}}^L, \quad (6.5)$$

where $\mathcal{K}_{\underline{\alpha}}^R$ are generators of right shifts on K corresponding to $\underline{\alpha} \in R_+$.

Suppose the operator (6.5) with $\exp 2B \equiv g_0 \in G_0$ acts on its eigenfunctions Ψ so that

$$\mathcal{K}_{\underline{\alpha}}^R \Psi = 0, \quad (6.6_1)$$

$$Z_{\underline{\alpha}}^L \Psi = \sum_{\alpha \in R_+^1} \delta_{\alpha\underline{\alpha}} v_\alpha c_\alpha \Psi. \quad (6.6_2)$$

Moreover, exclude from C_0 the terms associated with $\mathfrak{g}_0^0$. Here c_α denotes the nonzero coefficients of the decomposition of $J_+ \in A_1$ in terms of the basis of $\mathfrak{g}$, the element $H \in A_1$ performing the considered grading of $\mathfrak{g}$.

Since in what follows the Ψ will be interpreted as wave functions of a quantum-mechanical system and therefore should be normalized, they are transformed via an irreducible unitary representation of $G^{\mathbf{R}}$. Therefore, thanks to the above conditions, the functions Ψ are matrix elements of the principal continuous series of unitary representations between the states invariant with respect to right shifts on $\mathcal{K}_\alpha^R$. In turn the matrix elements of the transition between the states with vectors $\varphi_\pm(k)$ are determined in accordance with the realization (3.1) by the formula

$$\int \overset{*}{\varphi}_-(k) \prod_j [R_j(k,g)]^{\rho_j} \varphi_+(\tilde{k}) d\mu(k),$$

where for operator-irreducible components the functions $\varphi_\pm(k)$ should be transformed via unitary representations $\{l\}$ of the centralizer, see (3.2). Therefore, to get the desired wave functions Ψ one should construct an expression for $\varphi_+(k)$ which satisfies the above conditions and is scalar with respect to the transformations generated by $\mathfrak{g}_0^0$, with $\varphi_-(k)|_{K/S} = 1$ due to (6.6_1).

Let us first satisfy equations (6.6_2) formally with zero right-hand side for any $\underline{\alpha} \in R_+$ and then take the nonhomogeneity into account. Obviously, for a given representation $\{\rho, l\}$ of $G^{\mathbf{R}}$ such the functions φ_+^0 are matrix elements $\langle |k| \rangle$ between the highest states $|j\rangle$, $1 \le j \le r$, of the fundamental representations of $\mathfrak{g}$ and also between all the states of these representations annihilated under $\mathcal{X}_{\underline{\alpha}}$, $\underline{\alpha} \in R_+$. Therefore, the φ_+^0 are restrictions onto K of the matrix elements $\Xi^{\{\rho,l\}}$ of the fundamental representations of $G^{\mathbb{R}}$ which are highest with respect to $\mathfrak{g}/\mathfrak{g}_0$. The nonhomogeneous part of (6.6_2) can be recovered if $\Xi^{\{\rho,l\}}$ is multiplied by $\exp(J_-^L \log \xi')$. Here J_-^L is given by $J_-^L = \sum_\alpha c_{-\alpha} X_{-\alpha}^L$, where $X_{-\alpha}^L$ are generators of the left shifts on $\mathcal{K}$ corresponding to negative roots $-\alpha$ and $\xi' \equiv \prod_{1 \le i \le r} \xi^{\{i\}}$, $i \notin I_0$, where I_0 is the subset of $\{1,\ldots,r\}$, corresponding to $i_0 \in I_0$ with $h_{i_0} \in \mathfrak{g}_0^0$. The final expression

$$\varphi_+ \equiv W^{\{\rho,l\}} = \Xi^{\{\rho,l\}} \exp(J_-^L \ln \xi') \qquad (6.7)$$

obviously turns into a Whittaker vector (6.4) for the normal real form $G^{\mathbb{N}}$ in the canonical grading. Therefore it is natural to call the function $W^{\{\rho,l\}}$ determined by (6.7) a *generalized Whittaker vector*.

Chapter 3
A general method of integrating two-dimensional nonlinear systems

In this chapter we will show that to each $\mathbb{Z}$-graded Lie algebra there is associated a whole series of systems of equations in two-dimensional space which are completely integrable in the sense of the Goursát problem with initial values determined on characteristics. The explicit form of the nonlinear systems arising in this way depends on the structure of the Lie algebra and its grading. The integration of the system is closely related to the representation theory of the Lie algebra and the corresponding group.

The criterion for complete integrability of the systems with nonlinearities of a particular form takes the form of a condition imposed on the Lie-Bäcklund algebra of these equations and is equivalent in some sense to the solution of a classification problem for Lie algebras.

The solutions of the corresponding one-dimensional systems (of ordinary differential equations) are obtained from the general solution of the two-dimensional case by imposing certain symmetry conditions on the arguments z_+ and z_-, e.g., the solutions depend on $z_+ \pm z_-$ or $z_+ \cdot z_-$, etc. Hereafter z_+ and z_- represent independent coordinates in the two-dimensional space on which the dynamical systems are defined. The integration method developed here admits a direct extension for the case of supersymmetric systems associated with the corresponding Lie superalgebras.

§ 3.1 General method*)

3.1.1 Lax-type representation. The method is based on general properties of graded Lie algebras $\mathfrak{g}$ and Lax-type representation

$$[\partial/\partial z_+ + A_+, \, \partial/\partial z_- + A_-] \equiv A_{-,z_+} - A_{+,z_-} + [A_+, A_-] = 0, \qquad (1.1)$$

---

*) [67–69, 72, 73, 79, 126]

realized by a pair of operators $A_\pm$ with values in $\mathfrak{g}$. To this end, let us define $A_\pm$ as follows:

$$A_\pm = E_\pm^0 + \sum_{1 \le a \le m_\pm} E_\pm^a \equiv (e_0 \cdot u_\pm) + \sum_{1 \le a \le m_\pm} (e_\pm^a \cdot f_\pm^a), \qquad (1.2)$$

where $E_\pm^0$ and $E_\pm^a$ take values in the finite-dimensional subspaces $\mathfrak{g}_0$ and $\mathfrak{g}_\pm$, $1 \le a \le m_\pm$, respectively, of the decomposition (I.2.1) of $\mathfrak{g}$, and e_0 and $e_\pm^a$ are basis vectors of these subspaces. The summation of the components of the scalar products $(\cdot \cdot)$ of e_0, $e_\pm^a$ and vector-functions $u_\pm(z_+, z_-)$, $f_\pm^a(z_+, z_-)$ depending on $z_\pm$ runs from 1 to dim $\mathfrak{g}_a$, i.e., over the inner index of the a-th space.

Substituting (1.2) into (1.1) and using property (I.2.2) we get the defining system of equations

$$E_-^0, z_+ - E_+^0, z_- + [E_+^0, E_-^0] + \sum_{1 \le a \le \min(m_+, m_-)} [E_+^a, E_-^a] = 0,$$

$$\sum_{1 \le a \le m_\pm} \{ E_\pm^a, z_\mp + [E_\mp^0, E_\pm^a] \} - \sum_{1 \le b < a \le m_\pm} [E_\pm^a, E_\mp^b] = 0. \qquad (1.3)$$

In (1.3) there are three types of subsystems with values in the subspaces of zero, positive and negative indices, respectively. Equating to zero the coefficients of every linearly independent element of $\mathfrak{g}$ we get the following system of differential equations on the two-dimensional space for $u_{\pm i}(z_+, z_-)$ and $f_{\pm a}^a(z_+, z_-)$:

$$u_{-s, z_+} - u_{+s, z_-}$$

$$+ \sum_{1 \le i,j \le d_0} N_{ij}^s u_{+i} u_{-j} + \sum_{\substack{1 \le a \le \min(m_+, m_-) \\ 1 \le \alpha, \beta \le d_{\pm a}}} N_{\alpha, -\beta}^s f_{+\alpha}^a f_{-\beta}^a = 0,$$

$$(1 \le s \le d_0),$$

$$f_{\pm \alpha, z_\mp}^a + \sum_{\substack{1 \le i \le d_0 \\ 1 \le \beta \le d_{\pm a}}} N_{i, \pm \beta}^{\pm \alpha} u_{\mp i} f_{\pm \beta}^a \qquad (1.4)$$

$$\mp \sum_{\substack{1 \le b \le m_\pm \\ (b \ge a+1)}} \sum_{\substack{1 \le \nu \le d_b \\ 1 \le \beta \le d_{-(b-a)}}} N_{\nu, -\beta}^{\pm \alpha} f_{+\nu}^b f_{-\beta}^{b-a} = 0$$

$$(1 \le \alpha \le d_{\pm a}, \ 1 \le a \le m_\pm),$$

where the N_{AB}^G are the structure constants of $\mathfrak{g}$, $[e_A^a, e_B^b] = N_{AB}^G e_G^c$.

As we will show, the system (1.4) is completely integrable; its solutions are determined by an appropriate number of arbitrary functions depending on z_+ for the characteristic $z_- = $ const or z_- for $z_+ = $ const, i.e.,

$$\varphi^a_{+\alpha}(z_+) \text{ and } \varphi^b_{-\beta}(z_-), \ 1 \le a \le m_+, \ 1 \le b \le m_-,$$

$$1 \le \alpha \le d_{+a}, \ 1 \le \beta \le d_{-b}.$$

The form of representation (1.1) is invariant with respect to the gauge transformations generated by g_0:

$$A_\pm \to g_v^{-1} A_\pm g_v + g_v^{-1} g_{v,z_\pm}, \tag{1.5}$$

where $g_v = \exp(e_0 \cdot v(z_+, z_-))$.

This property is called *gauge invariance*. The algebraic structure of $A_\pm$ (in the sense of the set of subspaces $g_{\pm a}$, $0 \le a \le m_\pm$, on which $A_\pm$ take values), called in what follows as spectrum of $A_\pm$, does not change under such transformation due to the grading property (1.2.2). So, due to this property, the spectrum of (1.3) also does not change, nor do the equations for the gauge invariant quantities constructed with the functions $u_\pm$ and $f^a_\pm$.

The majority of the two-dimensional dynamical systems describing non-linear effects that we encounter in theoretical and mathematical physics correspond to the case $m_+ = m_- = 1$ of our construction.

In this case equations (1.4) simplify to the form

$$u_{-s,z_+} - u_{+s,z_-} + \sum_{ij} \mathcal{N}^s_{ij} u_{+i} u_{-j} + \sum_{\alpha,\beta} \mathcal{N}^s_{\alpha,-\beta} f_{+\alpha} f_{-\beta} = 0,$$

$$f_{\pm\alpha,z_\mp} + \sum_{i,\beta} \mathcal{N}^{\pm\alpha}_{i,\pm\beta} u_{\mp i} f_{\pm\beta} = 0. \tag{1.6}$$

In particular, for simple Lie algebras of rank r in the principal grading (see 1.2.3) for which

$$A_\pm = (\boldsymbol{h} \cdot \boldsymbol{u}_\pm) + (\boldsymbol{X}_\pm \cdot \boldsymbol{f}_\pm) \equiv \sum_{1 \le j \le r} (h_j u_{\pm j} + X_{\pm j} f_{\pm j}), \tag{1.7}$$

the system (1.6) with relations (I.2.8) being taken into account takes the form

$$u_{+j,z_-} - u_{-j,z_+} = f_{+j} f_{-j}, \ (\ln f_{\pm j})_{,z_\mp} = \mp(k u_\mp)_j, \tag{1.8}$$

where k is Cartan's matrix of g and $(k u_\pm)_j \equiv \sum_{1 \le i \le r} k_{ji} u_{\pm i}$. Introducing gauge invariant (in the sense of (1.5) with $e_0 \equiv h$) functions $\rho_j \equiv \ln f_{+j} f_{-j}$ we derive from (1.8) the system

$$\rho_{j,z_+z_-} = \sum_{1 \le i \le r} k_{ji} \exp \rho_i, \ 1 \le j \le r, \tag{1.9}$$

which for finite-dimensional g (with an invertible Cartan matrix) can be written in an equivalent form

$$x_{i,z_+z_-} = \exp(kx)_i, \ x_i \equiv (k^{-1}\rho)_i. \tag{1.10}$$

Notice that the systems (1.3) and (1.4) were deduced for an arbitrary, not necessarily finite-dimensional Lie algebra. Chapter 5 deals with infinite-dimensional case where unlike in the finite dimensional case, the integration of the equations that arise is not possible in a finite form; the solution of the corresponding Goursát problem is given by infinite formal series. An investigation of convergence properties requires additional reasoning using properties of the algebras such as the property of finite growth. Similar reasoning also holds for supersymmetric dynamical systems with the operators A taking values in the corresponding Lie superalgebra. Here, as in the case of Lie algebras, the systems of equations associated with finite dimensional Lie superalgebras are integrable in finite polynomials, while for infinite-dimensional Lie superalgebras they are integrable in formal series.

3.1.2 Examples. To illustrate the above general algebraic construction let us apply it to derive several concrete two-dimensional nonlinear equations. These examples clearly show how the specifics of the structure of the different types of Lie algebras and the choice of gradings manifest in the types of the arising equations. For this, for every system we will write the associated Lie algebra and explicit form for the operators $A_\pm$ with values in it substituting which into (1.1) yields the equations in question.

1) The Liouville equation

$$x_{,z_+z_-} = \exp 2x \qquad (1.11)$$

is associated with sl(2) (see (1.2.8) for $r = 1$, $k = (2)$), where

$$A^\pm = u_\pm h + f_\pm X_\pm, \quad x = (\ln f_+ f_-)/2.$$

2) The sine-Gordon ($\varepsilon = 0$) and Bulough-Dodd ($\varepsilon = 1$) equations

$$x_{,z_+z_-} = (1 + \varepsilon) \exp((2 - \varepsilon)x) - \exp(-2x) \qquad (1.12_\varepsilon)$$

are associated with infinite-dimensional affine Lie algebras $A_1^{(1)}$ and $A_2^{(2)}$ of finite growth, and $k = \begin{pmatrix} 2 & -(2-\varepsilon) \\ -2(1+\varepsilon) & 2 \end{pmatrix}$ respectively for $A_\pm = \sum_{j=1,2} (u_{\pm j} h_j + f_{\pm j} X_{\pm j})$. Here $((1 + \varepsilon)\rho_1 + \rho_2)_{,z_+z_-} = 0$ with $\rho_j = \ln f_{+j} f_{-j}$, so that

$$\rho_2 = -(1 + \varepsilon)\rho_1 + (2 + \varepsilon) \ln(\varphi_+(z_+)\varphi_-(z_-)),$$

where φ_+ and φ_- are arbitrary functions. Setting $\rho_1 \equiv (2 - \varepsilon)x + \ln \varphi_+\varphi_-$ and performing the corresponding conformal transformation in the equation for x we get (1.12).

3) The system

$$-x_{,z_+z_-} = \exp 2x + y^+y^- \exp x, \quad y_{,z_\mp}^\pm = \mp(3/2)y^\mp \exp x, \qquad (1.13)$$

is associated with the Virasoro algebra (1.2.19) which is infinite-dimensional Lie algebra of finite growth. Here

$$A_\pm = u_\pm X_0 + f_{\pm1} X_{\pm1} + f_{\pm2} X_{\pm2}, \quad x = \frac{1}{2} \ln 4 f_{+2} f_{-2}, \quad y_\pm = f_{\pm1}(f_{\pm2})^{-1/2}.$$

4) The system

$$x_{,z_+ z_-} = \exp 2x, \quad y_{,z_+ z_-} = l(l+1) y \exp 2x, \tag{1.14}$$

is associated with a nonsemisimple Lie algebra $A_1 \,\dot{\ni}\, W^{(l)}$ where the highest weight of the A_1-module $W^{(l)}$ is even in the realization $A_1 \simeq sl(2)$ or integer in the realization $A_1 \simeq O(3)$, i.e.

$$[h, R_m] = 2m R_m, \quad [X_\pm, R_m] = [(l \mp m)(l \pm m + 1)]^{1/2} R_{m\pm1};$$

with

$$A_\pm = u_\pm h + f_\pm X_\pm + f_\pm^0 R_0 + f_\pm^1 R_{\pm1}$$

and

$$x \equiv (1/2) \ln f_+ f_-, \quad y \equiv f_+^1 (f_+)^{-1} - f_-^1 (f_-)^{-1}.$$

5a) The system

$$x_{1,z_+ z_-} = \exp(2x_1 - 2x_2), \quad x_{2,z_+ z_-} = \exp(-x_1 + 2x_2) \tag{1.15a}$$

is associated with the principal embedding of A_1 into B_2 (the Cartan element of the embedding is $h = 3h_1 + 4h_2$ and $\mathfrak{g}_0 = \{h_1, h_2\}$) with

$$A_\pm = \sum_{j=1,2} (u_{\pm j} h_j + f_{\pm j} X_{\pm j}); \quad x_j \equiv (1/2) \ln f_{+j} f_{-j}.$$

5b) The system

$$x_{j,z_+ z_-} = \exp 2x_j + (-1)^j \operatorname{th}[(x_1 - x_2)/2] \cdot \operatorname{ch}^{-2}[(x_1 - x_2)/2] y_{,z_+} y_{,z_-} ,$$

$$(\operatorname{th}^2[(x_1 - x_2)/2] \cdot y_{,z_-})_{,z_+} + (\operatorname{th}^2[(x_1 - x_2)/2] \cdot y_{,z_+})_{,z_-} = 0 \tag{1.15b}$$

is associated with a nonprincipal embedding of A_1 into B_2 (with the Cartan element $h = 2h_1 + h_2$ and $\mathfrak{g}_0 = \{h_1; h_2, X_{\pm2}\}$). In an appropriate gauge we have

$$A_\pm = \sum_{j=1,2} (u_{\pm j} h_j + f_{\pm j} X_{\pm j}) + f_{\pm3} X_{\pm(\pi_1 + 2\pi_2)} + f_{\pm4} X_{\mp2};$$

$$x_1 \equiv (1/2) \ln f_{+1} f_{-1}, \quad x_2 \equiv (1/2) \ln f_{+3} f_{-3}; \quad y_{,z_\pm}$$

$$\equiv 2 \exp(x_2 - x_1) \cdot \operatorname{ch}^2[(x_2 - x_1)/2] \cdot f_{\pm2} f_{\pm1}^{1/2} f_{\pm3}^{-1/2}.$$

To deduce (1.15b) from (1.1) we use commutation relations of the form (1.2.8) for $r = 2$ and $k = \begin{pmatrix} 2 & -2 \\ -1 & 2 \end{pmatrix}$, i.e.,

$$B_2 \equiv \{X_{\pm\pi_j}, h_{\pm\pi_j}, j = 1, 2; X_{\pm(\pi_1+\pi_2)}, X_{\pm(\pi_1+2\pi_2)}\}$$

$$[X_{-\pi_1}, X_{-\pi_2}] = X_{-(\pi_1+\pi_2)}, [X_{\pi_1+2\pi_2}, X_{-\pi_2}] = -X_{\pi_1+\pi_2},$$

$$[X_{\pi_1}, X_{\pi_2}] = X_{\pi_1+\pi_2}, [X_{\pi_2}, X_{-(\pi_1+2\pi_2)}] = -X_{-(\pi_1+\pi_2)},$$

$$[h_{\pi_i}, X_{\pm(\pi_1+2\pi_2)}] = \pm 2\delta_{i2} X_{\pm(\pi_1+2\pi_2)}, [X_{\pi_1+2\pi_2}, X_{-(\pi_1+2\pi_2)}] = h_{\pi_1} + h_{\pi_2}.$$

The last two examples show how strongly the choice of the grading of the same algebra influences the form of the resulting system.

Naturally, equations (1.11), (1.12$_\varepsilon$) and (1.15a) are particular cases of (1.9) for $k = (2)$; $\begin{pmatrix} 2 & -2 \\ -2 & 2 \end{pmatrix}$, $\begin{pmatrix} 2 & -1 \\ -4 & 2 \end{pmatrix}$ and $\begin{pmatrix} 2 & -2 \\ -1 & 2 \end{pmatrix}$ respectively.

6) Supersymmetric Liouville equation.

$$x_{,z_+z_-} = \exp 2x + w^+ w^- \exp x, w^\pm_{,z_\mp} = w^\mp \exp x \qquad (1.16_1)$$

is related with the finite-dimensional simple Lie superalgebra $osp(1/2) = \{h, X_\pm, Y_\pm; [h, Y_\pm]_- = \pm Y_\pm, [Y_+, Y_-]_+ = h, [X_\pm, Y_\pm]_- = 0 [X_\pm, Y_\mp]_- = Y_\pm, [Y_\pm, Y_\pm]_+ = \mp 2X_\pm\}$, $A_\pm = u_\pm h + f_\pm X_\pm + f'_\pm Y_\pm$, $x \equiv \frac{1}{2}\ln f_+ f_-$, $w^\pm \equiv (f_\pm)^{-1/2} f'_\pm$. Here $w^\pm(z_+z_-)$ and $f'_\pm(z_+, z_-)$ are Majorana spinors of functions with anticommuting values. System (1.16$_1$) is a component form of the supersymmetric extension of the Liouville equation [142]

$$\mathcal{D}_-\mathcal{D}_+\varphi = \exp\varphi \qquad (1.16_2)$$

which corresponds to the action $\int dz_+ dz_- d\theta_+ d\theta_- (-\frac{1}{2}\varphi\mathcal{D}_-\mathcal{D}_+\varphi + \exp\varphi)$.

Here $\varphi = \varphi(z_\pm, \theta_\pm) = x(z_\pm) - \bar{\theta}w(z_\pm) - \frac{1}{2}\bar{\theta}\theta F(z_\pm)$ is a superscalar field (containing two scalars x and F, and a Majorana spinor $w^\pm$) which depends on the coordinates $z_\pm$ of a two-dimensional space. The Grassmann variables $\theta \equiv \begin{pmatrix} \theta_+ \\ \theta_- \end{pmatrix}$; $\bar{\theta} \equiv (-\theta_-, \theta_+)$; $\mathcal{D}_\pm \equiv \mp\frac{\partial}{\partial\theta_\pm} + \theta_\pm\frac{\partial}{\partial z_\pm}$ are covariant superderivatives, $\mathcal{D}^2_\pm = \mp\frac{\partial}{\partial z_\pm}$, $\mathcal{D}_+\mathcal{D}_- = -\mathcal{D}_-\mathcal{D}_+$. In terms of the superfield φ components ($F = \exp x$) equation (1.16$_2$) takes form (1.16$_1$) and reduces to the usual Liouville equation (1.11) for $w^\pm = 0$.

3.1.3 Construction of solutions.
The representation (1.1) expresses the fact that the vector with components A_+ and A_- is a gradient, i.e.,

$$A_\pm = g^{-1}g_{,z_\pm}, \qquad (1.17)$$

where g is an element of the complex hull of the Lie group G with $\text{Lie}(G) = \mathfrak{g}$. At this stage, the crucial difference between infinite- and finite-dimensional

Lie algebras is that the former cannot always be integrated to a group while the latter can.

First consider the construction of general solutions of system (1.3) or (1.4) in the finite-dimensional case and then discuss the modification of the method under the passage to infinite-dimensional algebras. In accordance with (1.17), solving these equations is the same as finding a group element g such that the spectra of the elements from $\mathfrak{g}$ in (1.17) and (1.2) are the same. In other words, it is necessary for A_+ (resp. A_-) in (1.17) not to contain elements from the subspaces $\mathfrak{g}_a$ in the decomposition (1.2.1) with $a < 0$ and $a > m_+$ (resp. $a > 0$ and $a < -m_-$); symbolically:

$$A_+ \cap \mathfrak{g}_{\substack{a<0 \\ a>m_+}} = 0, \quad A_- \cap \mathfrak{g}_{\substack{a>0 \\ a<-m_-}} = 0. \tag{1.18}$$

Making use of the modified Gauss decomposition for g in the form (1.5.3) and substituting it into (1.17) we get

$$\begin{aligned} A_+ &= g_{0-}^{-1}\mathcal{N}_+^{-1}(\mathcal{M}_-^{-1}\mathcal{M}_{-,z_+})\mathcal{N}_+g_{0-} \\ &\quad + g_{0-}^{-1}(\mathcal{N}_+^{-1}\mathcal{N}_{+,z_+})g_{0-} + g_{0-}^{-1}g_{0-,z_+}, \end{aligned} \tag{1.19_1}$$

$$\begin{aligned} A_- &= g_{0+}^{-1}\mathcal{N}_-^{-1}(\mathcal{M}_+^{-1}\mathcal{M}_{+,z_-})\mathcal{N}_-g_{0+} \\ &\quad + g_{0+}^{-1}(\mathcal{N}_-^{-1}\mathcal{N}_{-,z_-})g_{0+} + g_{0+}^{-1}g_{0+,z_-}. \end{aligned} \tag{1.19_2}$$

Now recall that $\mathcal{M}_+$ and $\mathcal{N}_+$ (resp. $\mathcal{N}_-$ and $\mathcal{M}_-$) are only generated by the elements from $\mathfrak{g}_a$ with $a > 0$ (resp. $a < 0$) and g_{0+} and g_{0-} belong to $\mathfrak{g}_0$, and the gauge transformations of $A_\pm$ does not affect their spectrum. Therefore, from (1.19) we derive, thanks to (1.2), (1.18) and (1.2.2) that $\mathcal{M}_-^{-1}\mathcal{M}_{-,z_+} = 0$, $\mathcal{M}_+^{-1}\mathcal{M}_{+,z_-} = 0$ which implies

$$\mathcal{M}_{-,z_+} = 0, \quad \mathcal{M}_{+,z_-} = 0, \quad \text{i.e. } \mathcal{M}_- = \mathcal{M}_-(z_-), \quad \mathcal{M}_+ = \mathcal{M}_+(z_+). \tag{1.20}$$

Therefore,

$$A_+ = g_{0-}^{-1}(\mathcal{N}_+^{-1}\mathcal{N}_{+,z_+})g_{0-} + g_{0-}^{-1}g_{0-,z_+}, \tag{1.21_1}$$

$$A_- = g_{0+}^{-1}(\mathcal{N}_-^{-1}\mathcal{N}_{-,z_-})g_{0+} + g_{0+}^{-1}g_{0+,z_-}. \tag{1.21_2}$$

Thanks to (1.5.4) the group parameters of the elements $\mathcal{N}_-$, $\mathcal{N}_+$ and $g_0 \equiv g_{0+}g_{0-}^{-1}$ are (because of the uniqueness of the decomposition (1.5.3) for regular g) functions in parameters of the element $\mathcal{M}_+^{-1}\mathcal{M}_-$. This enables us to reformulate the conditions of the absence of terms corresponding to the subspaces $\mathfrak{g}_a$ with $a > m_+$ and $a < -m_-$ in (1.21_1) and (1.21_2), respectively, in terms of $\mathcal{M}_+$ and $\mathcal{M}_-$. This finally reduces our problem to defining the $\mathcal{M}_\pm$.

If, in accordance with (1.2), we impose the conditions $A_+ \cap \mathfrak{g}_{a>m_+} = 0$ and $A_- \cap \mathfrak{g}_{a<-m_-} = 0$, then we get $\sum\limits_{m_++1\leq a\leq M_+} d_a$ differential equations relating the parameters of $\mathcal{N}_+$ and $\displaystyle\sum_{m_-+1\leq a\leq M_-} d_{-a}$ equations for the parameters of $\mathcal{N}_-$ which should hold for arbitrary values of z_- and z_+. (Here M_+ and M_- are maximal and minimal, respectively, values of the grading index a in the decomposition (1.2.1) for a finite-dimensional Lie algebra $\mathfrak{g}$ under consideration.) In particular, the realization of the relation

$$\mathcal{N}_+^{-1}\mathcal{N}_{+,z_+}\cap\mathfrak{g}_{a>m_+}=0 \quad (*)$$

in (1.21_1) does not depend on $\mathcal{M}_-$ and we may take $\mathcal{M}_-$ to be the identity. Then the identity (1.5.4) takes the form $\mathcal{M}_+^{-1}=\mathcal{N}_-g_0\mathcal{N}_+^{-1}$ (at $\mathcal{M}_-=\mathrm{id}$) and the elements $\mathcal{N}_-$ and g_0 in it also become the identity and therefore we may replace $\mathcal{N}_+$ by $\mathcal{M}_+$ in $(*)$. Similarly, $\mathcal{N}_-$ is replaced by $\mathcal{M}_-$ in $\mathcal{N}_-^{-1}\mathcal{N}_{-,z_-}\cap\mathfrak{g}_{a<-m_-}=0$.

As a result, the conditions (1.18) take an equivalent form

$$\begin{aligned}
\mathcal{M}_+^{-1}(z_+)(\mathcal{M}_+(z_+))_{,z_+}\cap\mathfrak{g}_{a>m_+}&=0,\\
\mathcal{M}_-^{-1}(z_-)(\mathcal{M}_-(z_-))_{,z_-}\cap\mathfrak{g}_{a<-m_-}&=0.
\end{aligned} \qquad (1.22)$$

Therefore, all the $\displaystyle\sum_{1\leq a\leq M_+} d_a$ (resp. $\displaystyle\sum_{1\leq a\leq M_-} d_{-a}$) parameters of $\mathcal{M}_+$ (resp. $\mathcal{M}_-$) are related by $\displaystyle\sum_{m_++1\leq a\leq M_+} d_a$ (resp. $\displaystyle\sum_{m_-+1\leq a\leq M_-} d_{-a}$) equations and hence each of $\mathcal{M}_+$ and $\mathcal{M}_-$ depends on $n_+\equiv\displaystyle\sum_{1\leq a\leq m_+} d_a$ and $n_-\equiv\displaystyle\sum_{1\leq a\leq m_-} d_{-a}$ arbitrary functions $\varphi_{+\alpha}^a(z_+)$ and $\varphi_{-\beta}^b(z_-)$, where $1\leq a\leq m_+$, $1\leq\alpha\leq d_a$, $1\leq b\leq m_-$ and $1\leq\beta\leq d_{-b}$. In other words, we see that $\mathcal{M}_+$ and $\mathcal{M}_-$ satisfy the following equations

$$\begin{aligned}
(\mathcal{M}_+(z_+))_{,z_+}&=\mathcal{M}_+(z_+)\sum_{1\leq a\leq m_+}(e_+^a\cdot\varphi_+^a(z_+))\equiv\mathcal{M}_+L_+,\\
(\mathcal{M}_-(z_-))_{,z_-}&=\mathcal{M}_-(z_-)\sum_{1\leq a\leq m_-}(e_-^a\cdot\varphi_-^a(z_-))\equiv\mathcal{M}_-L_-.
\end{aligned} \qquad (1.23)$$

A solution of (1.23), like a solution of the equation $dS/dt=S(t)L(t)$ for the scattering matrix in the quantum field theory (see, e.g. [9, 10]), is representable as a multiplicative integral, i.e. an infinite series of multiple integrals in retarded products

$$S=1+\sum_{n\geq 1}\frac{(-1)^n}{n!}\int\cdots\int\prod_{j=1}^n dt_j\theta(t_j-t_{j-1})L(t_1)\ldots L(t_n)$$

or as a symbolically anti-$\mathcal{Z}_\pm$-ordered exponentials with Lagrangians $L_\pm$:

$$\mathcal{M}_+=\tilde{\mathcal{Z}}_+\exp\int_{a_+}^{z_+} dz_+'L_+(z_+'),\quad \mathcal{M}_-=\tilde{\mathcal{Z}}_-\exp\int_{a_-}^{z_-} dz_-'L_-(z_-'). \qquad (1.24)$$

This expression can be written uniformly with the exponent being the sum of suitably ordered retarded commutators

$$R_m^\pm = \sum_{\{i_1,\ldots,i_m\}} \left[L_{\pm i_1}[\ldots[L_{\pm i_{m-1}}, L_{\pm i_m}]\ldots]\right]$$

$$\equiv \sum_\omega (-1)^{m+\varepsilon_\omega} \frac{\varepsilon_\omega!(m-\varepsilon_\omega-1)!}{m} \left[\ldots[[L_{\pm\omega(1)}, L_{\pm\omega(2)}]\ldots L_{\pm\omega(m)}]\right];$$

$$L_{\pm i} \equiv L_\pm(z_\pm^{(i)}), \quad z_\pm^{(0)} \equiv z_\pm,$$

$$\mathcal{M}_\pm = \exp \sum_{1 \leq m \leq \infty} (-1)^m/m! \int \ldots \int$$

$$\prod_{1 \leq i \leq m} dz_\pm^{(i)} \theta(z_\pm^{(i)} - z_\pm^{(i-1)}) R_m^\pm(z_\pm^{(1)}, \ldots, z_\pm^{(m)})$$

(1.25)

where sum runs over all permutations ω of $\{1,\ldots,m\}$ with ε_ω errors in ordering in $\{\omega(1),\ldots,\omega(m)\}$.

In particular,

$$R_1^\pm = L_{\pm 1}; \quad R_2^\pm = [L_{\pm 1}, L_{\pm 2}];$$

$$R_3^\pm = [L_{\pm 1}[L_{\pm 2}, L_{\pm 3}]] + [L_{\pm 3}[L_{\pm 2}, L_{\pm 1}]];$$

$$R_4^\pm = 2[L_{\pm 1}[L_{\pm 2}[L_{\pm 3}, L_{\pm 4}]]] + 2[L_{\pm 3}[L_{\pm 2}[L_{\pm 4}, L_{\pm 1}]]]$$

$$+ 2[L_{\pm 4}[L_{\pm 1}[L_{\pm 3}, L_{\pm 2}]]] + 2[L_{\pm 4}[L_{\pm 3}[L_{\pm 1}, L_{\pm 2}]]].$$

Thanks to the finite-dimensionality of the Lie algebras considered, the series in the exponent (1.25) contains only a finite number of terms.

Notice also that the realization of the action of elements of the form (1.24) on a basis of a finite-dimensional representation is not an infinite sum but a polynomial, i.e., only finitely many terms contribute to the matrix elements.

Therefore, the formulas (1.24) and (1.25) enable us to recover g and therefore to find an explicit form of the operators $A_\pm$ satisfying (1.18) with the help of equations (1.17). Indeed, knowing $\mathcal{M}_+$ and $\mathcal{M}_-$, we may use them to express the group parameters of $\mathcal{N}_+$, $\mathcal{N}_-$ and g_0 via (1.5.4) and therefore find the desired regular element g up to a gauge $g_{0+} \to g_{0+}g_v$, $g_{0-} \to g_{0-}g_v$, $g_v \in G_0$.

Let us give one more deduction of the spectral content of a group element in $g^{-1}g_{,z_\pm}$ corresponding to (1.2) directly from (1.5.4). Differentiating the latter identity first with respect to, say, z_- and taking the conditions (1.20) on $\mathcal{M}_+$ into account we get

$$\mathcal{N}_{-,z_-} g_0 \mathcal{N}_+^{-1} + \mathcal{N}_- g_{0,z_-} \mathcal{N}_+^{-1} + \mathcal{N}_- g_0 \mathcal{N}_{+,z_-}^{-1} = \mathcal{M}_+^{-1}\mathcal{M}_{-,z_-}.$$

Multiplying this relation by $\mathcal{N}_+ g_0^{-1} \mathcal{N}_-^{-1} \equiv \mathcal{M}_-^{-1} \mathcal{M}_+$ from the left we get

$$\mathcal{N}_+ g_0^{-1} (\mathcal{N}_-^{-1} \mathcal{N}_{-,z_-}) g_0 \mathcal{N}_+^{-1} + \mathcal{N}_+ (g_0^{-1} g_{0,z_-}) \mathcal{N}_+^{-1} + \mathcal{N}_+ \mathcal{N}_{+,z_-}^{-1}$$

$$= \mathcal{M}_-^{-1} \mathcal{M}_{-,z_-} \equiv L_-,$$

or, finally

$$g_0^{-1} (\mathcal{N}_-^{-1} \mathcal{N}_{-,z_-}) g_0 + g_0^{-1} g_{0,z_-} - \mathcal{N}_+^{-1} \mathcal{N}_{+,z_-} = \mathcal{N}_+^{-1} L_- \mathcal{N}_+. \tag{1.26$_1$}$$

The elements from the subspaces $\mathfrak{g}_a$ with $a < 0$ appear only contained in the first term of (1.26_1) and in its right-hand side. Since the grading index in L_- is more than $(-m_- - 1)$ due to (1.23), $g_0^{-1}(\mathcal{N}_-^{-1} \mathcal{N}_{-,z_-}) g_0$ also does not contain elements from $\mathfrak{g}_{a<-m_-}$. Therefore, an operator A_- of the form (1.21_2) satisfies the necessary conditions (1.18). A similar proof for A_+ follows from the formula

$$g_0 (\mathcal{N}_+^{-1} \mathcal{N}_{+,z_+}) g_0^{-1} - g_{0,z_+} g_0^{-1} - \mathcal{N}_-^{-1} \mathcal{N}_{-,z_+} = \mathcal{N}_-^{-1} L_+ \mathcal{N}_-. \tag{1.26$_2$}$$

Thus, the problem of integrating differential equations (1.4) is solved in three steps. First, we find the elements $\mathcal{M}_+$ and $\mathcal{M}_-$ in the form (1.24) or (1.25) as solutions of (1.23); then with the help of $(1.5.4)$ we recover the elements $\mathcal{N}_+$, $\mathcal{N}_-$ and g_0 parametrizing $A_\pm$, see (1.21); finally, comparing (1.21) and (1.2) we get the desired dependences of the vector-functions $u_\pm(z_+, z_-)$, $f_\pm^a(z_+, z_-)$ on the complete set of arbitrary functions $\varphi_\pm^a(z_\pm)$ parametrizing the operators $L_\pm$ of the form (1.23). Therefore, we get a solution of the Goursát problem for (1.4) from the initial values on characteristics determined by the functions $\varphi_\pm^a$.

Generally, for an arbitrary Lie group there is no explicit formula which expresses the group parameters of $\mathcal{N}_\pm$ and g_0 in terms of $\mathcal{M}_\pm$. However, their calculation is computable (by repeated use of Campbell-Hausdorff-type formulas), which enables one to perform the corresponding calculations in each concrete case. In many cases the representation theory of the Lie groups G enables one to express these elements in terms of matrix elements of $\mathcal{M}_+^{-1} \mathcal{M}_- \in G$ in some representation of G.

§ 3.2 Systems generated by the local part of an arbitrary graded Lie algebra*)

In the preceding section we gave the general scheme for constructing integrable dynamical systems in two-dimensional space connected with an arbitrary graded Lie algebra or superalgebra and developed a method for finding

---

*) [79]

their solutions. This method enables us to get closed expressions for solutions; however, since there is no general description of "generic" Lie algebras (i.e., of their structure constants) the equations themselves can not always be expressed in a form more explicit than (1.4). Besides, the form of the equations essentially depends on the choice of gauge conditions.

On the contrary, any nonlinear dynamical system generated by the local part of an arbitrary Lie algebra whose grading is compatible with an integer embedding of A_1 into the algebra may be expressed in a compact and explicit form. In what follows we will consider only such systems.

3.2.1 Exactly integrable systems.
Considering such systems we start with the equation of the form

$$[\partial/\partial z_+ + E_0^{0+} + E_0^{f+} + E_1^+, \; \partial/\partial z_- + E_0^{0-} + E_0^{f-} + E_1^-] = 0, \qquad (2.1)$$

where $E_0^{0\pm}$, $E_0^{f\pm}$ and $E_1^{\pm}(\equiv E_{\pm 1})$ take values in the subspaces $\mathfrak{g}_0^0$, $\mathfrak{g}_0^f$ and $\mathfrak{g}_{\pm 1}$ of $\mathfrak{g}$, respectively, and $E_a^{\pm} = \sum\limits_{1 \leq \alpha \leq d_a} \varphi_{\pm \alpha}^a(z_+, z_-) X_\alpha^a$, where $X_\alpha^a \in \mathfrak{g}_a$.

The decomposition of the local part $\hat{\mathfrak{g}} = \mathfrak{g}_{-1} \oplus (\mathfrak{g}_0^0 \oplus \mathfrak{g}_0^f) \oplus \mathfrak{g}_1$, with (1.2.2) being taken into account, induces the splitting of (2.1) into the subsystems

$$[\partial/\partial z_- + E_0^{0-} + E_0^{f-}, E_1] = 0, \; [\partial/\partial z_+ + E_0^{0+} + E_0^{f+}, E_{-1}] = 0,$$

$$[\partial/\partial z_+ + E_0^{0+} + E_0^{f+}, \partial/\partial z_- + E_0^{0-} + E_0^{f-}] + [E_1, E_{-1}] = 0. \qquad (2.2)$$

Thanks to the properties of embeddings A_1 into $\mathfrak{g}$ described in 1.3, the dimensions of $\mathfrak{g}_{\pm 1}$ and $\mathfrak{g}_0^f$ coincide and on the subspaces $\mathfrak{g}_{\pm 1}$ a representation of $\mathfrak{g}_0^f$ arises. The whole space $\mathfrak{g}_1(\mathfrak{g}_{-1})$ can be recovered from one element J_+ (resp. J_-) by conjugations by elements from $G_0 = \exp \mathfrak{g}_0$, namely $\mathfrak{g}_{\pm 1} = \{g_0^{-1} J_{\pm} g_0 : g_0 \in G_0\}$.

Since $J_{\pm}$ commute with transformations from $G_0^0 = \exp \mathfrak{g}_0^0$, the parameters of g_0 associated with $\mathfrak{g}_0^0$ are, in fact, absent in the $g_0^{-1} J_{\pm} g_0$, which guarantees the identity $d_{\pm 1} = d_0^f$. In other words, since G_0^0 is the stationary subgroup for $J_{\pm}$, then the elements $q \equiv g_0^{-1} g_0^0 g_0$, where $g_0^0 \in G_0^0$, constitute the stationary subgroup for "points" of $\mathfrak{g}_{\pm 1} = q \mathfrak{g}_{\pm 1} q^{-1}$. Therefore, by an appropriate gauge transformation we may reduce E_{+1} to J_+ and then the first of equations (2.2) takes the form

$$[E_0^{0-} + E_0^{f-}, J_+] = 0, \quad \text{or} \quad [E_0^{f-}, J_+] = 0.$$

Since any element of $\mathfrak{g}_{+1}$ can be obtained from the corresponding element of $\mathfrak{g}_0^f$ by an action of the raising operator J_+, this last relation implies $E_0^{f-} = 0$. Therefore (2.2) can be rewritten in the form

$$E_{-1,z_+} + [E_0^{0+} + E_0^{f+}, E_{-1}] = 0,$$

$$E_{0,z_-}^{f+} + [E_{-1}, J_+] = 0, \qquad (2.3)$$

$$E_{0,z_-}^{0+} - E_{0,z_+}^{0-} + [E_0^{0-}, E_0^{0+}] \equiv [\partial/\partial z_- + E_0^{0-}, \partial/\partial z_+ + E_0^{0+}] = 0.$$

The last equation in (2.3) means that $E_0^{0\pm}$ is a gradient, i.e. $E_0^{0\pm} = (g_0^0)^{-1} g_{0,z_\pm}^0$ and therefore can be excluded from the remaining equations by an appropriate gauge transformation. Thus, we arrive at the following finite form of expression of the initial system (2.2):

$$E_{-1,z_+} + [E_0^{f+}, E_{-1}] = 0, \quad E_{0,z_-}^{f+} + [E_{-1}, J_+] = 0. \qquad (2.4)$$

Expressing E_0^{f+} from the first equation and substituting it into the second one we get the desired system of second order nonlinear differential equations in functions $\varphi^1{}_\alpha(z_+, z_-)$. Since $d_{-1} = d_0^f$, the number of functions $\varphi_{-\alpha}^1$ and $\varphi_{+\alpha}^f$ coincides.

Explicitly introducing the matrix function $R_{\gamma\alpha} \equiv \sum_\beta C_{\alpha\beta}^\gamma \varphi_{-\beta}^1$, substitute the bracket $[E_0^{f+}, E_{-1}] = \sum_{\alpha,\beta,\gamma} C_{\alpha\beta}^\gamma \varphi_{+\alpha}^{0f} \varphi_{-\beta}^1 X_\gamma^-$, where $C_{\alpha\beta}^\gamma$ is the matrix of the representation of $\mathfrak{g}_0^f$ in $\mathfrak{g}_{-1}$, i.e., $[X_\alpha^{0f}, X_\beta^-] = \sum_\gamma C_{\alpha\beta}^\gamma X_\gamma^-$, into the first of equations (2.4). Then simplifying we get

$$\varphi_{+\alpha}^{0f} = -\sum_\beta R_{\alpha\beta}^{-1} \varphi_{-\beta,z_+}^1. \qquad (2.5)$$

Taking (2.5) into account we reduce (2.4) to the form

$$\sum_\beta (R_{\alpha\beta}^{-1} \varphi_{-\beta,z_+}^1)_{,z_-} = \sum_\beta A_{\alpha\beta} \varphi_{-\beta}^1, \qquad (2.6_1)$$

where $A_{\alpha\beta}$ is expressed from $[X_\beta^-, J_+] \equiv \sum_\alpha A_{\alpha\beta} X_\alpha^{0f}$. Finally,

$$\varphi_{-\alpha,z_+z_-}^1 = \sum_{\beta,\gamma} [R_{\alpha\beta,z_-} R_{\beta\gamma}^{-1} \varphi_{-\gamma,z_+}^1 + R_{\alpha\beta} A_{\beta\gamma} \varphi_{-\gamma}^1]. \qquad (2.6_2)$$

Since R is invertible, (2.5) and (2.6) are well-defined.

The system (2.4) can be represented as one second-order operator equation in $g_0 \in G_0$. Indeed, $E_{-1} = g_0 J_- g_0^{-1}$ by definition and the first of equations (2.4) takes the form $[g_0^{-1} g_{0,z_+} + g_0^{-1} E_0^{f+} g_0, J_-] = 0$ implying that $g_0^{-1} g_{0,z_+} + g_0^{-1} E_0^{f+} g_0 \equiv E_0^0$ takes values in $\mathfrak{g}_0^0$ and $E_0^{f+} = g_0 E_0^0 g_0^{-1} - g_{0,z_+} g_0^{-1}$. By a gauge transformation $g_0 \mapsto g_0 g_0^0$ with $g_0^0 \in G_0^0$ which does not affect the form of E_{-1} we may equate E_0^0 to zero getting $E_0^{f+} = -g_{0,z_+} g_0^{-1}$ (the last equality means that there are d_0^0 relations between the parameters of g_0 and their first derivatives with respect to z_+ which play the role of constraints). Substituting this equality into the second of equations (2.4) we get

$$(g_{0,z_+} g_0^{-1})_{,z_-} + [g_0 J_- g_0^{-1}, J_+] = 0; \qquad (2.7_1)$$

$$g_{0,z_+} g_0^{-1} \in \mathfrak{g}_0^f. \qquad (2.7_2)$$

Here (2.7_2) may be considered as an additional (or initial) condition on the characteristic since (2.7_1) implies that the components of $g_{0,z_+} g_0^{-1}$ belonging to $\mathfrak{g}_0^0$ depend only on z_+. Obviously, this condition is trivial if $\mathfrak{g}_0^0 = 0$, i.e., if $E_0^{0+} \equiv 0$, and in this case the systems (2.4) and (2.7) are completely equivalent. Clearly, (2.7) may be also rewritten in the form

$$(g_0^{-1} g_{0,z_-})_{,z_+} + [g_0^{-1} J_+ g_0, J_-] = 0;$$

$$g_0^{-1} g_{0,z_-} \in \mathfrak{g}_0^f. \tag{2.8}$$

3.2.2 Systems associated with infinite-dimensional Lie algebras.

In accordance with the construction from 1.3.2 the local part $\hat{\tilde{\mathfrak{g}}}$ of an infinite-dimensional Lie algebra $\tilde{\mathfrak{g}}$ contains (in addition to the generators $X_\alpha^\pm$, $1 \leq \alpha \leq d_{\pm 1}$, and X_a^0, $1 \leq a \leq d_0$, of the local part $\hat{\mathfrak{g}}$ of the initial finite-dimensional Lie algebra $\mathfrak{g}$) the elements Y_i^0 and $Y_\nu^\pm$ satisfying relations (1.3.2)–(1.3.5). The arguments leading to a suitably modified version of equations (2.4) remain essentially the same.

As in the deduction of equations (2.4) we here deduce from (2.2) thanks to the relations (1.2.4), (1.3.2)–(1.3.5):

$$E_{-1,z_+} + [E_0^{f+}, E_{-1}] = 0, \quad \tilde{E}_{-1,z_+} + [E_0^{f+}, \tilde{E}_{-1}] = 0,$$

$$E_{0,z_-}^{f+} + [E_{-1}, J_+] - [\tilde{E}_1, \tilde{E}_{-1}]' = 0, \tag{2.9}$$

$$\tilde{E}_{0,z_-}^{0+} - [\tilde{E}_1, \tilde{E}_{-1}]'' = 0, \qquad \tilde{E}_{1,z_-} = 0,$$

where

$$E_{0,\pm 1} \in \mathfrak{g}_{0,\pm 1}; \quad \tilde{E}_{0,\pm 1} \in \tilde{\mathfrak{g}}_{0,\pm 1}/\mathfrak{g}_{0,\pm 1};$$

$$[\tilde{E}_1, \tilde{E}_{-1}] \equiv [\tilde{E}_1, \tilde{E}_{-1}]'(\in \mathfrak{g}_0^f) + [\tilde{E}_1, \tilde{E}_{-1}]''(\in \tilde{\mathfrak{g}}_0^0/\mathfrak{g}_0^0);$$

$$E_{0,\pm 1} \equiv \sum_\alpha \varphi_\alpha^{0,\pm 1}(z_+, z_-) X_\alpha^{0,\pm 1}, \quad \tilde{E}_0^0 \equiv \sum_i \tilde{\varphi}_i^0(z_+, z_-) Y_i^0;$$

$$\tilde{E}_{\pm 1} \equiv \sum_\mu \tilde{\varphi}_{\pm\mu}^1(z_+, z_-) Y_\mu^\pm.$$

In the matrix form system (2.9) is as follows:

$$\varphi_{-\alpha,z_+}^1 + \sum_{\beta,\gamma} C_{\beta\gamma}^{-\alpha} \varphi_\beta^0 \varphi_{-\gamma}^1 = 0, \tag{2.10$_1$}$$

$$\tilde{\varphi}_{-\mu,z_+}^1 + \sum_{\beta,\nu} Q_{\beta\nu}^{+\mu} \varphi_\beta^0 \tilde{\varphi}_{-\nu}^1 = 0, \tag{2.10$_2$}$$

$$\varphi_{\alpha,z_-}^0 + \sum_\beta A_{\alpha\beta} \varphi_{-\beta}^1 - \sum_\nu \mathfrak{a}_{\alpha\nu} \tilde{\varphi}_{-\nu}^1 = 0, \tag{2.10$_3$}$$

$$\left(\mathfrak{a}_{\alpha\nu} \equiv \sum_{i,\mu} R_{\mu\nu}^i b_{i\alpha} \tilde{\varphi}_{+\mu}^1 \right).$$

$$\tilde{\varphi}^0_{i,z_-} - \sum_\nu \mathfrak{b}^i_\nu \tilde{\varphi}^1_{-\nu} = 0, \; (\mathfrak{b}^i_\nu \equiv \sum_\mu R^i_{\mu\nu}\tilde{\varphi}^1_{+\mu}), \qquad (2.10_4)$$

$$\tilde{\varphi}^1_{+\mu,z_-} = 0. \qquad (2.10_5)$$

In what follows we will assume that $1 \le \alpha, \; \beta \le d_{-1}, \; 1 \le \mu, \; \nu \le d_L$. The subsystem (2.10_1) coincides with the corresponding equations from 3.2.1 and solving it with respect to φ^0_β we get (2.5). The equation (2.10_2) implies that the $\tilde{\varphi}^1_{+\mu}$ are arbitrary functions depending only on z_+ and by appropriate transformations they can be excluded from the remaining equations.

Therefore, inserting (2.5) into (2.10_3) we get the desired system of second-order equations generalizing (2.6) to the case of infinite-dimensional algebras:

$$\varphi^1_{-\alpha,z_+z_-} = \sum_{\beta,\gamma}[R_{\alpha\beta,z_-}R^{-1}_{\beta\gamma}\varphi^1_{-\gamma,z_+} + R_{\alpha\beta}A_{\beta\gamma}\varphi^1_{-\gamma}] - \sum_{\beta,\nu}R_{\alpha\beta}\mathfrak{a}_{\beta\nu}\tilde{\varphi}^1_{-\nu}, \quad (2.11)$$

where the functions $\tilde{\varphi}^1_{-\mu}$ are related with $\varphi^1_{-\beta}$ in accordance to (2.10_2) by the equations

$$\tilde{\varphi}^1_{-\mu,z_+} = \sum_\alpha\left(\sum_\beta \mathcal{R}_{\mu\beta}R^{-1}_{\beta\alpha}\right)\varphi^1_{-\alpha,z_+}, \qquad (2.12)$$

$$\mathcal{R}_{\mu\beta} \equiv \sum_\nu Q^{+\mu}_{\beta\nu}\tilde{\varphi}^1_{-\nu}.$$

3.2.3 Hamiltonian formalism.
Here we give the Hamiltonian formulation of the main equations (2.8) (or (2.7)) above constructed in the one-dimensional case when all the unknown functions depend on one variable $t \equiv z_+ + z_-$ and these equations take the form

$$(g_0^{-1}g_{0,t})_{,t} + [g_0^{-1}J_+g_0, J_-] = 0 \qquad (2.13_1)$$

with additional condition

$$g_0^{-1}g_{0,t} \in \mathfrak{g}_0^f. \qquad (2.13_2)$$

Consider

$$\mathfrak{h} = (1/2)C_2 + \mathrm{tr}(g_0^{-1}J_+g_0J_-), \qquad (2.14)$$

where $C_2 \equiv \sum_a L^2_a$ is the quadratic Casimir operator constructed from infinitesimal left shifts L_a, $1 \le a \le d_0$, on G_0. The Poisson bracket of $\mathfrak{h}$ with the group element g_0 in a representation of G_0 with generators M_a is of the form $\{\mathfrak{h}, g_0\} = \sum_a L_a M_a g_0$ or $\{\mathfrak{h}, g_0\}g_0^{-1} = \sum_a L_a M_a$. Taking now the Poisson bracket of $\mathfrak{h}$ with the last element we get

$$\left\{\mathfrak{h}, \sum_a L_a M_a\right\} = \mathrm{tr}\left\{g_0^{-1}J_+g_0J_-, \sum_a L_a M_a\right\}$$

$$= -\sum_a M_a \mathrm{tr}\{M_a, [g_0^{-1}J_+g_0, J_-]\} = -[g_0^{-1}J_+g_0, J_-].$$

Here for simplicity we have made use of an orthonormal basis in the space of representation of G_0, i.e. $\operatorname{tr} M_a M_b = \delta_{ab}$, and made use of the fact that C_2 commutes with all the generators. Therefore, if we identify $\mathfrak{h}$ with the Hamiltonian of a dynamical system then the equations which describe this system coincide with (2.13_1), and (2.13_2) holds if we require that the generalized momenta M_a (right shifts) vanish for the elements $g_0^{-1} g_{0,t} = \sum_a L_a M_a$

taking values in $\mathfrak{g}_0^0$.

The existence of a Hamiltonian formalism for the dynamical system (2.8) considered enables one to apply the standard apparatus of the perturbation theory of classical and quantum mechanics. Then the role of the free Hamiltonian is played by the first term in (2.14), whereas the second term describes the interaction in the system. Multiplying this term by a constant λ we find out that the series of the perturbation theory in one- and two-dimensional cases are finite polynomials in λ and reproduce the exact solutions of the corresponding systems in a different but completely equivalent form.

3.2.4 Solutions of exactly integrable systems (Goursát problem). The representation (2.1) means that the operators

$$A_\pm \equiv E_0^{0\pm} + E_0^{f\pm} + E_1^\pm = g^{-1} g_{,z_\pm}, \qquad (2.15)$$

where $g \in G = \exp \mathfrak{g}$, are parametrized by the modified Gauss decomposition $g = \mathcal{M}_+ \mathcal{N}_- g_{0+} = \mathcal{M}_- \mathcal{N}_+ g_{0-}$. Here $\mathcal{M}_\pm$, $\mathcal{N}_\pm$ belong to the nilpotent subgroups of G generated by $\mathfrak{g}_{\pm a}$, $a \geq 1$ and $g_{0\pm} \in G_0$. In accordance with the general method let us assume that $\mathcal{M}_\pm$ satisfy equations of type (1.23):

$$\mathcal{M}_{\pm,z_\pm} = \mathcal{M}_\pm \sum_{1 \leq \alpha \leq d_{\pm 1}} \varphi_{\pm\alpha}(z_\pm) X_\alpha^\pm \equiv \mathcal{M}_\pm L_\pm, \qquad (2.16)$$

where $\varphi_{+\alpha}(z_+)$, $\varphi_{-\alpha}(z_-)$ are arbitrary functions. Set

$$K \equiv \mathcal{M}_+^{-1} \mathcal{M}_- = \mathcal{N}_- g_{0+} g_{0-}^{-1} \mathcal{N}_+^{-1} \qquad (2.17)$$

which (for regular g) uniquely determines $\mathcal{N}_+$, $\mathcal{N}_-$ and $g_0 \equiv g_{0+} g_{0-}^{-1}$. The parameters of these elements are functions of the group parameters of K, i.e., $\mathcal{M}_+(z_+)$, $\mathcal{M}_-(z_-)$.

The elements $\mathcal{M}_\pm$, $\mathcal{N}_\pm$ and g_0 define the general solutions of (2.2). To prove this let us make use of the relations $K_{,z_-} = KL_-$, $K_{,z_+} = -L_+K$, which follow from (2.17) and (2.16) or, equivalently,

$$\mathcal{N}_-^{-1} \mathcal{N}_{-,z_-} = g_0 L_- g_0^{-1}; \quad g_0^{-1} g_{0,z_-} - \mathcal{N}_+^{-1} \mathcal{N}_{+,z_-} = \mathcal{N}_+^{-1} L_- \mathcal{N}_+ - L_-;$$

$$\mathcal{N}_+^{-1} \mathcal{N}_{+,z_+} = g_0^{-1} L_+ g_0; \quad g_{0,z_+} g_0^{-1} + \mathcal{N}_-^{-1} \mathcal{N}_{-,z_+} = -\mathcal{N}_-^{-1} L_+ \mathcal{N}_- + L_+.$$

$$(2.18)$$

For the operators $A_\pm$, taking (2.16) into account, we get formula (1.23). Comparing (1.23) with (2.15) we see that A_+ and A_- take values in the subspaces $\mathfrak{g}_0 \oplus \mathfrak{g}_1$ and $\mathfrak{g}_0 \oplus \mathfrak{g}_{-1}$, respectively. Therefore, (2.17) and (2.16) enable us to solve the dynamical system completely in operator form (2.2). Notice that relations (2.18) automatically imply that the gauge-invariant element g_0 satisfies (and therefore is the explicit solution of) equation (2.7).

To construct solutions of the second-order nonlinear system (2.6) with operator form (2.4) containing $d_{\pm 1}$ unknown functions we have, in accordance with the results of 3.1, passed to the gauge

$$A_+ = E_0^{f+} + J_+, \quad A_- = E_{-1}. \tag{2.19}$$

Comparing (2.18) and (2.19) we see that $g_{0+,z_-} = 0$, i.e., $g_{0+} = g_{0+}(z_+)$, $\mathcal{N}_+^{-1}\mathcal{N}_{+,z_+} = g_{0-}J_+ g_{0-}^{-1}$ and $g_{0-}^{-1}g_{0-,z_+}$ takes values in $\mathfrak{g}_0^f$. Taking this into account we derive from (2.18)

$$L_+ = g_{0-}J_+ g_{0+}^{-1}, \tag{2.20}$$

which determines g_{0+} up to right shifts from G_0^0, which commute with $J_\pm$. Therefore, let us assume that g_{0+} does not depend on the parameters related to G_0^0, i.e. $g_{0+} \equiv g_0^{f+}$ depends only on d_{-1} parameters which, in turn, define a functional dependence of L_+. Similarly, we get $g_{0-} = g_0^{-1}g_0^{f+}$.

Equations (2.18) and (2.19) imply

$$A_- = E_{-1} = g_{0+}^{-1}\mathcal{N}_-^{-1}\mathcal{N}_{-,z_-}g_{0+} = g_{0-}^{-1}L_- g_{0-}.$$

Making use of (2.18) and (2.20) we find out that

$$-g_{0-}^{-1}g_{0-,z_+} + (\mathcal{N}_- g_{0+})^{-1}(\mathcal{N}_- g_{0+})_{,z_+} = -(g_{0+}^{-1}\mathcal{N}_- g_{0+})^{-1}J_+(g_{0+}^{-1}\mathcal{N}_- g_{0+}) + J_+,$$

implying $g_{0-}^{-1}g_{0-,z_+} \in \mathfrak{g}_0^f$. Therefore, the considered gauge leads to correct expressions for the operators $A_\pm$ from (2.19). The coefficient functions f_α in the decomposition of $A_- = g_{0+}^{-1}g_0 L_- g_0^{-1}g_{0+} = \sum\limits_{1 \le \alpha \le d_{-1}} f_\alpha X_\alpha^-$ satisfy the equations (2.6) which interest us.

Thus, the general scheme for constructing solutions of a dynamical system described by equations (2.6) is as follows.

1) Introduce two arbitrary elements $g_0^\pm(z_\pm) \in G_0 = \exp \mathfrak{g}_0$.

2) Construct operators $L_\pm = g_0^\pm J_\pm (g_0^\pm)^{-1}$ which depend on $d_{\pm 1}$ arbitrary functions $\varphi_{\pm\alpha}(z_\pm)$ and recover from them solutions of two S-matrix-type equations $\mathcal{M}_{\pm,z_\pm} = \mathcal{M}_\pm L_\pm$, in the form of multiplicative integrals, see e.g. (1.24) or (1.25).

3) Define an element $\hat{g}_0$ from the equation

$$(\mathcal{M}_+ g_0^+)^{-1}(\mathcal{M}_- g_0^-) = \hat{\mathcal{N}}_- \hat{g}_0 \hat{\mathcal{N}}_+^{-1},$$

$$\hat{\mathcal{N}}_\pm \equiv (g_0^\pm)^{-1}\mathcal{N}_\pm g_0^\pm, \quad \hat{g}_0 \equiv (g_0^+)^{-1}g_0 g_0^-.$$

4) The coefficients of the decomposition of the $\mathfrak{g}_{-1}$-valued element $E_{-1} = \hat{g}_0 J_- \hat{g}_0^{-1}$ with respect to its generators $X_{-\alpha}$ constitute the complete solution of the dynamical system considered.

Notice that in the deduction of the above relations we never used the simplicity of $\mathfrak{g}$ and therefore the formulas thus obtained give solutions of dynamical systems associated with the local part of an "arbitrary" Lie algebra containing a subalgebra sl(2) with respect to which $\mathfrak{g}$ splits into multiplets with integer values of angular momentum.

§ 3.3   Generalization for systems with fermionic fields*)

In the previous section we confined ourselves to non-linear systems related to integer embeddings of sl(2) into an arbitrary Lie algebra. However, half-integer embeddings are also of certain interest since the corresponding systems can be interpreted as models of non-linear bosonic and fermionic "fields". In particular, this consideration makes it possible to interpret Grassmann anticommuting variables as the limit case of a quantized fermionic field $\varphi(x)$ in a two-dimensional space at "zero value" of Planck's constant $\hbar$ since $\lim_{\hbar \to 0} \{\varphi(x), \bar{\varphi}(x')\} = 0$.

A Lax-type representation (3.1.1) realized by the operators $A_\pm$ taking values in subspaces $\mathfrak{g}_0 \oplus \mathfrak{g}_{\pm 1/2} \oplus \mathfrak{g}_{\pm 1}$ of a graded Lie algebra is

$$[\partial/\partial z_+ + E_0^+ + E_{1/2}^+ + E_1^+, \ \partial/\partial z_- + E_0^- + E_{1/2}^- + E_1^-] = 0. \qquad (3.1)$$

(Hereafter we use the notation of § 3.2.) The grading of $\mathfrak{g}$ is set by an embedding of sl(2)= 0(3) in $\mathfrak{g}$, the degree of the subspace $\mathfrak{g}_a$ is $a = m/2$ (with respect to the image of the Cartan element H of sl(2), cf. 1.3.1) and its parity is m mod 2. In accordance with (1.2.2), representation (3.1) can be rewritten in the form

$$E_{0,z_+}^- - E_{0,z_-}^+ + [E_{1/2}^+, E_{1/2}^-] + [E_1^+, E_1^-] = 0, \qquad (3.2_1)$$

$$E_{1/2,z_+}^- + [E_0^+, E_{1/2}^-] + [E_{1/2}^+, E_1^-] = 0,$$
$$E_{1/2,z_-}^+ + [E_0^-, E_{1/2}^+] + [E_{1/2}^-, E_1^+] = 0, \qquad (3.2_2)$$

$$E_{1,z_+}^- + [E_0^+, E_1^-] = 0, \quad E_{1,z_-}^+ + [E_0^-, E_1^+] = 0. \qquad (3.2_3)$$

A solution of (3.2_3) (with gauge $E_0^+ = 0$) is given as in § 3.2 by the formulas

$$E_1^- = J_-, \ E_1^+ = (g_0^0 g_0)^{-1} J_+ (g_0^0 g_0),$$

$$E_0^- = g_0^{-1} g_{0,z_-} + g_0^{-1} E_0^{0-} g_0,$$

---

*) [71, 76]

or, since the subalgebra $\mathfrak{g}_0^0$ commutes with the image of sl(2),

$$E_0^- = (g_0^0 g_0)^{-1}(g_0^0)_{,z_-} .$$

Then (3.2_2) can be represented as

$$E_{1/2,z_+}^- + [E_{1/2}^+, J_-] = 0,$$

$$E_{1/2,z_-}^+ + [g_0^{-1}g_{0,z_-}, E_{1/2}^+] + [E_{1/2}^-, g_0^{-1}J_+ g_0] = 0,$$

or, after the replacement $E_{1/2}^+ \to g_0^{-1}E_{1/2}^+ g_0$, in the symmetric form

$$E_{1/2,z_+}^- + [g_0^{-1}E_{1/2}^+ g_0, J_-] = 0,$$
$$E_{1/2,z_-}^+ + [g_0 E_{1/2}^- g_0^{-1}, J_+] = 0. \tag{3.3_2}$$

Similarly, from (3.2_1) we get

$$(g_0^{-1}g_{0,z_-})_{,z_+} + [g_0^{-1}E_{1/2}^+ g_0, E_{1/2}^-] + [g_0^{-1}J_+ g_0, J_-] = 0. \tag{3.3_1}$$

Thus (3.3) comprises equations for the element g_0 of the complex span of the group G_0 containing components of bosonic fields and for the two elements $E_{1/2}^\pm$ taking values in the subspaces $\mathfrak{g}_{\pm 1/2}$ of the initial Lie algebra $\mathfrak{g}$. When subspaces with half-integer values of the grading index correspond to odd elements of a Lie superalgebra, the coefficient functions $f_\pm^{1/2}(z_-, z_-)$ in the expansion of $E_{1/2}^\pm$ (3.1.2) should be regarded as taking values in an auxiliary supercommutative superalgebra C, cf. §1.7.

For example, for the Lie superalgebra osp(1/2) with the subspaces $\mathfrak{g}_0 \equiv \{h\}$, $\mathfrak{g}_{\pm 1/2} \equiv \{Y_\pm\}$ and $\mathfrak{g}_{\pm 1} \equiv \{X_\pm\}$, system (3.3) with $g_0 = \exp hx$, $E_{1/2}^\pm = \omega^\pm Y_\pm$ leads to the component form of the supersymmetric Liouville equation (1.16).

System (3.3) can be rewritten again in the form of (3.1.1) this time by a method invariant in a certain gauge as follows:

$$\left[\partial/\partial z_+ + g_0^{-1}E_{1/2}^+ g_0 + g_0^{-1}J_+ g_0, \ \partial/\partial z_- + g_0^{-1}g_{0,z_-} + E_{1/2}^- + J_-\right] = 0 \tag{3.4_1}$$

or, after an evident gauge transformation, in the equivalent form:

$$\left[\partial/\partial z_+ - g_{0,z_+}g_0^{-1} + E_{1/2}^+ + J_+, \ \partial/\partial z_- + g_0 E_{1/2}^- g_0^{-1} + g_0 J_- g_0^{-1}\right] = 0. \tag{3.4_2}$$

This clearly implies the invariance of (3.3) with respect to the replacement $z_+ \rightleftarrows z_-$, $g_0 \rightleftarrows g_0^{-1}$, $J_+ \rightleftarrows J_-$, $E_{1/2}^+ \rightleftarrows E_{1/2}^-$. For the zero value of spinor fields (coefficients in $E_{1/2}^\pm$), at the initial moment of time, (3.3) naturally transforms into (3.2.8) associated only with the integer embeddings of sl(2) in $\mathfrak{g}$.

The form (3.4) of expressing a class of nonlinear systems discussed above makes it possible to construct explicitly a solution of the Goursát problem in a way similar to that described in § 3.2.4 for (3.3) (in the form of finite polynomials of multiple integrals of the corresponding number of arbitrary functions) and a formal solution of the Cauchy problem for the same system as an integral ordered "chronologically" over a contour. All the calculations related to this approach are given in § 5.4.

§ 3.4 Lax-type representation as a realization of self-duality of cylindrically-symmetric gauge fields*)

The gauge field theory is an ideal test-field for an application of the general algebraic approach developed in the first sections of this chapter. In the framework of this theory we manage to give a unified constructive description of cylindrically-symmetric instanton and spherically-symmetric monopole (dyon) configurations of Yang-Mills fields (for an arbitrary embedding of A_1 into the complexified Lie algebra of a compact gauge group).

The investigations in this field were the starting point and provided for a stimulus for creating the approach we advocate for the study of nonlinear dynamical systems. Later on the method acquired greater generality. Therefore, it seems useful to trace how representations of type (1.1) actually arise for the applications to physics under consideration.

A Yang-Mills field over a Euclidean space $\mathbb{R}^4$ with gauge group G is given by its components $\mathcal{A}_\mu(x)$, $0 \leq \mu \leq 3$ which take values in $\mathfrak{g} = (\text{Lie }(G))^{\mathbb{C}}$ and which are C^∞ differentiable functions in $x \in \mathbb{R}^4$. Making use of the Bianci identity it is not difficult to show that the dynamical equations $[\partial/\partial x_\mu + \mathcal{A}_\mu, \mathcal{F}_{\mu\nu}] = 0$ which follow from the stationarity of the action

$$S = -(1/2)e^2 \int \text{tr } \mathcal{F}_{\mu\nu}\mathcal{F}_{\mu\nu}d^4x$$

are satisfied if the field is (anti) self-dual, i.e., if $\pm\mathcal{F}_{\mu\nu} = {}^*\mathcal{F}_{\mu\nu}(\equiv (1/2)\varepsilon_{\mu\nu\rho\tau} \mathcal{F}_{\rho\tau})$. Here $\mathcal{F}_{\mu\nu} \equiv [\partial/\partial x_\mu + \mathcal{A}_\mu, \partial/\partial x_\nu + \mathcal{A}_\nu]$ is the strength tensor of the Yang-Mills field, $\varepsilon_{\mu\nu\rho\tau}$ is tensor totally antisymmetric with respect to its indices with $\varepsilon_{1230} \equiv 1$, and e is the coupling constant which will be omitted in what follows. Among the solutions of duality equations a certain subclass of so-called (anti)instanton solutions is distinguished. These solutions are regular at all the points of $x \in \mathbb{R}^4$ and at infinity, i.e., on the whole 1-point compactification S^4 of $\mathbb{R}^4$, and they have finite action. The action is in this case proportional to the topological charge $Q \equiv \frac{1}{16\pi^2} \int \text{tr } {}^*\mathcal{F}_{\mu\nu}\mathcal{F}_{\mu\nu}d^4x$. (Obviously, anti-instantons are obtained from instantons by changing orientation of the space.)

---

*) [5, 22, 25, 26, 29, 63, 65, 66, 69, 70, 72, 92, 137, 138, 140, 143, 153, 156, 157, 159

The problem of the description of all instantons for an arbitrary compact classical Lie group was solved mathematically using methods of algebraic geometry [5].

It would be very interesting to get general solutions (not only parametric ones of instanton type) solutions of Yang-Mills equations determined by a set of arbitrary functions sufficient to formulate a Cauchy (or Goursát) problem, at least for the (anti) self-dual subclass. We have managed to do that under additional symmetry constraints which simplify the study of the considered system, reducing the total number of degrees of freedom to those which are invariant with respect to a subgroup of the conformal group of coordinate transformations. (Recall that the Yang-Mills theory is invariant with respect to the direct product of this group and the gauge group.) This requirement — of the cylindric symmetry in $\mathbb{R}^4$ — enables us to solve the considered problem completely and at the same time preserve a number of the main physical features of the theory. Let us describe this in detail.

First, complexify the picture, i.e., consider a Yang-Mills field $A_\mu(x)$ on $\mathbb{C}^4$ with values in $\mathfrak{g}^\mathbb{C} = (\text{Lie}\,G)^\mathbb{C}$ cylindrically symmetric with respect to the operator of the total momentum $T = J + L$, where $L = -ix \times \partial/\partial x$ are the operators of the spatial rotations and J are elements of an arbitrary embedded subalgebra $\text{sl}(2) \subset \mathfrak{g}^\mathbb{C}$. The components of A_0 and $\vec{A}$ of the field are transformed as a scalar and a vector, respectively, under the transformations generated by the operators T_i (intertwining the inner (gauge) and spatial indices of the field) and depend only on $\hat{r} \equiv (x^2)^{1/2}$ and $t \equiv x_0$.

Where it won't cause a misunderstanding we will simply write $\mathfrak{g}$ instead of $\mathfrak{g}^\mathbb{C}$. Now consider the subalgebra $\mathfrak{g}_0$ of $\mathfrak{g}$ spanned by the elements commuting with T independent on t and preserving the cylindrical symmetry of the fields. This subalgebra defines a Yang-Mills field on the two-dimensional space with coordinates $(\hat{r}, t)$ and is obviously determined by an embedding $\text{sl}(2) \subset \mathfrak{g}$. In accordance with the multiplet structure of $\mathfrak{g}$ with respect to T (see 1.3) the elements of $\mathfrak{g}_0$ are of the form $W^{l,\nu_l} = \sum\limits_M Y_m^{l*}(n) F_m^{l,\nu_l}$, where $Y_m^l(n)$ are spherical functions in $n \equiv x/\hat{r}$. Using the components of every irreducible multiplet W^{l,ν_l} and a unit vector n (transformed with respect to the vector, i.e., three-dimensional, representation of $\text{SL}(2) \simeq 0(3)$) we can, thanks to the rule of moment's summation, construct three types of operator vector structures namely:

$$W_{1,2,3}^{l,\nu_l} \equiv \{nW^{l,\nu_l}, LW^{l,\nu_l}, n \times LW^{l,\nu_l}\}.$$

For $G = SU(2)$ the role of $Y_m^l(n)$, W^l, nW^l, LW^l and $n \times LW^l$ are played by n, $(n\sigma)$, $n(n\sigma)$, $n \times \sigma$ and $n \times n \times \sigma$, respectively, where σ are Pauli matrices.

Notice that we consider here only integer embeddings of $A_1 \simeq 0(3)$ into G, i.e. those which do not lead to spinorial multiplets. The reason is that using the components of spinorial multiplets and vector n, it is impossible to construct a scalar with respect to T.

Therefore, the components $\mathcal{A}_\mu(x)$ of the field are parametrized by four operator structures $\mathcal{W}_\mu(\hat{r}, t)$ $(= \sum_{\{l\}} \varphi^l_\mu(\hat{r}, t) \mathcal{W}^{l, \nu_l})$:

$$\mathcal{A}_0 = \mathcal{W}_0, \ \mathbf{A} = \mathbf{n}\mathcal{W}_1 + \mathbf{L}\mathcal{W}_2 + \mathbf{n} \times \mathbf{L}\mathcal{W}_3 \tag{4.1}$$

where the $\mathcal{W}_\mu$ are scalars with respect to the total momentum, i.e. $[\mathbf{T}, \mathcal{W}_\mu] = 0$.

The electric strength $\boldsymbol{\mathcal{E}}$ and magnetic strength $\boldsymbol{\mathcal{H}}$ of the gauge field are calculated according to the usual rules and, in the coordinate system where $\mathbf{n} = (0, 0, 1)$, are represented in the following form, where $\mathcal{E}_i \equiv \mathcal{F}_{i0}$, $\mathcal{H}_i \equiv (1/2)\varepsilon_{ijk}\mathcal{F}_{jk}$:

$$\mathcal{E}_0 = (1/2)[\mathcal{D}_+, \mathcal{D}_-] \equiv (1/2)\mathcal{F}_{z_+ z_-},$$

$$\mathcal{H}_0 = -(1/2)(z_+ + z_-)^{-2}\{[\mathcal{W}^+, \mathcal{W}^-] + H\} \equiv -(1/2)V,$$

$$\mathcal{E}_\pm - \mathcal{H}_\pm = -(z_+ + z_-)^{-1}(\mathcal{D}_\mp \mathcal{W}^\pm), \tag{4.2}$$

$$\mathcal{E}_\pm + \mathcal{H}_\pm = (z_+ + z_-)^{-1}(\mathcal{D}_\pm \mathcal{W}^\pm),$$

$$\mathcal{W}^\pm \equiv \mp(z_+ + z_-)\{[J_\pm, \pm i\mathcal{W}_2 - \mathcal{W}_3] + (z_+ + z_-)^{-1}J_\pm\},$$

$$\mathcal{W}^0_\pm \equiv -\mathcal{W}_1 \mp i\mathcal{W}_0;$$

$(\mathcal{D}_\pm \mathcal{W}) \equiv \mathcal{W}_{,z_\pm} + [\mathcal{W}^0_\pm, \mathcal{W}]$ are covariant derivatives, $\{B_0, B_\pm\} \equiv \{iB_3, B_1 \pm iB_2\}$, $\mathbf{B} \equiv \{\vec{\mathcal{E}}, \vec{\mathcal{H}}, \mathbf{J}\}$, $J_0 \equiv H$ is the Cartan element of the embedding and $2z_\pm \equiv \hat{r} \mp it$.

In accordance with (4.2) the densities of the action $S \equiv -\pi \int s\, dz_+ dz_-$ and the topological charge $Q \equiv -(1/16\pi) \int q\, dz_+ dz_-$ are of the form

$$s = \mathrm{tr}\{(1/2)(z_+ + z_-)^2(\mathcal{F}^2_{z_+ z_-} + V^2) - (\mathcal{D}_+ \mathcal{W}^-)(\mathcal{D}_- \mathcal{W}^+)$$

$$\qquad - (\mathcal{D}_+ \mathcal{W}^+)(\mathcal{D}_- \mathcal{W}^-)\}; \tag{4.3}$$

$$q = \mathrm{tr}\{-(z_+ + z_-)^2 \mathcal{F}_{z_+ z_-} V + (\mathcal{D}_+ \mathcal{W}^-)(\mathcal{D}_- \mathcal{W}^+) - (\mathcal{D}_+ \mathcal{W}^+)(\mathcal{D}_- \mathcal{W}^-)\}. \tag{4.4}$$

Therefore, the Lagrangian of the system describes the interaction of the charged fields $\mathcal{W}^\pm$ and the gauge field $\mathcal{F}_{z_+ z_-}$ with the self-action in a curved 2-dimensional space.

In terms of these fields we may get the equations of motion by varying the functional of the action (4.3):

$$[\mathcal{D}_+, \mathcal{D}_-]_+ \mathcal{W}^\pm = \pm[\mathcal{W}^\pm, V],$$

$$[\mathcal{D}_+, \mathcal{F}_{z_+ z_-}] = [\mathcal{W}^+, (\mathcal{D}_+ \mathcal{W}^-)] - [\mathcal{W}^-, (\mathcal{D}_+ \mathcal{W}^+)], \tag{4.5}$$

where $[\cdot, \cdot]_+$ stands for the anticommutator.

The self-duality equations constitute a particular subclass of this general system

$$(\mathcal{D}_-\mathcal{W}^+) = 0, \ (\mathcal{D}_+\mathcal{W}^-) = 0, \ [\mathcal{D}_+, \mathcal{D}_-] + [\mathcal{W}^+, \mathcal{W}^-] = 0, \qquad (4.6)$$

and follow from (4.2) by equating the components of the electric and magnetic strengths of the field. (Their expression in the form (4.6) requires an additional change

$$\mathcal{W}_\pm^0 \to \mathcal{W}_\pm^0 \pm H/2(z_+ + z_-),$$

$$\mathcal{W}^\pm \to (z_+ + z_-)\mathcal{W}^\pm$$

which "rectifies" the space and cancels the term $-(1/2)(z_+ + z_-)^{-2}H$ in the expression for $\mathcal{H}_0$.) The corresponding equations for anti-selfdual fields are obtained from (4.6) by an obvious substitution $\mathcal{W}^+ \rightleftarrows \mathcal{W}^-$.

The left-hand sides of (4.6) take values in different subspaces of $\mathfrak{g}$ ($\mathfrak{g}_{+1}$, $\mathfrak{g}_{-1}$ and $\mathfrak{g}_0$, respectively) whose grading is determined by $H \in A_1$. Taking this into account we can rewrite (4.6) in the Lax form (1.1) with the operators $A_\pm = \partial/\partial z_\pm + \mathcal{W}_\pm^0 + \mathcal{W}^\pm$ (cf. (2.1)) taking values in the local part of $\mathfrak{g}$ for an arbitrary integer embedding of A_1 into $\mathfrak{g}$.

Therefore, the general procedure for constructing general solutions developed in §1 enables us to get the complete description of cylindrically-symmetric self-dual Yang-Mills fields.

Notice once again that we have not made any assumptions on $\mathfrak{g}$ such as compactness, simplicity, etc., except for the existence of a subalgebra isomorphic to A_1.

In the simplest case of the principal embedding of A_1 into a simple Lie algebra $\mathfrak{g}$ this system, in accordance with the results from 3.1, reduces to equations (1.9) -- the generalized two-dimensional Toda lattice — to be studied in detail in the next chapter.

Let us briefly discuss the problem of calculating the topological charge (4.4) of a self-dual system or, equivalently, of its action (4.3). For this, rewrite the system with the help of simple algebraic operations in the form

$$q = \text{tr} H\{[(1/2)\partial^2/\partial z_+ \partial z_- (z_+ + z_-)^2 - 1][\mathcal{W}_{-,z_+}^0 - \mathcal{W}_{+,z_-}^0 + H(z_+ + z_-)^{-2}]\}. \tag{4.7}$$

Since $\text{tr}\, H\mathcal{W}^\pm \equiv 0$, $\mathcal{W}_\pm^0$ in (4.7) can be equivalently replaced by $A_\pm$ of the form (1.21) implying

$$\text{tr}\, H(\mathcal{W}_{-,z_+}^0 - \mathcal{W}_{+,z_-}^0) = \text{tr} H(g_0^{-1}g_{0,z_+}),_{z_-}$$

$$= \partial^2/\partial z_+ \partial z_- \sum_j H_j \, \text{tr}(h_j H), \tag{4.8}$$

where H_j are the parameters of the element $\exp \sum_j h_j H_j$ of the Cartan subgroup in the Gauss decomposition for $\mathfrak{g}_0$. According to formulas from Chapter 1 the H_j are expressed in terms of the highest matrix elements $\xi^{\{j\}}$ of the fundamental representations of G, i.e. $H_j = \xi^{\{j\}}((\mathcal{M}_+)^{-1}\mathcal{M}_-)$. Taking this formula into account and substituting (4.8) into (4.7) we get the following finite expression for the density of the topological charge of a self-dual configuration:

$$q = -\frac{1}{2}\left[\frac{1}{2}\frac{\partial^2}{\partial z_+\partial z_-}(z_+ + z_-)^2 - 1\right]\frac{\partial^2}{\partial z_+\partial z_-}\sum_{i,j}k_j v_i k_{ij}\ln\frac{\xi^{\{i\}}(\mathcal{M}_+^{-1}\mathcal{M}_-)}{(z_z + z_-)^{c_i}},$$

(4.9)

where c_i are the coefficients of the decomposition of a Cartan element H with respect to generators of $\mathfrak{h}$ and $\mathrm{tr}(h_i h_j) = (1/2)v_i k_{ij}$.

Notice that for the principal embedding $(c_j = k_j \ (\equiv \delta_j))$, after summation over i and j, this formula takes the form

$$q = -[(1/2)\partial^2/\partial z_+\partial z_-(z_+ + z_-)^2$$
$$- 1]\partial^2/\partial z_+\partial z_-\ln[\xi^{\{v\}}(\mathcal{M}_+^{-1}\mathcal{M})/(z_+ + z_-)^{(vk)}]$$

$$\equiv (1/2)\sum_j v_j k_j[\exp\hat{\rho}_j - \hat{\rho}_j]_{,z_+z_-}, \quad \hat{\rho}_j \equiv \rho_j + 2\ln(z_+ + z_-) - \ln k_j.$$

(4.10)

To conclude this section consider briefly the problem of description of spherically-symmetric monopoles (dyons) on Minkowski space (with Higgs field in the adjoint representation of the gauge group) in the Bogomolny-Prasad-Sommerfeld limit. In this case the density of the Hamiltonian $\hat{\mathfrak{h}}$ for purely magnetic, time-independent solutions is given by the expression

$$\hat{\mathfrak{h}} = (1/2)\mathrm{tr}\hat{\mathcal{H}}^2 + (1/2)\mathrm{tr}(\boldsymbol{D}\chi)^2$$

$$\equiv (1/2)\mathrm{tr}(\hat{\mathcal{H}} \mp \boldsymbol{D}\chi)^2 \pm \mathrm{tr}(\hat{\mathcal{H}}\hat{\boldsymbol{D}}\chi),$$

where $\hat{\mathcal{H}}$ is the magnetic strength of the Yang-Mills field $\hat{\boldsymbol{A}}_\mu = (0, \hat{\boldsymbol{A}})$ parametrized by $\hat{\boldsymbol{W}}(\hat{r})$ (see (4.1)) and $(\boldsymbol{D}\chi) \equiv n\chi_{,\hat{r}} -[\hat{\boldsymbol{A}}\chi]$. Any solution of the differential equations $\hat{\mathcal{H}} = \boldsymbol{D}\chi (\hat{\mathcal{H}} = -\boldsymbol{D}\chi)$ realizes an energy minimum $\mathcal{J} = \int \hat{\mathfrak{h}}d^3 x \geq \int \mathrm{tr}(\hat{\mathcal{H}}\boldsymbol{D}\chi)d^3 x$ determined by the formula

$$2\pi\int\limits_0^\infty d\hat{r}\mathrm{tr}\{\hat{r}^2(\mathcal{D}_{\hat{r}}\chi)^2 - [\hat{W}^+, \chi][\hat{W}^-, \chi]\},$$

(4.11)

where $(\mathcal{D}_{\hat{r}}\chi) \equiv \chi_{,\hat{r}} -[\hat{W}^1, \chi]$. Comparing (4.11) with the integral of s over $\hat{r}$ given by (4.3), where $\mathcal{W}_0^0$ and $\mathcal{W}^\pm$ satisfy the self-duality equations (4.6), and passing to the static limit we see that the two expressions coincide for $\mathcal{W}_0 = \chi$ and $\hat{\boldsymbol{W}} = \boldsymbol{W}$.

Therefore, the preimages $\hat{\boldsymbol{A}} = \boldsymbol{A}$ and $\chi = \mathcal{A}_0$ of static self-dual fields $(\mathcal{A}_0, \boldsymbol{A})$ in $\mathbb{R}^4$ are monopole solutions in the Minkowski space with the Higgs field from the adjoint representation of G. To describe nonsingular magnetic monopoles it is necessary to pass to the static limit in the general solutions of equations (4.6) constructed in § 3.1 and ensure the finiteness of the energy functional (4.11) imposing appropriate boundary conditions. Then the solutions determined by the number of arbitrary functions in z_+, z_- necessary for formulation of the Cauchy problem turn into parametric solutions in complete analogy with the corresponding instanton problem (for details, see 4.1.6). The charge of a magnetic monopole is also given by the static limit of (4.10). Notice that dyon solutions can be obtained from $\boldsymbol{\mathcal{W}}$ and χ with the help of an obvious change $\boldsymbol{\mathcal{W}}' = \hat{\boldsymbol{\mathcal{W}}}$, $\mathcal{W}'_0 = \mathrm{sh}\theta\chi$; $\chi' = \mathrm{ch}\theta\chi$, where θ is an arbitrary constant.

Chapter 4

Integration of nonlinear dynamical systems associated with finite-dimensional Lie algebras

In this chapter we will explicitly construct general solutions for a number of concrete two-dimensional classical nonlinear systems of the type (3.1.4) associated with finite-dimensional Lie algebras. Moreover, in the cases where the one-dimensional (or parametric) solutions are important in applications we perform the necessary reduction.

§ 4.1 The generalized (finite nonperiodic) Toda lattice

4.1.1 Preliminaries*). Consider a two-dimensional ·dynamical system described by differential equations with exponential nonlinearities of the form

$$\rho_{i,z_+z_-} = \sum_j k_{ij} \exp \rho_j, \tag{1.1}$$

where k is an arbitrary numerical matrix. Equations (1.1) result from the representation (3.1.1) in which the operators $A_\pm = \sum_j [u_{\pm j}(z_+, z_-)h_j + f_{\pm j}(z_+, z_-)X_{\pm j}]$ are $\mathbb{C}$-spanned by the generators h_j, $X_{\pm j}$ of a contragredient algebra satisfying (1.2.8) with an arbitrary numerical matrix k and $\rho_j \equiv \ln f_{+j} f_{-j}$. Notice that for the Cartan matrices k of simple finite-dimensional Lie algebras $\mathfrak{g}$ the system (1.1) obviously follows either from (3.2.6) for the principal grading of $\mathfrak{g}$:

$$\mathfrak{g}_0 = \{h_j\}, \ \mathfrak{g}_{\pm 1} = \{X_{\pm j}\}, \ 1 \le j \le r, \ J_\pm = \sum_j k_j^{1/2} X_{\pm j}, \ k_i \equiv 2 \sum_j (k^{-1})_{ij},$$

---

*) [11, 15, 56, 63, 67–70, 72, 88, 90–92, 100, 104, 117, 136, 144–146, 149, 154, 155, 158, 160]

where $R_{ij} = -k_{ij}\varphi^1_{-j}$, $\varphi^1_{-j} \equiv \exp\rho_j$, $A_{ij} = -\delta_{ij}$ or from (3.2.8) with $g_0 \equiv \exp[-\sum_{i,j} h_i(k^{-1})_{ij}(\rho_j - \ln k_j)]$. Note that if we consider embeddings which are not principal then the corresponding equations (3.2.8) describe nonabelian versions of the Toda systems.

The dynamical systems of this type are encountered in the description of various essentially nonlinear physical phenomena. The attempts to take such effects into account systematically for such phenomena go back perhaps to works of Fermi and his collaborators on the heat equilibrium problem. The one-dimensional case

$$d^2\rho_i(t)/dt^2 \equiv \ddot{\rho}_i = \pm \sum_j k_{ij} \exp\rho_j \qquad (1.2)$$

(equations (1.1) with the matrix $k_{ij} = 2\delta_{ij} - \delta_{i+1j} - \delta_{ij+1}$) was first introduced and studied by Toda in the description of the exponential interaction between the nearest neighbouring particles on the one-dimensional lattice with fixed end-points $1 \leq i, j \leq r$, $\rho_0 = \rho_{r+1} = -\infty$.

More precisely, he considered the dynamical model of interacting particles of the same mass m on the line with the potential $V(\rho_i) = (a/b)\exp(-b\rho_i) + a\rho_i$, $a, b > 0$, where the first term corresponds to the repulsive force and the second one to the attractive force, ρ_i being the relative displacement of neighbouring particles (the i-th and $(i-1)$-th from the equilibrium position. This gives rise to the system of equations of motion

$$m\ddot{\rho}_i = a[2\exp(-b\rho_i) - \exp(-b\rho_{i-1}) - \exp(-b\rho_{i+1})],$$

which reduces to (1.2) with the minus sign in the right-hand side (e.g. under substitution $\rho_i \to -\rho_i$, $a = b = m = 1$).

Later on both physicists and mathematicians devoted many papers to general theorems proving complete integrability and existence of an additional series of integrals of motion for such one-dimensional systems. In particular. a relation has been established between the system (1.2) and the classical series A_r, and the so-called *generalized Toda lattice with fixed end-points* has been constructed using the root system of an arbitrary finite-dimensional simple complex Lie algebra. Finally, in the study of cylindrically symmetric config-urations of Yang-Mills fields with an arbitrary simple compact gauge group $\mathcal{K}$ and the principal embedding of A_1 into $(\text{Lie}(\mathcal{K}))^\mathbb{C}$ the system (3.1.9), the generalized two-dimensional Toda lattice, had been derived.

In this section following the general method we will deduce explicitly the general solutions (in the sense of Goursát's problem) for (3.1.9), i.e., for equations (1.1) where k is Cartan's matrix of an arbitrary simple finite-dimensional complex Lie algebra. Therefore, we will completely solve the two-dimensional generalized (finite nonperiodic) Toda lattice.

Another important particular case of system (1.1), which arises when k coincides with the generalized Cartan matrix of an infinite dimensional affine Lie algebra associated with the periodic Toda lattice, is considered in Chapters 5 and 6.

Notice that, though at the first glance it may look paradoxical, the two-dimensional variant of the generalized Toda lattice in the classical and quantum cases (the latter is considered in Chapter 7) admits a much simpler method of solution, i.e., explicit integration as compared with the corresponding ordinary differential equations. The general solutions of one-dimensional classical systems are obtained from the two-dimensional ones comparatively easy whereas for the quantum systems the situation is much more complicated.

4.1.2 Construction of exact solutions on the base of the general scheme of Chapter 3*). According to the general method, a solution of (3.1.9) for a finite-dimensional simple Lie algebra $\mathfrak{g}$ describing two-dimensional generalized Toda lattice with fixed end-points ($\rho_0 = \rho_{r+1} = -\infty$) is completely determined by the element K, see (3.2.17). In the principal grading of $\mathfrak{g}$ the equations (3.2.16) for $\mathcal{M}_\pm(z_\pm)$ take the form

$$\mathcal{M}_{+,z_+} = \mathcal{M}_+ \sum_{1 \le j \le r} \varphi_{+j}(z_+) X_{+j}, \quad \mathcal{M}_{-,z_-} = \mathcal{M}_- \sum_{1 \le j \le r} \varphi_{-j}(z_-) X_{-j} \quad (1.3)$$

and are expressed by the formulas (3.1.26) (or (3.1.27)) with operators

$$L_+ = -K_{,z_+} K^{-1} = \sum_{1 \le j \le r} \varphi_{+j}(z_+) X_{+j}, \quad L_- = K^{-1} K_{,z_-} = \sum_{1 \le j \le r} \varphi_{-j}(z_-) X_{-j}.$$
$$(1.4)$$

Here $\mathcal{N}_\pm$ and $\mathcal{M}_\pm$ denote the elements from the maximal nilpotent subgroups of $G = \exp \mathfrak{g}$ corresponding to positive and negative roots, respectively, and the elements $g_{0\pm}$ from (3.2.17) are parametrized by the generators of a Cartan subgroup of G in the form

$$g_{0\pm} = \exp H_\pm, \quad H_\pm = \sum_{1 \le j \le r} H_j^\pm h_j. \quad (1.5)$$

As we have already mentioned, the identity (3.2.17) enables one to define $\mathcal{N}_\pm$ and $\exp H$, $\hat{H} \equiv H_+ - H_- \equiv \sum_{1 \le j \le r} H_j h_j$ and therefore find the $f_{\pm j}$ from (3.1.7) and express the solutions $\rho_j \equiv \ln f_{+j} f_{-j}$ of (3.1.9) or, equivalently, of (3.1.10) in terms of the $f_{\pm j}$. However, even in the considered case of finite-dimensional simple Lie groups the explicit calculation of $\mathcal{N}_\pm$ is cumbersome and laborious. On the other hand, as we will show, it turns out that the solutions are determined by the functions $H_j = \mathrm{tr}(h^j \hat{H})$.

---

*) [67–69, 72]

Comparing (3.1.7) with (3.1.21) we see that $(hu_+) = H_{-,z_+}$, $(hu_-) = H_{+,z_-}$, i.e., in accordance with (3.1.8)

$$(hu_{+,z_-}) - (hu_{-,z_+}) = \sum_j h_j \exp \rho_j = -\hat{H}_{,z_+,z_-} , \qquad (1.6)$$

whereas thanks to (3.2.17), definitions (1.4.19) and (1.6.1) of the highest matrix elements $\xi^{\{j\}}(g)$ of the j-th fundamental representation of G we have

$$\exp H_j = \xi^{\{j\}}(K). \qquad (1.7)$$

Therefore, (1.6) and (1.7) imply

$$\exp \rho_j = -[\ln \xi^{\{j\}}(\mathcal{M}_+^{-1}\mathcal{M}_-)]_{,z_+z_-} . \qquad (1.8)$$

With the help of (1.3) and (1.6.2) we can show that the highest matrix elements $\xi^{\{L\}}(K)$ of an irreducible representation of G with the highest weight L satisfies

$$[\xi^{\{L\}}(K)]_{,z_+} = -\sum_j \varphi_{+j}(z_+)\langle L|X_{+j}\mathcal{M}_+^{-1}\mathcal{M}_-|L\rangle$$

$$= -\sum_j \varphi_{+j}X_j^L \xi^{\{L\}}(K), \ [\xi^{\{L\}}(K)]_{,z_-} = \sum_j \varphi_j(z_-)X_{-j}^R \xi^{\{L\}}(K). \qquad (1.9)$$

Taking the highest matrix elements of the fundamental representation $\xi^{\{i\}}(\mathcal{M}_+^{-1}\mathcal{M}_-)$ for $\xi^{\{L\}}$ in (1.9) and taking into account that $X_{-j}^R \xi^{\{i\}} = X_j^L \xi^{\{i\}} = 0$ thanks to (1.4.19) for all $i \neq j$ we get

$$[\xi^{\{j\}}(K)]_{,z_+} = -\varphi_{+j}X_j^L \xi^{\{j\}}(K), \ [\xi^{\{j\}}(K)]_{,z_-} = \varphi_{-j}X_{-j}^R \xi^{\{j\}}(K)$$

and therefore the right-hand side of (1.8)

$$[\ln \xi^{\{j\}}]_{,z_+z_-} \equiv [\xi^{\{j\}}\xi^{\{j\}}_{,z_+z_-} - \xi^{\{j\}}_{,z_+} \xi^{\{j\}}_{,z_-}](\xi^{\{j\}})^{-2}$$

$$\equiv (\xi^{\{j\}})^{-2} \det \begin{pmatrix} \xi^{\{j\}} & \xi^{\{j\}}_{,z_+} \\ \xi^{\{j\}}_{,z_-} & \xi^{\{j\}}_{,z_+z_-} \end{pmatrix}$$

takes the form

$$-(\xi^{\{j\}})^{-2}\varphi_{+j}\varphi_{-j} \det \begin{pmatrix} \xi^{\{j\}} & X_{+j}^L \xi^{\{j\}} \\ X_{-j}^R \xi^{\{j\}} & X_{-j}^R X_{+j}^L \xi^{\{j\}} \end{pmatrix}$$

and thanks to (1.6.16) is equal to

$$-\varphi_{+j}\varphi_{-j}(\xi^{\{j\}}(K))^{-2} \prod_{j \neq i}[\xi^{\{i\}}(K)]^{-k_{ji}} .$$

Therefore, the final expression for a solution of (3.1.9) is of the form

$$\exp \rho_i(z_+, z_-) = \varphi_{+i}(z_+)\varphi_{-i}(z_-) \prod_{1 \leq j \leq r}[\xi^{\{j\}}(\mathcal{M}_+^{-1}\mathcal{M}_-)]^{-k_{ij}}, \qquad (1.10)$$

or in terms of $x_i \equiv \sum\limits_{1 \leq j \leq r} (k^{-1})_{ij} \rho_j$, it takes the form

$$\exp(-x_i, (z_+, z_-)] = \exp\left\{ -\sum_{1 \leq j \leq r} k_{ij}^{-1} \ln[\varphi_{+j}\varphi_{-j}] \right\} \cdot \xi^{\{i\}}(\mathcal{M}_+^{-1}\mathcal{M}_-), \quad (1.11)$$

and is determined by $2r$ arbitrary functions $\varphi_{\pm j}(z_\pm)$.

These formulas enable one to get a solution of the Goursát problem for the two-dimensional generalized finite nonperidic Toda lattice. Indeed, suppose the initial conditions on $\mathcal{M}_\pm$ are $\mathcal{M}_+(a_+) = 1$, $\mathcal{M}_-(a_-) = 1$, where a_+ and a_- are arbitrary numbers, see (3.1.24). Then (1.10) implies

$$\exp \rho_i(a_+, z_-) = \varphi_{+i}(a_+)\varphi_i(z_-), \ \exp \rho_i(z_+, a_-) = \varphi_{+i}(z_+)\varphi_{-i}(a_-)$$

and therefore the general solution (1.10) expresses in terms of the initial values $\exp \rho_i(a_+, z_-)$ and $\exp \rho_i(z_+, a_-)$ on the characteristics $z_+ = a_+$ and $z_- = a_-$.

In order to compare with the case of infinite-dimensional Lie algebras of finite growth considered in what follows, recall that the highest vectors $\xi^{\{j\}}(\mathcal{M}_+^{-1}\mathcal{M}_-)$ of the fundamental representations of G can be expressed thanks to (1.6.11) as the scalar product

$$\begin{aligned}
\xi^{\{j\}}(\mathcal{M}_+^{-1}\mathcal{M}_-) &= \operatorname{tr}(\mathcal{M}_+^\cdot R_j^- \mathcal{M}_+^{-1} \cdot \mathcal{M}_- R_j^+ \mathcal{M}_-^{-1}) \\
&\equiv (\mathcal{M}_+ R_j^- \mathcal{M}_+^{-1}, \mathcal{M}_- R_j^+ \mathcal{M}_-^{-1}).
\end{aligned} \quad (1.12)$$

Let us give one more method for deducing solutions (1.11) based directly on relations (3.2.18) in our particular case. For this introduce the current-like operators $\mathcal{J}_+ \equiv N_+^{-1} N_{+,z_+} = \exp(-\hat{H})L_+ \exp \hat{H}$ and $\mathcal{J}_- \equiv N_+^{-1} N_{+,z_-} = \hat{H}_{,z_-} + L_- - N_+^{-1}L_- N_+$. Their compatibility condition in the form $[\partial/\partial z_+ + \mathcal{J}_+, \partial/\partial z_- + \mathcal{J}_-] = 0$ implies

$$\hat{H}_{,z_+z_-} = [L_-, \exp(-\hat{H})L_+ \exp \hat{H}], \quad (1.13)$$

which is a particular case of (3.2.8) for $g_0 = \exp \hat{H}$. Making use of (1.4), (1.2.8) and decomposing $\hat{H}$ with respect to generators h_j from $\mathfrak{h}$ we get

$$\begin{aligned}
\hat{H}_{,z_+z_-} &= \sum_{i,j} \varphi_{+j}\varphi_{-i}\left[X_{-i}\exp\left(-\sum_p H_p h_p \right) X_j \exp\left(\sum_q H_q h_q \right) \right] \\
&= \sum_{i,j} \varphi_{+j}\varphi_{-i}\exp[-(kH)_j][X_{-i}, X_j] = -\sum_i \varphi_{+i}\varphi_{-i}\exp[-(k\hat{H})_i]h_i,
\end{aligned}$$

i.e.

$$H_{i,z_+z_-} = -\varphi_{+i}(z_+)\varphi_{-i}(z_-) \exp[-(k\hat{H})_i].$$

Setting

$$x_j \equiv -H_j + \sum_{1 \leq i \leq r} (k^{-1})_{ji} \ln[\varphi_{+i}(z_+)\varphi_{-i}(z_-)], \qquad (1.14)$$

we see that the x_j satisfy (3.1.10) and their exponents are presentable in the form (1.11).

4.1.3 Examples. To illustrate the structure of general solutions constructed in the above subsection consider the algebras A_1 and A_2. The first case is absolutely obvious since the exponents (3.1.25) contain only one term, namely

$$\mathcal{M}_+ = \exp \int_{a_+}^{z_+} \varphi_+(z'_+)dz'_+ X_+, \ \ \mathcal{M}_- = \exp \int_{a_-}^{z_-} \varphi_-(z'_-)dz'_- X_-$$

and thanks to identity (3.2.17)

$$\mathcal{N}_\pm = \exp \left[\int^{z_\pm} \varphi_\pm(z'_\pm)dz'_\pm X_\pm \exp(-H_1) \right], \ \ \exp \hat{H} = \exp H_1 h,$$

$$H_1 = \ln \left[1 - \int^{z_+} \varphi_+(z'_+)dz'_+ \int^{z_-} \varphi_-(z'_-)dz'_- \right].$$

Therefore, due to (1.14)

$$\exp(-x) = [\varphi_+(z_+)\varphi_-(z_-)]^{-1/2} \left[1 - \int^{z_+} \varphi_+(z'_+)dz'_+ \int^{z_-} \varphi_-(z'_-)dz'_- \right],$$
$$\qquad (1.15)$$

which coincides with the known general solution of Liouville's equation (3.1.11) integrated in the middle of the last century.

The case of A_2 is less trivial. In accordance with (3.1.27) set

$$\mathcal{M}_\pm = \exp[a_{\pm 1}X_{\pm\pi_1} + a_{\pm 2}X_{\pm\pi_2} + a_{\pm 12}X_{\pm(\pi_1+\pi_2)}],$$

where

$$a_{\pm 1} \equiv \int^{z_\pm} \varphi_{\pm 1}(z'_\pm)dz'_\pm, \ \ a_{\pm 2} = \int^{z_\pm} \varphi_{\pm 2}(z'_\pm)dz'_\pm,$$

$$a_{\pm 12} \equiv \frac{1}{2} \int^{z_\pm} dz'_\pm \int^{z'_\pm} dz''_\pm [\varphi_{\pm 1}(z'_\pm)\varphi_{\pm 2}(z''_\pm) - \varphi_{\pm 1}(z''_\pm)\varphi_{\pm 2}(z'_\pm)].$$

The coefficient functions $b_\pm$ and H_j in the exponents of the elements

$$\mathcal{N}_\pm = \exp \left[\sum_{j=1,2} b_\pm X_{\pm\pi_j} + b_{\pm 12}X_{\pm(\pi_1+\pi_2)} \right] \ \text{and} \ \exp \hat{H} = \exp \sum_{j=1,2} H_j h_{\pi_j}$$

are calculated with the help of (3.2.17) and are of the form

$$b_{\pm 1} = (a_{\pm 1} - a_{\pm 3}a_{\mp 2})\exp(-H_1),$$

$$b_{\pm 2} = [a_{\pm 2} + (a_{\pm 3} - a_{\pm 1}a_{\pm 2})a_{\mp 1}]\exp(-H_2),$$

$$b_{\pm 3} = a_{\pm 3}\exp(-H_1),$$

$$H_1 = \ln(1 - a_1 a_{-1} + a_3 a_{-3}), \quad H_2 = \ln(1 - a_2 a_{-2} + a'_3 a'_{-3}),$$

where

$$a_{\pm 3} \equiv \mp a_{\pm 12} + (1/2)a_{\pm 1}a_{\pm 2} = \begin{cases} \int\limits_{z_+}^{z_+} \int\limits^{z'_+} \varphi_{+1}(z''_+)\varphi_{+2}(z'_+), dz'_+ dz''_+, \\ \int\limits_{z_-}^{z'_-} \int\limits \varphi_{-1}(z'_-)\varphi_{-2}(z''_-), dz'_- dz''_- \end{cases}$$

$$a'_{\pm 3} \equiv \pm a_{\pm 12} + (1/2)a_{\pm 1}a_{\pm 2}.$$

Therefore, from (1.14) we deduce

$$\exp(-x_1) = [\varphi_{+1}(z_+)\varphi_{-1}(z_-)]^{-2/3}[\varphi_{+2}(z_+)\varphi_{-2}(z_-)]^{-1/3}$$

$$\left[1 - \int\limits^{z_+} \varphi_{+1}(z'_+)dz'_+ \int\limits^{z_-} \varphi_{-1}(z'_-)dz'_- \right.$$

$$+ \int\limits^{z_+} dz'_+ \int\limits^{z'_+} dz''_+ \varphi_{+1}(z''_+)\varphi_{+2}(z'_+) \tag{1.16}$$

$$\left. \int\limits^{z_-} dz'_- \int\limits^{z'_-} dz''_- \varphi_{-1}(z'_-)\varphi_{-2}(z''_-) \right],$$

$$\exp(-x_2) = \exp(-x_1(\varphi_{\pm 1} \rightleftarrows \varphi_{\pm 2})).$$

These expressions show that though the dependence of the group parameters of $\mathcal{N}_\pm$ on the parameters of $\mathcal{M}_\pm$ is of a rather complicated form even in this very simple case, and therefore the desired element g determining the functions $u_{\pm j}$ and $f_{\pm j}$ via (3.1.17) is defined rather complicatedly, the solutions connected with the element of Cartan subgroup in the Gauss decomposition we are ultimately interested in are quite constructible. Actually they are just a polynomial form of expression for the highest matrix elements of the fundamental representations in terms of parameters of a group element $\mathcal{M}_+^{-1}\mathcal{M}_-$ in accordance with (1.4.22).

4.1.4 Construction of solutions without appealing to the Lax-type representation*). In a number of cases, in particular for calculations in concrete physical applications associated with simple classical Lie algebras, it is more advisable to use explicit expressions for solutions of (3.1.9) or (3.1.10) in terms of the determinants of a certain matrix rather than the general formula (1.11) or (1.10). Therefore, we will give here a direct method for constructing general solutions which does not appeal to Lax-type representation.

4.1.4.1 Symmetry properties of the Toda lattice for the series A, B, C and the reduction procedure. We will need the symmetry properties of the considered system which follow from the structure of the corresponding Cartan matrices which intertwine nonlinearities. It is more convenient for us to start from the equations in the form (3.1.10).

The Cartan matrices of the classical series A_r, B_r and C_r are identical: $k_{ij} = 2\delta_{ij} - \delta_{ij+1} - \delta_{i+1j}$, $1 \le i, j \le r - 2$, except for the elements of the lower right-hand 2×2 corner $\hat{k}$:

$$\hat{k}_A = \begin{pmatrix} 2 & -1 \\ -1 & 2 \end{pmatrix}, \quad \hat{k}_B = \begin{pmatrix} 2 & -2 \\ -1 & 2 \end{pmatrix}, \quad \hat{k}_C = \begin{pmatrix} 2 & -1 \\ -2 & 2 \end{pmatrix}.$$

Explicitly substituting these matrices into (3.1.10) we get

$$\begin{cases} x_{1,z_+z_-} = \exp(2x_1 - x_2), \\ x_{2,z_+z_-} = \exp(-x_1 + 2x_2 - x_3), \\ \qquad \cdots \cdots \cdots \cdots \cdots \cdots \cdots \cdots \cdots \cdots \cdots \\ x_{j,z_+z_-} = \exp(-x_{j-1} + 2x_j - x_{j+1}), \\ \qquad \cdots \cdots \cdots \cdots \cdots \cdots \cdots \cdots \cdots \cdots \cdots \\ x_{r-2,z_+z_-} = \exp(-x_{r-3} + 2x_{r-2} - x_{r-1}); \end{cases} \tag{1.17}$$

$$x_{r-1,z_+z_-} = \begin{cases} \exp(-x_{r-2} + 2x_{r-1} - x_r) & \text{for } A_r, C_r, \\ \exp(-x_{r-2} + 2x_{r-1} - 2x_r) & \text{for } B_r; \end{cases} \tag{1.18}$$

$$x_{r,z_+z_-} = \begin{cases} \exp(-x_{r-1} + 2x_r) & \text{for } A_r, B_r, \\ \exp(-2x_{r-1} + 2x_r) & \text{for } C_r; \end{cases} \tag{1.19}$$

which enables us to express the unknown functions $\exp(-x_j)$, $2 \le j \le r$, in terms of one unknown function $X \equiv \exp(-x_1)$. Indeed, the first equation in (1.17) yields

$$\exp(-x_2) = -\det \begin{pmatrix} X & X_{,z_-} \\ X_{,z_+} & X_{,z_+z_-} \end{pmatrix} \equiv -\Delta_2(X);$$

---

*) [65–68, 70]

from the second one we get

$$\exp(-x_3) = -\det \begin{pmatrix} X & X_{,z_-} & X_{,z_-z_-} \\ X_{,z_+} & X_{,z_+z_-} & X_{,z_+z_-z_-} \\ X_{,z_+z_+} & X_{,z_+z_+z_-} & X_{,z_+z_+z_-z_-} \end{pmatrix} = -\Delta_3(X).$$

Continuing the reduction process to the $(r-2)$-th step we get

$$\exp(-x_j) = (-1)^{j(j+1)/2}\Delta_j(X), \qquad \begin{array}{l} 1 \le j \le r \text{ for } A_r, \\ 1 \le j \le r-1 \text{ for } B_r, C_r, \end{array} \tag{1.20}$$

where $\Delta_j(X)$ is the principal j-th order minor of the matrix $(\partial/\partial z_+)^{p-1} \times (\partial/\partial z_-)^{q-1}X$.

Notice that this recurrence follows from the easily verified identity

$$[\log \Delta_j(X)]_{,z_+z_-} = \Delta_{j-1}(X)\Delta_{j+1}(X)\Delta_j^{-2}(X). \tag{1.21}$$

The further stages of reduction are different for different series. From (1.18) we get

$$(-1)^{r(r-1)/2}\Delta_r(X) = \begin{cases} \exp(-x_r) & \text{for } A_r, C_r, \\ \exp(-2x_r) & \text{for } B_r. \end{cases} \tag{1.22}$$

The r-th and last step for A_r gives

$$\Delta_{r+1}(X) = (-1)^{r(r+1)/2}, \tag{1.23}$$

which is a $2r$-th order nonlinear equation defining one unknown function X in terms of which all the other unknown functions $\exp(-x_j)$, $2 \le j \le r$ are expressed via (1.20).

For the series B_r and C_r the equations to define X are more complicated and even more complicated are the cases with D_r and the exceptional Lie algebras G_2, F_4, E_6, E_7 and E_8.

Therefore, the above reduction scheme gives an algorithm to get explicit formulas for $\exp(-x_j)$, $2 \le j \le r$, in terms of one unknown function X defining solutions of (3.1.10). The last stage of the reduction defining X is constructible in this approach only for the series A. However, the detailed analysis of the root structure of other Lie algebras enables one to deduce from the above scheme the answer in some cases (B and C series) making use of outer automorphisms of the algebra $\mathfrak{g}$ for which the solution is known.

Indeed, the system (3.1.10) for A_r is symmetric with respect to the change $x_j \rightleftharpoons x_{r-j+1}$, $1 \le j \le r$, and therefore possesses solutions with $x_j = x_{r-j+1}$. But (for $r = 2s$) equating $x_j = x_{r-j+1}$ we get a system for B_s (under the additional replacement of x_s by $2x_s$). For $r = 2s-1$ we similarly get C_s from A_{2s-1}. Similarly, equating $x_1 = x_6$, $x_3 = x_5$ and with the change $x_1 \rightarrow x_4$, $x_2 \rightarrow x_1$, $x_3 \rightarrow x_3$, $x_4 \rightarrow x_2$ in the equations (3.1.10) for E_6 we get a system corresponding to F_4, while equating $x_1 = x_3$ in the system for B_3 gives a system for G_2.

4.1.4.2 Direct solution of the system (3.1.10) for the series A. In accordance with the above reduction scheme the complete integration of the system (3.1.10) for the algebra A_r reduces to that of the nonlinear equation (1.23) for one unknown function $X(\equiv \exp(-x_1))$, in terms of which all the other functions $\exp(-x_j)$, $2 \le j \le r$, are expressed via (1.20). Let us seek the solutions of (1.23) in the form

$$X = \sum_{0 \le a \le r} P_+^a(z_+)P_-^a(z_-), \qquad (1.24)$$

where $P_\pm^a(z_\pm)$ are arbitrary functions. Notice that the form of this substitution is suggested by the general solution (1.15) of the Liouville equation (3.1.11) which is, as we have already mentioned, the particular case of our system corresponding to A_1. The substitution of (1.24) into (1.23) factors the dependence on z_+ and z_-; namely, we get

$$\Delta_{r+1}(X) = \det P_+ \cdot \det P_- = (-1)^{r(r+1)/2} \qquad (1.25)$$
$$\text{where} \quad P_{\pm b}^a = P_{\pm, \underbrace{z_\pm \ldots z_\pm}_{b}}^a, \ 0 \le a, \ b \le r.$$

Therefore, since P_+ (P_-) does not depend on z_- (z_+), the essentially nonlinear system of r equations (3.1.10) of second-order partial derivatives for the series A, i.e., (1.17) with $1 \le j \le r$, finally reduces to two ordinary r-th order differential equations determining the unknown functions $P_+^r(z_+)$ and $P_-^r(z_-)$ from the known functions $P_+^j(z_+)$ and $P_-^j(z_-)$, $0 \le j \le r-1$, respectively. Without loss of generality we may rewrite (1.25) in the form

$$\det P_+ = 1, \quad \det P_- = (-1)^{r(r+1)/2}, \qquad (1.26)$$

which is easy to solve by induction.

Indeed, for $r = 1$ (the algebra A_1) we get from (1.26)

$$P_+^0 P_{+,z_+}^1 - P_{+,z_+}^0 P_+^1 = 1, \ P_-^0 P_{-,z_-}^1 - P_{-,z_-}^0 P_-^1 = -1, \qquad (1.27)$$

with a particular solutions $P_\pm^0 = z_\pm$, $P_\pm^1 = \mp 1$.

Thanks to the conformal invariance of the Liouville equation, the function X with

$$P_\pm^0 = u_\pm u_{\pm,z_\pm}^{-1/2}, \ P_\pm^1 = \mp u_{\pm,z_\pm}^{-1/2}, \qquad (1.28)$$

where $u_+(z_+)$ and $u_-(z_-)$ are arbitrary functions, also satisfies the equation as is clear from direct substitution of (1.28) into (1.27).

It is also possible to solve (1.27) directly without making use of the conformal invariance of the initial equation by reducing them to the form $(P_\pm^1/P_\pm^0)_{,z_\pm} = \pm (P_\pm^0)^{-2}$ which implies

$$P_\pm^1 = \pm P_\pm^0 \int^{z_\pm} (P_\pm^0(z_\pm'))^{-2} dz_\pm',$$

or, setting $P_\pm^0 \equiv \varphi_\pm^{-1/2}(z_\pm)$,

$$P_\pm^1 = \pm[\varphi_\pm(z_\pm)]^{-1/2} \int^{z_\pm} \varphi_\pm(z_\pm') dz_\pm'.$$

Therefore, starting from the substitution (1.24) we get the known general solution for the Liouville equation containing two arbitrary functions $\varphi_\pm(z_\pm)$.

Now use induction to study the general case A_r. Suppose we know the general solution for A_{r-1} determined by $2(r-1)$ arbitrary functions $\varphi_{\pm j}(z_\pm)$, $1 \leq j \leq r-1$, i.e., we know the functions $_{(r-1)}P_\pm^a$, $0 \leq a \leq r-1$, satisfying (1.26) for A_{r-1}. Then the following is a particular solution for (1.26) for A_r which also depends on $2(r-1)$ arbitrary functions:

$$_{(r)}P_\pm^a(z_\pm) = \int^{z_\pm} dz_\pm' {}_{(r-1)}P_\pm^a(z_\pm'), \ 0 \leq a \leq r-1; \ {}_{(r)}P_\pm^r(z_\pm) = \pm 1. \quad (1.29)$$

Obviously, the system (3.1.10) is conformally invariant with respect to the transformation

$$z_\pm \to \varphi_\pm(z_\pm), \ x_j \to x_j + \ln \left(\varphi_+(z_+)\varphi_-(z_-) \sum_{1 \leq i \leq r} (k^{-1})_{ji} \right).$$

Therefore, applying this transformation to the function (1.29) we get the complete solution of equations (1.26):

$$_{(r)}P_\pm^a = \int^{\varphi_\pm(z_\pm)} dz_\pm' {}_{(r-1)}P_\pm^a(z_\pm')\varphi_{\pm,z_\pm}^{-r/2}, \ {}_{(r)}P_\pm^r = \pm\varphi_{\pm,z_\pm}^{-r/2}, \ 0 \leq a \leq r-1$$

and therefore, the general solution of the system (3.1.10) for A_r. This solution is expressed in quadratures depending on $2r$ arbitrary functions. As follows from the reduction procedure, it is natural to represent the explicit dependence of $P_\pm^a(z_\pm)$ on these functions $\varphi_{\pm j}(z_\pm)$, $1 \leq j \leq r$, in the form

$$P_\pm^a(z_\pm) = (\pm 1)^a \varphi_0^\pm(z_\pm) \cdot (1 \ldots a)_\pm,$$

$$(\varphi_0^\pm)^{-1} \equiv \prod_{1 \leq j \leq r} \varphi_{\pm j}^{1/(r+1)}, \quad (1.30)$$

where

$$(1 \ldots a)_\pm \equiv \int^{z_\pm} dz_\pm^{(1)} \varphi_{\pm 1}(z_\pm^{(1)}) \int^{z_\pm^{(1)}} dz_\pm^{(2)} \varphi_{\pm 2}(z_\pm^{(2)}) \ldots \int^{z_\pm^{(a-1)}} dz_\pm^{(a)} \varphi_{\pm a}(z_\pm^{(a)}),$$

$$a \geq 1, \ (1 \ldots 0) \equiv 1,$$

$$(1.31)$$

and $P_\pm^a(z_\pm) =_{(r)} P_\pm^{r-a}(z_\pm)$. Substituting (1.30) into (1.24) we get the following expression for X:

$$X = \prod_{1 \leq i \leq r} [\varphi_{+i}(z_+)\varphi_{-i}(z_-)]^{-1/(r+1)} \left[1 + \sum_{1 \leq j \leq r} (-1)^j (1 \ldots j)_+ (1 \ldots j)_- \right].$$
(1.32)

In accordance with (1.20) the expressions for $\exp(-x_j)$, $2 \leq j \leq r$, are

$$\exp(-x_j) = (-1)^{j(j+1)/2} \sum \Delta_j(P_+)\Delta_j(P_-),$$
(1.33)

where sum runs over all the j-th order minors constituted by the first j rows of the matrices P_+, P_-.

In particular, $\exp(-x_r) = \sum_{0 \leq a \leq r} \bar{P}_+^a \bar{P}_-^a$ where $\bar{P}_\pm^a(z_\pm)$ are r-th order minors of the form (1.30) obtained from $P_\pm^a$ by replacing $\varphi_{\pm(r-a+1)}$ with $\varphi_{\pm a}$. (There is a similar relation between the functions $\exp(-x_j)$ and $\exp(-x_{r-j+1})$ which mirrors the symmetry of the system mentioned in 4.1.4.1 corresponding to outer automorphisms of $\mathfrak{g}$.)

Notice that any permutation of the functions P_+^a with the simultaneous identical permutation of P_-^a that preserves the sign in the right-hand side of (1.25) also leads to a complete solution of the considered system and reflects the invariance of the root system of A_r with respect to the Weyl group. Such an invariance of solutions of Liouville's equation has been noticed first by Bianci.

Naturally, the solutions constructed by the above reduction scheme coincide with the expressions obtained for A_r with the help of the Lax-type representation as a particular case of (1.11).

To see this one should make use of the explicit expression of the highest matrix elements of the fundamental representations of A_r as a polynomial in the corresponding multiple integrals.

In a number of cases, in particular for calculations connected with the explicit integration of Volterra equations (see 4.2) we will need a somewhat different ("local") form for solutions for A_r which does not contain multiple integrals. For this denote the multiple integral (1.31) by $\Phi_{\pm j}(z_\pm)$.

Then all the functions $\varphi_{\pm j}$ may be expressed via the linear combinations of the principal minors $\Delta_j(\Phi_\pm)$ of the matrices $\Phi_{i,\pm j} \equiv \Phi_{\pm j, \underbrace{z_\pm \ldots z_\pm}_{i}}$. We can verify by induction that the final formula for X is

$$X = [\Delta_r(\Phi_+)\Delta_r(\Phi_-)]^{-1/(r+1)} \left[1 + \sum_{1 \leq j \leq r} (-1)^j \Phi_{+j}(z_+)\Phi_{-j}(z_-) \right],$$
(1.34)

where $\Phi_{\pm j}(z_\pm)$ are new arbitrary functions. Notice that the form of (1.34) is the closest to the initial form of the solution of the Liouville equation and coincides with it for $r = 1$.

The above method can be generalized to B_r and C_r for which we derive from (1.18) and (1.19) the following equations for $X \equiv \exp(-x_1)$:

$$B_r : \Delta_{r+1}(X) = 2(-1)^r \Delta_r(X); \quad C_r : \Delta_{r+1}(X) = -\Delta_{r-1}(X).$$

However, these equations are rather complicated and for the series B and C it is more convenient to make use of the symmetry property of (3.1.10) established at the end of 4.1.4.1.

Relations $x_j = x_{r-j+1}$ for A_r leads to the relations $\varphi_{\pm j}(z_\pm) = \varphi_{\pm(r-j+1)}(z_\pm)$ between the functions defining the solution (1.33) and (1.32). Thus, we get for $r = 2s$ ($r = 2s - 1$) the solutions corresponding to B_s (C_s). Naturally, we thus get the solutions of these two series obtained from the general formula (1.11) with the help of explicit expressions for the highest matrix elements of their fundamental representations.

4.1.4.3 Invariant generalization of the reduction scheme for arbitrary simple Lie algebras.

In the method for integrating (3.1.10) for the classical series described above we use neither the properties of these algebras nor their representation theory. However, in much the same way as the form of the functions (1.30) is obviously connected with the corresponding root systems, all the determinants (1.33) which enter the final expression for the solution of (3.1.10) are actually multiplicative combinations of the highest matrix elements of the fundamental representations of these algebras. It was only the very simple structure of the system considered for the series A and the possibility of reducing this system in a rather elementary way to the cases corresponding to the algebras of types B and C that enabled us to perform all the calculations without referring directly to the representation theory. However, this method can be generalized to arbitrary simple Lie algebras. For this we will need recurrence relations (1.6.16) on the highest matrix elements of the fundamental representations which are more complicated for algebras other than A_r.

For this let us rewrite (3.1.10) in the form

$$\det \begin{pmatrix} X_j & X_{j,z_-} \\ X_{j,z_+} & X_{j,z_+z_-} \end{pmatrix} = \prod_{i \neq j} X_i^{-k_{ji}} \tag{1.35}$$

where in analogy with $X \equiv \exp(-x_1) \equiv X_1$, we have introduced the new functions

$$X_j \equiv \exp(-x_j), \ 2 \leq j \leq r; \ X_0 \equiv X_{r+1} \equiv 1.$$

Now, notice the complete analogy of (1.35) with the recurrence relation (1.6.16), more exactly with the formula

$$\det \begin{pmatrix} \xi^{\{j\}}(K) & (\xi^{\{j\}}(K))_{,z_-} \\ (\xi^{\{j\}}(K))_{,z_+} & (\xi^{\{j\}}(K))_{,z_+z_-} \end{pmatrix} = -\varphi_{+j}(z_+)\varphi_{-j}(z_-) \prod_{i\neq j} [\xi^{\{i\}}(K)]^{-k_{ji}},$$

(1.36)

obtained in 4.1.4.2 when we deduced the general solution of (1.11). By elementary transformations of (1.36) it is not difficult to show that the functions

$$Y_i \equiv \exp\left\{ -\sum_j (k^{-1})_{ij} \ln \varphi_{+j}\varphi_{-j} \right\} \xi^{\{i\}}(K)$$

satisfy the same system (1.35) as the functions X_i do. Therefore, the general solution of (3.1.10) depending on $2r$ arbitrary functions $\varphi_{\pm j}(z_\pm)$ is given by (1.11), i.e.

$$X_i = \exp\left\{ -\sum_j (k^{-1})_{ij} \ln \varphi_{+j}\varphi_{-j} \right\} \langle i|\mathcal{M}_+^{-1}\mathcal{M}_-|i\rangle.$$

(1.37)

Equations (1.3) satisfied by $\mathcal{M}_\pm(z_\pm)$ enable one to perform further transformations of the expression

$$\langle i|\mathcal{M}_+^{-1}\mathcal{M}_-|i\rangle = 1 + \sum_{\substack{p_1\cdots p_s \\ q_1\cdots q_s}} (-1)^s (p_s \ldots p_1)_+ (q_1 \ldots q_s)_-$$

(1.38)

$$\langle i|X_{p_s} \ldots X_{p_1} X_{-q_1} \ldots X_{-q_s}|i\rangle,$$

where $(p_1 \ldots p_s)_\pm$ are determined from (1.31). Taking (1.4.22) into account we rewrite the identity (1.38) for the matrix element $\langle i|X_{p_s} \ldots X_{p_1} X_{-q_1} \ldots X_{-q_s}|i\rangle$ in the form

$$\langle i|\mathcal{M}_+^{-1}\mathcal{M}_-|i\rangle = 1 + \sum_{p_1\cdots p_s} \sum_\omega (-1)^s \delta_{i,p_s}\delta_{i,p_{\omega(s)}} (p_s \ldots p_1)_+ (p_{\omega(1)} \ldots p_{\omega(s)})_-$$

$$\times \prod_{1\leq j\leq s} \left[\delta_{ip_s} - \sum_{t\in I_s^{(S_j)}} k_{p_{\omega(s)},p_j} + \delta_{i,p_j}\delta_{T_j,s-1} \right].$$

(1.39)

To prove the validity of (1.38) it suffices to make use of the representation of the solutions (1.3) entering (1.37) as the ordered exponentials (3.1.24) $\mathcal{M}_\pm = \tilde{Z}_\pm \exp \int^{z_\pm} L_\pm(z'_\pm)dz'_\pm$, and take into account the factor $(-1)^s$ which appears because we pass from $\mathcal{M}_+$ to $\mathcal{M}_+^{-1}$.

For finite-dimensional representations the series in (1.38) is obviously finite: having reduced the quadratic form $\langle \cdot|\cdot \rangle$ to the canonical form in all the orders with respect to the number of generators $X_{\pm j}$ we see that the total number of terms in (1.38) is equal to the dimension of the i-th fundamental representation. The final expression for these solutions of (3.1.10) is representable in the form

$$
\exp(-x_i) = \exp\left[-\sum_{1\leq j\leq r}(k^{-1})_{ij}\ln\varphi_{+j}\varphi_{-j}\right]
$$

$$
\times\left\{1+\sum_{p_1\ldots p_t}\sum_\omega(-1)^t\delta_{ip_t}\delta_{ip_{\omega(t)}}(p_t\ldots p_1)_+(p_{\omega(1)}\ldots p_{\omega(t)})_-\right.
$$

$$
\left.\times\prod_{1\leq j\leq t}\left[\delta_{ip_j}-\sum_{s\in I_t^{(S_j)}}k_{p_{\omega(s)}p_j}+\delta_{ip_j}\delta_{S_j,t-1}\right]\right\},\qquad(1.40)
$$

which is an equivalent form of expression (1.11).

4.1.5 The one-dimensional generalized Toda lattice*). Here we will construct and study properties of general solutions of the one-dimensional generalized Toda lattice with fixed end-points described by system of ordinary differential equations (1.2) or, equivalently

$$
\ddot{x}_i = \pm\exp(kx)_i,\qquad(1.41)
$$

where, as in (3.1.10), k is Cartan's matrix of a finite-dimensional simple Lie algebra $\mathfrak{g}$.

For definiteness consider the equations with the plus sign (the other possibility reduces to the first one under the change $t\to it$). The Lagrangean $\mathcal{L}$ and the Hamiltonian $\mathfrak{h}$ corresponding to a relative displacement of particles of this dynamical system in variables ρ_i or x_i take the form

$$
\mathcal{L} = (1/2)\sum_{i,j}(vk)_{ij}^{-1}\dot{\rho}_i\dot{\rho}_j + \sum_i v_i^{-1}\exp\rho_i;
$$

$$
\mathcal{L} = (1/2)\sum_{i,j}(vk)_{ij}\dot{x}_i\dot{x}_j + \sum_i v_i\exp(kx)_i;
$$

$$
\mathfrak{h} = -(1/2)\sum_{i,j}(vk)_{ij}p_ip_j + \sum_i v_i^{-1}\exp\rho_i;
$$

$$
\mathfrak{h} = -(1/2)\sum_{i,j}(vk)_{ij}^{-1}p_ip_j + \sum_i v_i\exp(kx)_i,
$$

(1.42)

where p_i are the generalized momenta corresponding to ρ_i or x_i. The system (1.41) (or (1.2)) possesses exactly r functionally independent integrals of

---

*) [64–70, 108]

motion one of which is the Hamiltonian $\mathfrak{h}$ (1.42). This follows automatically from the Lax-type representation for the system (1.41):

$$\dot{L} = -(i/2)[A, L];$$

$$A = \sum_j \exp(1/2)\,(kx)_j(X_j - X_{-j}),$$

$$L = \sum_j \exp(1/2)\,(kx)_j(X_j + X_{-j}) - i \sum_{i,j}(vk)_{ij}^{-1} p_j h_i; \; p_j \equiv \sum_i (vk)_{ji}\dot{x}_i.$$

$$(1.43)$$

(The latter follows from the representations (3.1.7) in the two-dimensional space expressed in the form free of the gauge arbitrariness, and the functions $f_{\pm j} = \exp(kx)_j$ and $u_{\pm j} = -(i/2)\dot{x}_j$ depend only on one variable $t \equiv -i(z_+ - z_-)$.)

Here $X_{\pm j}$ are root vectors and h_j are the elements of the Cartan subalgebra $\mathfrak{h}$ corresponding to simple roots of $\mathfrak{g}$. Therefore, in accordance with (1.43) we have $\frac{d}{dt}\operatorname{tr} L^q = 0$, i.e. $I_q(p, x) \equiv \operatorname{tr} L^q(t)$ are integrals of motion.

From an explicit realization of the (A, L)-pair we may indicate r functionally independent integrals $I_q(p, x)$ in involution, i.e., $\{I_q, I_l\} = 0$, and such that the spectrum of q's coincides with the values of orders of Casimir operators of $\mathfrak{g}$. The clearest proof follows from the explicit form of the asymptotics in t of the solutions of (1.41) depending on $2r$ independent parameters m_j, d_j, $1 \leq j \leq r$ constructed in what follows.

Indeed, since $I_q(p, x)$ and their Poisson brackets are integrals of motion and therefore do not depend on t, then it is possible to consider the asymptotic values of their constituents p_j and x_j. Then the terms in the expression for $L(t)$ as $t \to \infty$ connected with $X_{\pm j}$ vanish whereas p_i tend to $m_i v_i$ provided that certain inequalities between the parameters $\{m_i\}$ hold. The further calculations of asymptotics of these integrals are trivial and the final answer reproduces the corresponding expressions for the eigenvalues of Casimir operators of $\mathfrak{g}$.

Therefore, in accordance with the Liouville theorem the system (1.41) is completely integrable since it possesses r functionally independent global integrals in involution. In what follows we will give an independent constructive proof of this theorem and find explicitly the solutions of (1.41) defined by the necessary number $(2r)$ of arbitrary parameters.

There is another approach to the study of properties of equations (1.2) or (1.41) that is of particular importance in the quantization of the corresponding dynamical system. It is quite general and is also applicable to other completely integrable (in the sense of Liouville) one-dimensional dynamical systems.

For this consider a simple Lie algebra $\mathfrak{g}$ given by infinitesimal operators in the regular representation as a functional algebra (see 1.1.3). In other words, let us replace all the commutation relations between the elements of $\mathfrak{g}$ by the classical Poisson brackets, i.e., in physical terms, let us pass from the quantum case (the commutators) to the classical one when all the brackets tend to the Poisson brackets between the corresponding dynamical values as the Planck's constant $\hbar$ tends to zero.

Since the generators of the regular representation are linear with respect to derivatives of the parameters of $\mathfrak{g}$, such a passage does not cause any difficulties and in the finite-dimensional case the phase space of the dynamical system considered is described by $2(N-r)$ generalized positions and momenta and therefore r integrals in involution corresponding to r cyclic coordinates, where r is the rank and N is the dimension of $\mathfrak{g}$.

The second-order Casimir operator is given by the formula $(2.2.5_2)$ or setting $\partial/\partial\tau_i \equiv p_i$ in $(2.2.5_2)$ in the form

$$
C_2 = (1/2) \sum_{i,j} (vk)_{ij}^{-1} p_i p_j + \sum_{\alpha > 0} v_\alpha^{-1} [\exp(-\alpha(\tau)) K_\alpha^R X_\alpha^L \\
- \exp(-2\alpha(\tau)) X_\alpha^L X_\alpha^L]. \tag{1.44}
$$

Here we have taken into account the properties of the functional algebras. Indeed, as compared with the initial formula $(2.2.5)$ the symbols δ_i are absent here since in the physical interpretation they are proportional to $\hbar$ and therefore vanish in the classical limit. (In other words, the terms proportinal to δ_i cancel under the canonical transformation

$$
C_2 \rightarrow \exp((\delta\tau)/2) \cdot C_2 \cdot \exp(-(\delta\tau)/2), \ (\delta\tau) \equiv \sum_j \delta_j \tau_j.)
$$

The quantity C_2 of the form (1.44) may be considered as a Hamiltonian operator of a dynamical system described by r generalized positions τ_i which turn into cyclic ones as $\alpha(\tau) \rightarrow \infty$; r generalized momenta p_i; $(N-r)/2$ dynamical variables X_α^L; and $(N-r)/2$ quantities K_α^R with $\alpha > 0$. The Poisson brackets between them are given by the relations

$$
\{p_i, \tau_j\} = \delta_{ij}, \ \{p_i, X_\alpha^L\} = \{\tau_i, X_\alpha^L\} = \{p_i, K_\alpha^R\} = \{\tau_i, K_\alpha^R\} = 0;
$$

$$
\{X_\alpha^L, K_\beta^R\} = 0, \ \{X_\alpha^L, X_\beta^L\} = \mathcal{N}_{\alpha,\beta} X_{\alpha+\beta}^L; \tag{1.45}
$$

$$
\{K_\alpha^R, K_\beta^R\} = \mathcal{N}_{\alpha,\beta} K_{\alpha+\beta}^R - \mathcal{N}_{\alpha,-\beta} K_{\alpha-\beta}^R.
$$

Here the constants $\mathcal{N}_{\alpha,\beta}$ and $\mathcal{N}_{\alpha,-\beta}$ are nonzero only for $\alpha + \beta \in R_+$ and $\alpha - \beta \in R_+$, respectively, where $\alpha, \beta \in R_+$. The complete system of equations for these dynamical quantities $Q \equiv (p, \tau, X^L, K^R)$ with the Hamiltonian (1.44), i.e., $\dot{Q} = \{C_2, Q\}$, takes the form (with (1.45) being taken into account)

$$\dot{\tau}_i = ((vk)^{-1}p)_i,$$

$$\dot{p}_i = \sum_{\alpha>0} v_\alpha^{-1}\alpha_i[\exp(-\alpha(\tau))K_\alpha^R X_\alpha^L + 2\exp(-2\alpha(\tau))(X_\alpha^L)^2],$$

$$\dot{X}_\beta^L = \sum_{\alpha>0} v_\alpha^{-1}\mathcal{N}_{\alpha,\beta}[\exp(-\alpha(\tau))K_\alpha^R X_{\alpha+\beta}^L \qquad (1.46)$$

$$+ \exp(-2\alpha(\tau))(X_\alpha^L X_{\alpha+\beta}^L + X_{\alpha+\beta}^L X_\alpha^L)],$$

$$\dot{K}_\beta^R = \sum_{\alpha>0} v_\alpha^{-1}\exp(-\alpha(\tau))X_\alpha^L(\mathcal{N}_{\alpha,\beta}K_{\alpha+\beta}^R - \mathcal{N}_{\alpha,-\beta}K_{\alpha-\beta}^R).$$

It follows from (1.46) that imposing the initial conditions $K_\alpha^R = 0$ for all $\alpha \in R_+$ and $X_\alpha^L = 0$ for all α, except for the simple roots π_i, we get

$$K_\alpha(t) = 0; \; X_\alpha^L(t) = c_\alpha \delta_{\alpha\pi_i}, \; 1 \le i \le r,$$

where c_α are arbitrary constants which without loss of generality we may set equal to $v_\alpha/2$. Then the remaining equations of (1.46) take the form

$$\dot{\tau}_i = ((vk)^{-1}p)_i; \; \dot{p}_i = \sum_j v_j k_{ji}\exp(-2(k\tau)_j), \qquad (1.47)$$

which under the change $\tau_i \to -\tau_i/2$ implies (1.41) which describes the finite unperiodic Toda lattice.

Naturally, (1.47) are Hamiltonian equations $\dot{p}_i = -\partial\mathfrak{h}/\partial\tau_i, \; \dot{\tau}_i = \partial\mathfrak{h}/\partial p_i$ with the Hamiltonian $\mathfrak{h} = C_2$ for $K_\alpha^R = 0, \; X_\alpha^L = (1/2)v_\alpha\delta_{\alpha\pi_i}, 1 \le i \le r$.

The Casimir operators of the functional algebra $\mathfrak{g}^f$ are expressed in terms of generalized momenta corresponding to r cyclic variables of the phase space. In other words the r integrals of motion of (1.46) in involution, one of which is the Hamiltonian (1.44), are invariant polynomials on $\mathfrak{g}^f$. It is not difficult to see that the choice of initial conditions above does not affect these statements.

Therefore, the dynamical system (1.41) has r integrals in involution coinciding in their form with the invariant polynomials on $\mathfrak{g}^f$, where $K_\alpha^R = 0$, $X_\alpha^L = (1/2)v_\alpha\delta_{\alpha\pi_i}$.

The simplest way to get the explicit expressions for the solutions of a one-dimensional generalized Toda lattice is to impose an additional constraint of functional independence onto general solutions (1.11) of the two-dimensional system (3.1.10), say, in the form of dependence only on $\hat{r} = z_+ + z_-$ (or t or z_+z_-). For this it suffuces to select arbitrary functions $\varphi_{\pm j}(z_\pm)$ in the form $\varphi_{\pm j} = c_{\pm j}\exp m_j z_\pm$ which makes it possible to calculate explicitly the integrals of the type (1.31) and express the final formulas as functions of $z_+ + z_-$ only. After such a substitution the general solutions of (3.1.10) depending on $2r$ arbitrary functions turn into $2r$-parametric solutions of (1.41) and the parameters $d_j \equiv c_{+j}c_{-j}$ and m_j, $1 \leq j \leq r$ play the role of constants of integration for these ordinary differential equations.

Integrating term-wise the expressions $(p_1 \ldots p_t)_\pm$ entering the general formula (1.40) and using the fact that the scalar products of basic vectors in it are nonzero only for the same sets of indices $\{p_t, \ldots, p_1\}$ and $\{q_1, \ldots, q_t\}$ we get in the exponential parametrization of functions $\varphi_{\pm j}$ the following result:

$$
\exp(-x_j(\hat{r})) = \prod_{1 \leq j \leq r} d_j^{-(k^{-1})_{ij}} \exp(-\hat{r}(k^{-1}m)_i)
$$

$$
\times \left\{ 1 + \sum_{\substack{p_1 \cdots p_t \\ q_1 \cdots q_t}} (-1)^t \prod_{1 \leq s \leq t} d_{p_s} \exp\left(-\hat{r} \sum_{1 \leq l \leq t} m_{p_l} \right) \right.
$$

$$
\times \left. \frac{\langle i | X_{p_t} \ldots X_{p_1} X_{-q_1} \ldots X_{-q_t} | i \rangle}{\displaystyle\prod_{1 \leq \theta \leq t} \left(\sum_{1 \leq a \leq \theta} m_{p_a} \right) \left(\sum_{\theta \leq b \leq t} m_{q_b} \right)} \right\}
$$

$$
= \prod_{1 \leq j \leq r} d_j^{-(k^{-1})_{ij}} \exp(-\hat{r}(k^{-1}m)_i) \left\{ 1 + \sum_{p_1, \ldots, p_t} \sum_{\omega} (-1)^t \delta_{ip_t} \delta_{ip_{\omega(t)}} \right.
$$

$$
\times \prod_{1 \leq e \leq t} \left[\delta_{ip_e} - \sum_{s \in I_t^{(S_e)}} k_{p_{\omega(s)}} p_e + \delta_{ip_e} \delta_{S_e, t-1} \right] d_{p_e}
$$

$$
\times \exp\left(\hat{r} \sum_{1 \leq l \leq t} m_{p_l} \right) \prod_{1 \leq \theta \leq t} \left(\sum_{1 \leq a \leq \theta} m_{p_a} \right)^{-1} \left. \left(\sum_{\theta \leq b \leq t} m_{p_{\omega(b)}} \right)^{-1} \right\}.
$$

$$(1.48)$$

Formula (1.48) solves the integration problem for the classical one-dimensional generalized Toda lattice with fixed end-points i.e., a system (1.41) with a plus sign. To solve (1.41) with a minus sign the functions $\varphi_{\pm j}(z_\pm)$ should be expressed in the form $\varphi_{\pm j}(z_\pm) = c_{\pm j} \exp(\pm m_j z_\pm)$. Then formula (1.48) retains its form but $\hat{r}$ is replaced by t and $(-1)^t$ becomes 1.

Let us illustrate the structure of the general formula (1.48) for the simple Lie algebras of rank 1 and 2, see Table 4.1, choosing for brevity the form which depends on $R \equiv 2(z_+ z_-)^{1/2}$.

Table 4.1

A_1 :
$$\exp(-x) = c^{-1} R^{-\frac{m}{2}+1}\left(R^m - \frac{c^2}{m^2}\right);$$

A_2 :
$$\exp(-x_1) = \left(\frac{c_1}{2}\right)^{-1/3}\left(\frac{c_2}{2}\right)^{-2/3} R^{\frac{-2m_1-m_2+6}{3}}\left(R^{m_1+m_2} - \frac{c_2}{m_2^2}R^{m_1} + \frac{c_1 c_2}{m_1^2(m_1+m_2)^2}\right),$$

$$\exp(-x_2) = \left(\frac{c_1}{2}\right)^{-2/3}\left(\frac{c_2}{2}\right)^{-1/3} R^{\frac{-m_1-2m_2+6}{3}}\left(R^{m_1+m_2} - \frac{c_1}{m_1^2}R^{m_2} + \frac{c_1 c_2}{m_2^2(m_1+m_2)^2}\right);$$

B_2 :
$$\exp(-x_1) = \left(\frac{c_1}{3}\right)^{-1}\left(\frac{c_2}{4}\right)^{-1/2} R^{\frac{-2m_1-m_2+6}{2}}\left(R^{2m_1+m_2} - \frac{c_1}{m_1^2}R^{m_1+m_2} +\right.$$
$$\left. + \frac{c_1 c_2}{m_2^2(m_1+m_2)^2}R^{m_1} - \frac{c_1^2 c_2}{m_1^2(m_1+m_2)^2(2m_1+m_2)^2}\right),$$

$$\exp(-x_2) = \left(\frac{c_1}{3}\right)^{-1}\left(\frac{c_2}{4}\right)^{-1} R^{-m_1-m_2+4}\left(R^{2m_1+2m_2} - \frac{c_2}{m_2^2}R^{2m_1+m_2} +\right.$$
$$\left. + \frac{2c_1 c_2}{m_1^2(m_1+m_2)^2}R^{m_1+m_2} - \frac{c_1^2 c_2}{m_1^4(2m_1+m_2)^2}R^{m_2} + \frac{c_1^2 c_2^2}{m_2^2(m_1+m_2)^4(2m_1+m_2)^2}\right);$$

G_2 :
$$\exp(-x_1) = \left(\frac{c_1}{6}\right)^{-2}\left(\frac{c_2}{10}\right)^{-1} R^{-2m_1-m_2+6}\left(R^{4m_1+2m_2} - \frac{c_1}{m_1^2}R^{3m_1+2m_2} +\right.$$
$$+ \frac{c_1 c_2}{m_2^2(m_1+m_2)^2}R^{3m_1+m_2} - \frac{2c_1^2 c_2}{(2m_1+m_2)^2(m_1+m_2)^2 m_1^2}R^{2m_1+m_2} +$$
$$+ \frac{c_1^3 c_2}{m_2^2(m_1+m_2)^2 m_1^4}R^{m_1+m_2} - \frac{c_1^3 c_2^2}{(3m_1+2m_2)^2(2m_1+m_2)^4(m_1+m_2)^2 m_2^2}R^{m_1} +$$
$$\left. + \frac{c_1^4 c_2^2}{(2m_1+m_2)^4(3m_1+2m_2)^2(m_1+m_2)^2 m_1^2}\right).$$

$$\exp(-x_2) = \left(\frac{c_1}{6}\right)^{-3}\left(\frac{c_2}{10}\right)^{-2} R^{-3m_1-2m_2+10}\left(R^{6m_1+4m_2} - \frac{c_2}{m_2^2}R^{6m_1+3m_2} + \right.$$

$$+\frac{3c_1c_2}{m_1^2(m_1+m_2)^2}R^{5m_1+3m_2} - \frac{3c_1^2c_2}{m_1^4(2m_1+m_2)^2}R^{4m_1+3m_2}$$

$$+\frac{3c_1^2c_2^2}{m_2^2(m_1+m_2)^4(2m_1+m_2)^2}R^{4m_1+2m_2} + \frac{c_1^3c_2}{m_1^6(3m_1+m_2)^2}R^{3m_1+3m_2}-$$

$$-\frac{24c_1^3c_2^2(3m_1^2+3m_1m_2+m_2^2)}{m_1^2m_2^2(m_1+m_2)^2(2m_1+m_2)^2(3m_1+2m_2)^2}R^{3m_1+2m_2}+$$

$$+\frac{c_1^3c_2^3}{m_2^4(m_1+m_2)^6(3m_1+2m_2)^2}R^{3m_1+m_2}+$$

$$+\frac{3c_1^4c_2^2}{m_2^4(2m_1+m_2)^4(3m_1+m_2)^2(m_1+m_2)^2}R^{2m_1+2m_2}-$$

$$-\frac{3c_1^4c_2^3}{m_2^2(m_1+m_2)^4(2m_1+m_2)^4m_1^2(3m_1+2m_2)^2}R^{2m_1+m_2}+$$

$$+\frac{3c_1^5c_2^3}{m_1^4(m_1+m_2)^4(2m_1+m_2)^4(3m_1+2m_2)^2}R^{m_1+m_2}-$$

$$-\frac{c_1^6c_2^3}{m_2^2(2m_1+m_2)^6(3m_1+2m_2)^4(3m_1+m_2)^2}R^{m_2}+$$

$$\left.+\frac{c_1^6c_2^4}{m_2^2(m_1+m_2)^6(3m_1+2m_2)^4(3m_1+m_2)^2}R^{m_1+m_2}\right).$$

4.1.6 Boundary value problem (instantons and monopoles)*). The knowledge of explicit expressions for general solutions (4.1.11) of the generalized Toda lattice enables one to apply them effectively in the study of nonlinear effects of concrete physical processes described by equations (3.1.9) under certain boundary conditions. Here we confine ourselves to the study of the application of these solutions to the problem of distinguishing instanton and non-singular monopole configurations for the principal embedding of A_1 into the complexified Lie algebra of the gauge group G.

As has been noticed in 3.4, the cylindrically symmetric instantons constitute a subclass of $2r$-parameter solutions of (3.1.9) regular on the one-point compactification of $\mathbb{R}^4$ and with finite action or, equivalently, with topological charge (3.4.10). The conformal invariance enables us to pass from solutions $\rho_j^0(z)$ depending on one variable, say $z + \bar{z} \equiv \hat{r}$ or $z\bar{z}$, to $\rho_j(z) = \rho_j^0(g(z)) + \ln|dg/dz|^2$, which are solutions of (3.1.9) for an arbitrary analytic function $g(z)$ (Here for convenience z_+ and z_- are denoted by z and $\bar{z}$ respectively.)

To justify the requirement of finiteness at $\hat{r} = 0$ of the functions $\hat{\rho}_j$ entering the expression (3.4.10) for the density of the topological charge, we have to compensate the double zero at this point with the poles of the function $\exp \rho_j^0$ of corresponding order, i.e., impose the boundary conditions on the behaviour of the solution (4.1.48) at small "distances". It is not difficult to see that to justify these conditions it suffices that X ($\equiv \exp(-x_1)$) has a root of order k_1 at $\hat{r} = 0$. Then (3.1.9) automatically guarantees the roots of the necessary order ($= k_j$) at $\hat{r} = 0$ of the remaining functions $\exp(-x_j)$, $2 \le j \le r$.

The realization of this requirement in the form $(\partial/\partial\hat{r})^a X|_{\hat{r}=0} = 0$, $0 \le a \le k_1 - 1$, leads to the complete definition of the parameters d_j, $1 \le j \le r$, in terms of parameters m_j. Then the solutions $\exp(-x_j)$ can be expressed in the form

$$\exp(-x_j) = R^{-(k^{-1}m)_j + k_j}(R - 1)^{k_j} \Pi_j(R; m), \qquad (1.49)$$

where Π_j are finite positive definite nowhere vanishing polynomials, and moreover we may assume that $R \equiv 2[g(z)g^*(z)]^{1/2}|_{\hat{r}=0} = 1$ under an appropriate choice of g. Besides, to guarantee the analyticity of the solution we have to require that $m_j \in \mathbb{Z}$.

Now let us find the conditions imposed on g to guarantee nonsingularity of the solution of the self-duality equations obtained after a conformal transformation.

As we have mentioned, for this it is necessary that $|g(z)| = 1$ at $\hat{r} = 0$. Furthermore, we have to require that the functions in $\exp \hat{\rho}_j$ remaining after cancellation with $\hat{r}^2$ pole-bearing factors from $\exp x_j$ have neither poles nor

---

*) [15, 65, 66, 69, 70, 92, 137, 138, 140, 143, 153, 156, 157]

zeros in the right half-plane of the variable z. With (1.49) this implies that g should be the square of an analytic function $g(z) = B^2(z)$ with $|B(z)| = 1$ at $\hat{r} = 0$ and $|B(z)| < 1$ for $\hat{r} > 0$. The most general analytic function satisfying these conditions is the Blaschke function $B(z) = \prod_{1 \leq i \leq l} (a_i - z)(\overset{*}{a}_i + z)$, where a_i are arbitrary complex numbers with positive real part. The function B has l zeros and no pole in the right half-plane of the variable z. Thanks to Stokes' theorem the formula for the topological charge of the considered system with density (3.4.10) can be presented as the linear integral

$$Q = \frac{1}{4\pi i} \sum_{1 \leq j \leq r} v_j k_j \oint \partial_\mu \log B^{m_j - 1} \frac{dB}{dz} dx_\mu$$

(here $x_0 \equiv t$, $x_1 \equiv \hat{r}$). Since $B^{m_j - 1} dB/dz$ has $m_j l - 1$ zeros in the right z-half-plane, each of them contributing $2\pi i$ to the integral, we get the following final expression

$$Q = (1/2) \sum_{1 \leq j \leq r} v_j k_j (lm_j - 1). \tag{1.50}$$

Thus, the constructed instanton solutions correspond to the topological charge (1.50) and are characterized not only by l arbitrary complex parameters a_i, $\text{Re} a_i > 0$, but also by r "quantum" numbers which can be interpreted as the parameters of the inner degrees of freedom of instantons.

The correspondence mentioned in 3.4 between cylindrically-symmetric static self-dual fields in $\mathbb{R}^4$ and spherically symmetric-monopoles in the Minkowski space enables us to reconstruct directly the exact monopole solutions starting from the solutions (4.1.40) already constructed. The latter have no singularities in a finite interval of values of $\hat{r}$. To guarantee finiteness of the density of the Hamiltonian, it is necessary to satisfy certain boundary conditions at infinity and at small distances.

The procedure for distinguishing the subclass of solutions having the right behaviour as $\hat{r} \to 0$ is similar to the one performed in the instanton case, and as a result the monopole configurations are characterized by a set of r parameters m_j.

The energy of the system is given in the Bogomolny-Prasad-Sommerfield limit by (3.4.11), taking $\mathcal{D}\hat{\mathcal{H}} = 0$ into account. Through integration by parts it can be rewritten in the form

$$M = \lim_{\hat{r} \to \infty} \hat{r}^2 \int_{S^2} \text{tr}(\mathcal{H}_{\hat{r}} \chi) d\Omega.$$

The matrix of the magnetic charge $\hat{g}$ is determined, as usual, by the asymptotics of the magnetic field $\hat{\mathcal{H}}$, where

$$\mathcal{H}n \equiv \mathcal{H}_{\hat{r}} \to \frac{\hat{g}}{4\pi \hat{r}^2} \text{ as } \hat{r} \to \infty.$$

Therefore, to express these quantities we need an explicit form of the asymptotics of the Higgs field χ and $\mathcal{H}_{\hat{r}}$ as $\hat{r} \to \infty$. For our solutions they are of the form

$$\mathcal{H}_{\hat{r}} = -(1/2\hat{r}^2) \sum_j k_j h_j (\exp \hat{\rho}_j - 1), \quad \chi = (1/2) \sum_{i,j} (k^{-1})_{ji} h_j \hat{\rho}_{i,\hat{r}}.$$

Thanks to the boundary conditions at infinity the contribution of the Higgs field is completely determined by the exponents of prepolynomial factors of the corresponding solutions (1.48), i.e., by $\exp(-(k^{-1}m)_j \hat{r})$, or $R_j^{-(k^{-1}m)+k_j}$ in (1.49): indeed $\exp \hat{\rho}_j \to 0$ as $\hat{r} \to \infty$ and

$$\lim_{\hat{r} \to \infty} \hat{\rho}_{i,\hat{r}} = -\sum_j k_{ij} \lim \ [\ln \exp(-x_j)]_{,\hat{r}} = (1/2)m_i.$$

Finally, we get

$$\hat{g} = 2\pi \sum_j k_j h_j, \quad M = \pi \sum_{i,j} k_i (k^{-1})_{ij} m_j. \tag{1.51}$$

Let us illustrate formulas (1.49) by the corresponding expressions for simple Lie algebras of rank 2.

A_2 :	$\exp(-x_1) = R^{(-2m_1 - m_2 + 6)/3}(R_1)^2 \Pi_1^A;$ $\qquad\qquad d_1 = \dfrac{m_1^2(m_1 + m_2)}{m_2};$ $\exp(-x_2) = R^{(-m_1 - 2m_2 + 6)/3}((R - 1)^2 \Pi_2^A;$ $\qquad\qquad d_2 = \dfrac{m_2^2(m_1 + m_2)}{m_1};$
B_2 :	$\exp(-x_1) = R^{(-2m_1 - m_2 + 6)/2}(R - 1)^2 \Pi_1^B;$ $\qquad\qquad d_1 = \dfrac{m_1^2(2m_1 + m_2)}{m_2};$ $\exp(-x_2) = R^{-m_1 - m_2 + 4}(R - 1)^4 \Pi_2^B;$ $\qquad\qquad d_2 = \dfrac{m_2^2(m_1 + m_2)^2}{m_1^2};$
G_2 :	$\exp(-x_1) = R^{-2m_1 - m_2 + 6}(R - 1)^6 \Pi_1^G;$ $\qquad\qquad d_1 = \dfrac{m_1^2(2m_1 + m_2)(3m_1 + m_2)}{(m_1 + m_2)m_2};$ $\exp(-x_2) = R^{-3m_1 - 2m_2 + 10}(R - 1)^{10} \Pi_2^G;$ $\qquad\qquad d_2 = \dfrac{(m_1 + m_2)^3(3m_1 + 2m_2)m_2^2}{(3m_1 + m_2)m_1^3}.$

4.2

Expression (1.48) for the solutions of a monopole system can be rewritten from (1.37) in the form (cf. [15])

$$\exp(-x_j) = \langle j|\tilde{\mathcal{P}}(0)\exp[2(ph) - (mh)\hat{r}]\mathcal{P}(0)|j\rangle, \qquad (1.52)$$

where $\mathcal{P}(z)$ satisfies

$$d\mathcal{P}(z)/dz = \mathcal{P}(z)\sum_{1\le i\le r}\exp(m_i z)X_{-i}$$

with the boundary condition $\mathcal{P}(\infty) = 1$ and is related with $\mathcal{M}_-(z)$ via relation $\mathcal{M}_-(z) = \exp(-mh)\mathcal{P}(z)\exp(mh)$. Then

$$\mathcal{M}_+(z) = \exp(mh)\tilde{\mathcal{P}}^{-1}(z)\exp(-mh),$$
$$\tilde{\mathcal{P}} \equiv (-1)^H \mathcal{P}^\dagger(-1)^H; \ m_i < 0; \ p_i \equiv -(1/2)\ln d_i.$$

Introducing the complete set of orthonormal states of the i-th fundamental representation, i.e., the representation (with the highest weight f_i satisfying $f_i(h_j) = \delta_{ij}$) with weights $\lambda^{(i)}$ of multiplicities $\nu_{\lambda(i)}$, we get from (1.52)

$$\exp(-x_j) = \sum_{\{\lambda^{(j)},\nu_{\lambda(j)}\}}(-1)^{\eta_j}\exp[-(\lambda^{(j)}m)\hat{r} + 2(\lambda^{(j)}p)]|\langle\lambda^{(j)},\nu_{\lambda(j)}|\mathcal{P}(0)|j\rangle|^2,$$
$$(1.53)$$

where $\eta_j \equiv \eta(f_j - \lambda^{(j)})$ is determined from the relation $[H, X_\alpha] = \eta_\alpha X_\alpha$. The boundary condition at small distances, thanks to which the energy functional converges if the monopole solution is a regular function, establishes the following explicit relation among the parameters d_i and m_i. Namely,

$$\prod_i[(-m_i)^{-2/(\pi_i\pi_j)}d_i]^{\lambda_i^{(j)}} = 1; \ (\lambda_i^{(j)} \equiv \lambda^{(j)}(h_i)). \qquad (1.54)$$

The final solution defining a monopole system with finite energy is

$$\exp(-x_j) = \sum_{\{\lambda^{(j)},\nu_{\lambda(j)}\}}(-1)^{\eta_j}\exp[-(\lambda^{(j)}m)\hat{r}]\mathcal{Q}_{\lambda(j)}(m)\prod_i(-m_i)^{p+q}, \ (1.55)$$

where $\mathcal{Q}_{\lambda(j)}(m)$ is a polynomial in m, while p and q are the maximal and the minimal numbers from the string of roots through $\lambda^{(j)}$

$$\lambda^{(j)} - qf_j,\ldots,\lambda^{(j)} - f_j,\ \lambda^{(j)}, \lambda^{(j)} + f_j,\ldots,\lambda^{(j)} + pf_j.$$

§ 4.2 Complete integration of the two-dimensionalized system of Lotka-Volterra-type equations (difference KdV) as the Bäcklund transformation of the Toda lattice*)

The two-dimensional generalized Toda lattice discussed in the preceding section is closely connected with a number of differential-difference equations of first order in derivatives. In particular, the finite system of the form

$$N_{2i-1,z_+} = -N_{2i-1}(N_{2i} - N_{2i-2}), \quad N_{2i,z_-} = N_{2i}(N_{2i+1} - N_{2i-1}),$$
$$1 \leq i \leq r+1, \quad N_0 = N_{2r+2} = 0, \tag{2.1}$$

where the N_a, $1 \leq a \leq 2r+1$, depend on two independent variables z_+ and z_-, can be considered as the Bäcklund transformation of the system (3.1.9) with the Cartan matrix k of the Lie algebra A_r.

(In the one-dimensional case the Bäcklund transformation for the Toda lattice of infinite length has been studied in a number of papers, cf., e.g., [50, 84, 104, 117].

In what follows the system (2.1) will be called the *two-dimensionalized Volterra equations* because in the one-dimensional case ($t \equiv z_+ - z_-$, $N_a = N_a(t)$) it represents a particular class of equations introduced by Volterra in the study of ecology problems (in particular, the dynamics of coexistence of species) called the Lotka-Volterra system [14].

In the physical applications this system arises in the study of fine structure of the spectra of Langmuir waves in plasma and describes under boundary conditions $N_a \to$ const as $a \to \pm\infty$ the propagation of the spectral package of Langmuir oscillations with heat noise background [40].

In the one-dimensional case it is proved that a more general nonlinear system

$$\dot{N}_a = (a_0 + a_1 N_a + a_2 N_a^2)(N_{a+1} - N_{a-1}), \tag{2.2}$$

where a_0, a_1, a_2 are arbitrary constants, is completely integrable (in the Liouville sense) and the explicit solutions will be constructed here.

A detailed review with different applications of the chains of infinite length for $N_a(t)$ and associated nonlinear differential-difference systems, in particular, nonlinear induction-capacity chains of ladder type used in radiotechnical schemes, is contained in [117]. Also given in [117] are the known particular solutions of these equations, e.g., of the soliton-type, obtained either via Bäcklund transformation or numerically with the help of a computer.

The periodic problem for the one-dimensional Volterra chain (the difference Korteweg-de Vries equation) is considered in [30], [31].

---

*) [72, 74]

Moreover, in [84] with the help of the ISP method a scheme of integration of this chain is developed for rapidly decreasing initial conditions, and a Lax pair and integrals of motion are found.

The method for construction of general solutions of the system (2.1) depending on $2r + 1$ arbitrary functions that we propose is based on the fact that this system is a Bäcklund transformation for the two-dimensional (finite nonperiodic) Toda lattice.

Note, that the same takes place also for the periodic problem ($N_0 = N_{2r}$, $N_{2r+1} = N_1$); however, the recovery of the solutions here is much more complicated since the associated Lie algebras are infinite-dimensional and the corresponding equations are not integrable in quadratures.

The explicit form of the system (2.1) directly implies that the functions

$$\rho_i \equiv \ln(N_{2i-1}N_{2i}), \ \rho_i' \equiv \ln(N_{2i}N_{2i+1}), \ 1 \le i \le r, \tag{2.3}$$

satisfy the systems

$$\rho_{i,z_+z_-} = 2\exp\rho_i - \exp\rho_{i-1} - \exp\rho_{i+1}, \ \rho_0 = \rho_{r+1} = -\infty, \tag{2.4$_1$}$$

$$\rho_{i,z_+z_-}' = 2\exp\rho_i' - \exp\rho_{i-1}' - \exp\rho_{i+1}', \ \rho_0' = \rho_{r+1}' = -\infty, \tag{2.4$_2$}$$

which coincide with the equation (3.1.9) for the series A_r describing the two-dimensional Toda lattice. Therefore, the system (2.1) may be considered as a realization of a Bäcklund transformation relating the solutions of the systems (2.4$_1$) and (2.4$_2$). Since the general solutions of each of the systems (2.4) contain $2r$ arbitrary functions, whereas the general solutions of (2.1) depend on $2r + 1$ arbitrary functions, the Bäcklund transformation (2.1) for the systems (2.4) has a functional arbitrariness measured by one function.

Therefore, the scheme of integration of the system (2.1) consists in establishing the relation among $2r$ arbitrary functions $\Phi_{+i}(z_+)$ and $\Phi_{-i}(z_-)$, $1 \le i \le r$ (determining according to the preceding section a general solution for the system (2.4$_1$) for ρ_i, $1 \le i \le r$) and $2r$ functions $\Phi_{+i}'(z_+)$ and $\Phi_{-i}'(z_-)$ (defining the solutions of (2.4$_2$) for ρ_i').

It will be convenient to make use of the polynomial form of the solutions of (2.4), where thanks to (1.20) in the notations of 4.1.4.2

$$X_i \equiv \exp\left(-\sum_{1\le j\le r}(k^{-1})_{ij}\rho_j\right) = (-1)^{i(i+1)/2}\Delta_i(X), \ 1 \le i \le r, \tag{2.5}$$

and $X_1 \equiv X$ is given by the formula (1.34).

The solutions of the system (2.1) are completely determined by the relations

$$N_{2i-1}N_{2i} = \exp\rho_i(\Phi_\pm); \ N_{2i}N_{2i+1} = \exp\rho_i'(\Phi_\pm'), \ 1 \le i \le r, \tag{2.6}$$

and additionally by one of the equations (2.1), say, by

$$N_{1,z_+} = -N_1 N_2$$

$$(= -\exp\rho_1 = -\exp(kx)_1 = -x_{1,z_+z_-} \qquad (2.7)$$

$$\equiv -(\ln X)_{,z_+z_-}, \quad \text{see (1.2))},$$

which introduces the lacking arbitrary function after integration of (2.7). Indeed, from (2.7) we get

$$N_1 = (\ln X)_{,z_-} + g(z_-), \qquad (2.8)$$

where $g(z_-)$ is an arbitrary function realizing the above-mentioned arbitrariness of the Bäcklund transformation.

Our first illustration is the simplest example A_1, for which (2.1) takes the form

$$N_{1,z_+} = -N_1 N_2, \quad N_{2,z_-} = N_2(N_3 - N_1), \quad N_{3,z_+} = N_2 N_3. \qquad (2.9)$$

Substituting the known general solution

$$\exp\rho(= N_1 N_2) = \Phi_{+,z_+}\Phi_{-,z_-}(1 - \Phi_+\Phi_-)^{-2}$$

of the Liouville equation in the first of equations (2.9) and integrating them consecutively we get

$$N_1 = u^{-1}[1 - \Phi_+(\Phi_- + u\Phi_{-,z_-})](1 - \Phi_+\Phi_-)^{-1},$$

$$N_2 = u\Phi_{+,z_+}\Phi_{-,z_-}(1 - \Phi_+\Phi_-)^{-1}[1 - \Phi_+(\Phi_- + u\Phi_{-,z_-})]^{-1},$$

$$N_3 = u^{-1}\Phi_{-,z_-}^{-1}(\Phi_- + u\Phi_{-,z_-})_{,z_-}(1 - \Phi_+\Phi_-)[1 - \Phi_+(\Phi_- + u\Phi_{-,z_-})],$$
$$(2.10)$$

where $\Phi_+(z_+)$, $\Phi_-(z_-)$ and $u(z_-)$ are arbitrary functions. Therefore, the Bäcklund transformation realized by (2.9) and connecting the solutions ρ and ρ' of the Liouville equation leads to the replacement of an arbitrary function $\Phi_-(z_-)$ in $\exp\rho$ by an arbitrary function $\Phi_-(z_-) + u(z_-)(\Phi_-(z_-))_{,z_-}$ in the solution $\exp\rho'(= N_2 N_3)$.

Now pass to the general case. In accordance with formula (2.5) for X_2 we get the following expression for the function $\exp\rho_1 \equiv X_2 X_1^{-1}$:

$$\exp\rho_1 = \left[-\sum_i (-1)^i \Phi_{+,i,z_+}\Phi_{-i,z_-} + \sum_{i<j}(\Phi_{+i}\Phi_{+j,z_+} - \Phi_{+i,z_+}\Phi_{+j}) \right.$$

$$\left. \times (\Phi_{-i}\Phi_{-j,z_-} - \Phi_{-i,z_-}\Phi_{-j}) \right] Y^{-2}, \; Y \equiv 1 + \sum_j (-1)^j \Phi_{+j}\Phi_{-j}.$$

$$(2.11)$$

Then from (2.7) and with (2.8) we get

$$N_2 = \exp\rho_1[(\ln X)_{,z_-} + g(z_-)]^{-1},$$

which, together with the second of equations (2.1), namely,

$$N_{2,z_-} = N_2(N_3 - N_1) = \exp \rho_1' - \exp \rho_1$$

yields

$$
\exp \rho_1' = (Y')^{-2} \Bigg\{ - \sum_i (-1)^i \Phi_{+i,z_+} (\Phi_{-i} + u\Phi_{-i,z_-})_{,z_-}
$$

$$
+ \sum_{i<j} (\Phi_{+i}\Phi_{+j,z_+} - \Phi_{+i,z_+}\Phi_{+j})[(\Phi_i + u\Phi_{-i,z_-})(\Phi_{-j} + u\Phi_{-j,z_-})_{,z_-}
$$

$$
\tag{2.12}
$$

$$
- (\Phi_{-i} + u\Phi_{-i,z_-})_{,z_-} (\Phi_{-j} + u\Phi_{-j})_{,z_-}] \Bigg\};
$$

$$
Y' \equiv 1 + \sum_j (-1)^j (\Phi_{+j}(\Phi_{-j} + u\Phi_{-j,z_-}),
$$

where $u(z_-)$ is related with $g(z_-)$ via

$$
g - (1/(r+1))[\ln \Delta_r(\Phi_-)]_{,z_-} = u^{-1}. \tag{2.13}
$$

(The validity of (2.12) which actually already contains the result of the transformation $\Phi'_{+i} = \Phi_{+i}$, $\Phi'_{-i} = \Phi_{-i} + u\Phi_{-i,z_-}$ can be established directly by substituting

$$
\exp \rho_1 = -(\ln Y)_{,z_+ z_-}, \quad \exp \rho_1' = -(\ln Y')_{,z_+ z_-}
$$

and $N_1 \equiv Y'/(uY)$ in the equality

$$
\exp \rho_1' - \exp \rho_1 = N_{2,z_-} = [N_1^{-1} \exp \rho_1]_{,z_-}
$$

which implies the chain of relations

$$
\{ (\ln Y'/Y)_{,z_+} - (\ln Y)_{,z_+ z_-} uY/Y' \}_{,z_-}
$$

$$
= \{ Y^{-1}Y'^{-1}[YY'_{,z_+} - Y_{,z_+} Y' - u(YY_{,z_+ z_-} - Y_{,z_+} \cdot Y_{,z_-})] \}_{,z_-} = 0,
$$

which completes the verification.

Comparing (2.11) and (2.12) we see that the result of integration of the Bäcklund transformation for (2.4) can be represented in the form

$$
X_i = X_i(\Phi_+, \Phi_-), \quad X_i' = X_i(\Phi_+, \Phi_- + u\Phi_{-,z_-}). \tag{2.14}
$$

With the help of (2.6), i.e.,

$$
N_{2i-1} \cdot N_{2i} = \exp \rho_i(\Phi_+, \Phi_-), \quad N_{2i} \cdot N_{2i+1} = \exp \rho_i(\Phi_+, \Phi_- + u\Phi_{-,z_-}),
$$

and knowing the general solutions X_i (and therefore ρ_i) this enables us to recover the general solutions of (2.1) depending on $2r+1$ arbitrary functions Φ_{+i}, Φ_{-i} and $u(z_-)$, $1 \le i \le r$.

The final expressions are

$$N_{2i-1} = u^{-1}[\Delta_r(\Phi_- + u\Phi_{-,z_-})/\Delta_r(\Phi)]^{1/(r+1)}X_{i-1}X_i'(X_{i-1}'X_i)^{-1},$$

$$N_{2i} = u[\Delta_r(\Phi_-)/\Delta_r(\Phi_- + u\Phi_{-,z_-})]^{1/(r+1)}X_{i-1}'X_{i+1}(X_iX_i')^{-1}. \tag{2.15}$$

Notice that the equations (2.1) may also be considered in the case of an even number of functions N_a as Bäcklund transformation connecting the solutions of the system (2.4$_1$) for A_r with the solutions for A_{r-1}, which corresponds to the replacement of the boundary conditions $N_0 = N_{2r+2} = 0$ by $N_0 = N_{2r+1} = 0$. With the help of the above solutions (2.15) this procedure reduces to trivial operations: the condition $N_{2r+1} = 0$ leads to an ordinary differential equation $\Delta_r(\Phi_- + u\Phi_{-,z_-}) = 0$ relating $2r + 1$ functions Φ_{+i}, Φ_{-i} and u by the relation $u^{-1} = -[\ln(\bar{a}_0 + \sum_i \bar{a}_i\Phi_{-i})]_{,z_-}$, where $\bar{a}_0$ and $\bar{a}_i$ are arbitrary constants. The solutions obtained for the system (2.1) for an even number of functions N_a depend on $2r$ arbitrary functions.

Notice that the vanishing of $\Delta_r(\Phi'_-)$ which guarantees the verification of the boundary condition $N_{2r+1} = 0$ does not imply the vanishing of the remaining functions N_a, $1 \le a \le 2r$, since the common factor $\Delta_r(\Phi'_-)$ in (2.15) is canceled for N_a, $1 \le a \le 2r$, by the appropriate dependence of the functions X_i' on this factor.

As for the generalized Toda lattice, the solutions of (2.1) in the one-dimensional case may be obtained from the general solutions (2.15) constructed here of the two-dimensional Volterra equations by a special choice of the functions $\varphi_{\pm i}$ and u so that N_a depend only on one variable, say, $t \equiv z_+ - z_-$. This can be obtained by substituting $\varphi_{\pm i} = c_{\pm i}\exp(\pm m_i z_\pm)$ and $u = u_0 = \text{const}$ which guarantees that the general solutions of the corresponding one-dimensional equations of the form (2.2) with $a_0 = a_2 = 0$ are given by $2r + 1$ arbitrary numerical parameters: m_i, $d_i \equiv c_{+i}c_{-i}$, $1 \le i \le r$, and u_0. In such a parametrization the function X that, thanks to (2.5), determines the solutions X_i and is expressed by (1.34) takes the form

$$X = \left[\prod_{1\le j\le r} d_j m_j^2 W_r^2 \exp\left(t\sum_{1\le i\le r} m_i\right)\right]^{\frac{-1}{(r+1)}}\left[1 + \sum_{1\le\theta\le r}(-1)^\theta d_\theta\exp(tm_\theta)\right]$$

$$\equiv \sum_{0\le a\le r}(-1)^a\tilde{c}_a\exp(t\tilde{m}_a),$$

$$\tag{2.16}$$

where $W_r \equiv W_r(m_1\ldots m_r) = \prod_{i>j}(m_i - m_j)$ is the r-th order Vandermonde determinant, $\tilde{c}_i \equiv \tilde{c}_0d_i$, $\tilde{m}_i \equiv m_i + \tilde{m}_0$, $1 \le j)$ is the r-th order Vandermonde determinant, $\tilde{c}_i \equiv \tilde{c}_0 d_i$, $\tilde{m}_i \equiv m_i + \tilde{m}_0$, $1 \leq i \leq r$,

$$\tilde{c}_0 \equiv \left[\prod_{1 \leq j \leq r} d_j m_j^2 W^2(m_1, \ldots, m_r) \right]^{-1/(r+1)},$$

and $\tilde{m}_0 \equiv - \sum_{1 \leq j \leq r} m_j/(r+1)$. The definition of $\tilde{c}_i$ and $\tilde{m}_i$ obviously implies that $\sum_{0 \leq a \leq r} \tilde{m}_a = 0$, $\prod_{0 \leq a \leq r} \tilde{c}_a = [\prod_{1 \leq j \leq r} m_j W_r]^{-2}$, and there are only these two relations between $2(r+1)$ parameters $\tilde{c}_a$ and $\tilde{m}_a$, $0 \leq a \leq r$.

A formula similar to (2.16) holds also for X' with d_j replaced by $d_j(1 + u_0 m_j)$ because $\varphi'_{-i} = c_i(1 + u_0 m_i) \exp(-m_i z_-)$.

The remaining functions $X_j \equiv \exp(-x_j)$ are obtained by substituting (2.16) into (2.5):

$$X_j = (-1)^{j(j+1)/2} \sum_{s_1 > s_2 > \ldots > s_j} (-1)^{\Sigma_i s_i} W_j^2(\tilde{m}_{s_1} \ldots \tilde{m}_{s_j}) \tilde{c}_{s_1} \ldots \tilde{c}_{s_j} \exp\left[t \sum_{1 \leq i \leq j} \tilde{m}_{s_i} \right], \tag{2.17}$$

where the indices $s_1, \ldots, s_j$ run from 0 to $r + 1$.

Therefore, the desired solutions of the one-dimensional Volterra equations are characterized by $2r + 1$ arbitrary parameters: d_j, m_j, $1 \leq j \leq r$ and u_0 and can be expressed in the form

$$N_{2i-1} = u_0^{-1} \left[\prod_{1 \leq j \leq r} (1 + u_0 m_j) \right]^{1/(r+1)} . X_{i-1} X_i' (X_{i-1}' X_i)^{-1},$$

$$\tag{2.18}$$

$$N_{2i} = u_0 \left[\prod_{1 \leq j \leq r} (1 + u_0 m_j) \right]^{-1/(r+1)} X_{i-1}' X_{i+1} (X_i X_i')^{-1},$$

where the X_j are represented by formula (2.17) and the X_j' are obtained from them under the change $d_j \to d_j(1 + u_0 m_j)$.

If the number of N_a's is even the parameters d_j, m_j and u_0 cease to be independent.

One-dimensional Volterra equations of the type considered above are closely connected with the following nonlinear differential-difference system

$$\dot{T}_j = (1 + T_j^2)(T_{j+1} - T_{j-1}), \tag{2.19}$$

which can be made finite by imposing the boundary conditions $T_{-1} = -T_{r+1} = i$. In other words

$$\dot{Q}_j = (1 - Q_j^2)(Q_{j+1} - Q_{j-1}), \quad 0 \leq j \leq r, \tag{2.20}$$

where $Q_j \equiv -iT_j$, $Q_{-1} = -Q_{r+1} = 1$.

Notice that (2.2) reduces to (2.19) under a linear transformation in the functions and the argument. The relation between the solutions of the Volterra equations and (2.20) is given by a finite discrete variant of the Miura transformation [89]:

$$N_j = (1 + Q_j)(1 - Q_{j-1}), \quad 1 \le j \le r. \tag{2.21}$$

Therefore, if the solutions of Volterra equations are known it is not difficult to recover exact solutions of (2.20) via formulas (2.21) and one of equations (2.20), say, the first one (with $j = 0$). Indeed, the equation $\dot{Q}_0 = (1 - Q_0^2)(Q_1 - 1)$ and the relation $N_1 = (1 + Q_1)(1 - Q_0)$ enable one to exclude Q_1 and reduce the problem of finding the function $Q_0 \equiv f^{-1} - 1$ to the integration of the equation $\dot{f} + (N_1 - 4)f + 2 = 0$, whose solution is of the form

$$(1 + Q_0)^{-1} \equiv f = -2 \left[1 + \sum_j (-1)^j d_j \exp(m_j t) \right]^{-1}$$

$$\times \left\{ \lambda^{-1} + c_0 \exp(-\lambda t) + \sum_i (-1)^i \exp(m_i t) d_i / (m_i + \lambda) \right\}, \tag{2.22}$$

$$\lambda \equiv u_0^{-1} - 4, \quad c_0 = \text{const}.$$

Now the remaining unknown functions Q_j, $1 \le j \le r$, are expressed algebraically thanks to (2.21) via the known functions N_0 and Q_0. The final expression for the solutions is given in the form of continuous fractions

$$1 + Q_j = \cfrac{N_j}{2 - \cfrac{N_{j-1}}{2 - \cfrac{N_{j-2}}{2 - \cfrac{N_{j-3}}{\cfrac{\cdots}{\cfrac{N_3}{2 - \cfrac{N_2}{2 - \cfrac{N_1}{1 - Q_0}}}}}}}, \tag{2.23}$$

where Q_0 is given by (2.22). For an odd number of functions Q_j the general formula (2.23) is the same, only the parameters d_j, m_j and u_0 are now dependent, as above.

§ 4.3 String-type systems (nonabelian versions of the Toda system)*)

For an explicit realization of the nonlinear equations constructed in 3.2 we have to find an appropriate $\mathbb{Z}$-graded Lie algebra $\mathfrak{g}$ and a suitable parametrization of unknown functions $\varphi_{-\alpha}^1 \equiv f_\alpha(z_+, z_-)$. In other words, we have to define the explicit form of the matrices $R_{\alpha\beta}$ and $A_{\alpha\beta}$ entering (3.2.6) or for the system (3.2.7) or (3.2.8) define the operators $J_\pm$ and a convenient representation of the element g_0.

For the generalized Toda lattice solution to this problem is quite obvious (see 4.1.1) whereas for nonprincipal embeddings A_1 into $\mathfrak{g}$ it generally requires some calculations which will be illustrated here with the series B.

One of the classes of the systems arising describes the dynamics of the relativistic string and corresponds to the parametric formulation of 2-dimensional minimal surfaces in the N-dimensional Euclidean or pseudoeuclidean space (here $N = 2r + 1$ for B_r and $N = 2r$ for D_r). That is why these equations will be called of *string-type*.

Consider the gradings of B_r associated with the integer embeddings of A_1 into B_r so that $\mathfrak{g}_0 \equiv \mathfrak{g}_0^{(m)} = (\underset{1 \le j \le r-m}{\oplus} \mathfrak{gl}(1)_j) \oplus B_m$, where $\mathfrak{gl}(1)_j$ is the j-th copy of $\mathfrak{gl}(1)$. We will exclude the cases $m = 0$ and $m = r$ since the first one corresponds to the generalized Toda lattice whereas the second one is trivial.

Consider the simplest cases: $m = 1$ and $m = r - 1$, which also clarify the problem in general case.

1) $\mathfrak{g}_0^{(1)} = (\underset{1 \le j \le r-1}{\oplus} \mathfrak{gl}(1)_j) \oplus B_1$.

This embedding corresponds (see 1.3.1) to excluding the vertex corresponding to π_r from the extended Dynkin diagram, where $m \equiv 2\pi - \pi_1$, $\pi \equiv \sum_{1 \le j \le r} \pi_j$, and leads to the maximal π-system of B_r coinciding with the π-system D_r. For the further calculations it is convenient to perform a transformation with the help of $W(B_r)$ which transforms π into π_r. Then in accordance with the results from 1.3 we get the following structure of the local part of B_r:

$$\mathfrak{g}_0^0 = \{h_r\}, \quad \mathfrak{g}_0^f = \{h_j, \ 1 \le j \le r - 2, \ H, \ X_{\pm\pi_r}\};$$
$$\mathfrak{g}_{\pm 1} = \{X_{\pm\pi_j}, \ 1 \le j \le r - 1; \ X_{\pm(\pi_{r-1}+\pi_r)}, \ X_{\pm(\pi_{r-1}+2\pi_r)}\}, \tag{3.1}$$

with the subalgebra

$$A_1 = \{H = \sum_{1 \le j \le r} h_j + h_{r-1}, \ J_\pm = \sum_{1 \le j \le r-2} X_{\pm\pi_j} + X_{\pm(\pi_{r-1}+\pi_r)}\}.$$

---

*) [79, 129, 151]

The matrix functions $R_{\alpha\beta}$ and $A_{\alpha\beta}$ are of the form

$$R = -\mathrm{diag}(f_1 \ldots f_{r+1}) \cdot \begin{pmatrix} & & 0 & 0 \\ & k^{B_{r-1}} & 0 & 0 \\ & & -2f_r/f_{r-1} & 0 \\ 0 & -1 \quad 2 & f_{r+1}/f_r & -f_{r-1}/f_r \\ 0 & -1 \quad 2 & 0 & 2f_r/f_{r+1} \end{pmatrix},$$

$$A = -\begin{pmatrix} 1_{r-2} & 0 & 0 & 0 \\ 0 & 0 & 1 & 0 \\ 0 & -1 & 0 & 0 \\ 0 & 0 & 0 & 1 \end{pmatrix}.$$

Let us number the indices f_α, $1 \le \alpha \le r+1$, in accordance with the above ordering of root vectors of $\mathfrak{g}_{-1}$ and parametrize them via

$$f_j \equiv \exp\rho_j; \ 1 \le j \le r-2; \quad f_{r-1} \equiv x^{(1)} \exp\sigma,$$
$$f_r \equiv x_0 \exp\sigma, \qquad\qquad f_{r+1} \equiv x^{(2)} \exp\sigma;$$
$$x_0^2 + x^{(1)}x^{(2)} \equiv 1.$$

Taking this into account we rewrite (3.2.6) in the form

$$\rho_{i,z+z-} = \sum_{1 \le j \le r-2} k_{ij}^{A_{r-2}} \exp\rho_j, \ 1 \le i \le r-2; \ \sigma_{,z+z-}$$
$$= 2[1 - x^{(1)}x^{(2)}]^{1/2} \exp\sigma;$$
$$\{[1 - x^{(1)}x^{(2)}]^{-1/2}x_{,z+}^{(\varepsilon)}\}_{,z-} = 2x^{(\varepsilon)} \exp\sigma, \ \varepsilon = 1,2. \tag{3.2}$$

Setting

$$\sigma \equiv (\rho_{r-1} + \rho_r)/2, \ \delta \equiv (\rho_{r-1} - \rho_r)/2,$$
$$x^{(\varepsilon)} \equiv (-1)^{\varepsilon-1} \operatorname{sh}\delta \exp[(-1)^{\varepsilon-1}\theta],$$
$$\theta_{,z+} = \operatorname{ch}^{-2}(\delta/2) \operatorname{ch}\delta \cdot \omega_{,z+},$$
$$\theta_{,z-} = \operatorname{ch}^{-2}(\delta/2)\omega_{,z-},$$

we get

$$\rho_{i,z+z-} = \sum_{1 \le j \le r} k_{ij}^{D_r} \exp\rho_j - 2(\delta_{ir-1} - \delta_{ir})\omega_{,z+}\omega_{,z-} \operatorname{sh}((\rho_{r-1} - \rho_r)/2)$$
$$\times \operatorname{ch}^{-3}((\rho_{r-1} - \rho_r)/2),$$
$$[\operatorname{th}^2((\rho_{r-1} - \rho_r)/2)\omega_{,z+}]_{,z-} + [\operatorname{th}^2((\rho_{r-1} - \rho_r)/2)\omega_{,z-}]_{,z+} = 0. \tag{3.3}$$

These equations can be obtained by variation of the action with the Lagrangian

$$\mathcal{L} = \frac{1}{2} \sum_{1 \le i,j \le r} (k_{ij}^{D_r})^{-1} \rho_{i,z_+} \rho_{j,z_-} + \sum_{1 \le i \le r} \exp \rho_i - \omega_{,z_+} \omega_{,z_-} \operatorname{th}^2((\rho_{r-1} - \rho_r)/2) \tag{3.4}$$

and they turn into the equation of generalized Toda lattice for D_r for $\omega = $ const.

Notice that in order to construct (3.3) starting from (3.2.8) we should express g_0 in the form $g_0 = \exp(a^+ X_{\pi_r}) \exp(a^- X_{-\pi_r}) \exp \sum_{1 \le j \le r} a_j h_j$, where

$$\rho_j = - \sum_{1 \le i \le r} k_{ji}^{B_r} a_i, \ 1 \le j \le r-2; \ \rho_{r+1-\varepsilon}$$

$$= a_{r-2} - a_{r-1} + (-1)^\varepsilon \operatorname{Ar} \operatorname{sh}(a^+ a^-)^{1/2};$$

$$\begin{pmatrix} \omega_{,z_+} \\ \omega_{,z_-} \end{pmatrix} = \mp(1 + a^+ a^-) \begin{pmatrix} 1 \\ (1 + 2a^+ a^-)^{-1} \end{pmatrix} \tag{3.5}$$

$$\times \ [\ln(\exp(\mp a_{r-1} \pm 2a_r) a^\mp]_{,z_\pm} \pm \frac{1}{2} [\ln(a^+ a^-)]_{,z_\pm} \ .$$

The general solutions of (3.3) are obtained in complete agreement with 3.2.4 and depend on $2(r+1)$ arbitrary functions $\varphi_{+\alpha}(z_+)$, $\varphi_{-\alpha}(z_-)$. The functions ρ_j and $\omega_{,z_\pm}$ are expressed with the help of (3.5) via the group parameters $a_i(z_+, z_-)$ and $a^\pm(z_+, z_-)$ of g_0. The latter in their turn are determined by the matrix elements $\langle s_i | g | s_j \rangle$, $g \in B_r$, where the states $\{|s_i\rangle\}$ are annihilated by the action of the elements from $\mathfrak{g}_{+a}$, $a \ge 1$, i.e., $X_\alpha^a |s_i\rangle = \langle s_i | X_\alpha^{-a} = 0$.

Then according to the general scheme from 3.2.4 the desired solutions are given by the matrix elements

$$D_{s_i s_j} \equiv \langle s_i | \hat{\mathcal{K}} | s_j \rangle, \ \hat{\mathcal{K}} \equiv (\mathcal{M}_+ g_0^+)^{-1} (\mathcal{M}_- g_0^-).$$

Therefore, the problem reduces to constructing the states $\{|s_i\rangle\}$. It is not difficult to see that in the case which is of interest to us these are the highest states $|i\rangle$, $1 \le i \le r$, of the basis of the fundamental representations of B_r and the state $|r'\rangle \equiv X_{-\pi_r} |r\rangle$.

(Recall that for the generalized Toda lattice we have $\{|s_i\rangle\} \equiv \{|i\rangle\}$.)

Indeed, the relations (1.4.19) imply $\mathfrak{g}_1 |s_i\rangle = \langle s_i | \mathfrak{g}_{-1} = 0$, i.e., $X_\alpha |i\rangle = 0$, $1 \le i \le r$, $X_\alpha |r'\rangle = 0$, where $\alpha = \{\pi_i, 1 \le i \le r-1; \pi_{r-1} + \pi_r; \pi_{r-1} + 2\pi_r\}$. Thanks to (1.2.2) and since $\mathfrak{g}$ is generated by its local part, $X_\alpha^a |s_i\rangle = 0$ for any $a \ge 1$, $1 \le \alpha \le d_a$. Finally,

$$\exp a_j = \langle j | \hat{\mathcal{K}} | j \rangle, \ 1 \le j \le r-1; \ \exp a_r (1 + a^+ a^-) = \langle r | \hat{\mathcal{K}} | r \rangle;$$

$$a^+ \exp(-a_r) = \langle r | \hat{\mathcal{K}} | r' \rangle (\langle r-1 | \hat{\mathcal{K}} | r-1 \rangle)^{-1}; \tag{3.6}$$

$$a^- \exp a_r = \langle r' | \hat{\mathcal{K}} | r \rangle,$$

where the elements $\mathcal{M}_\pm$ and

$$g_0^\pm = \exp A_{(\pm)}^\pm X_{\pm\pi_r} \exp A_{(\pm)}^\mp X_{\mp\pi_r} \exp \sum_{1\le i \le r} A_i^{(\pm)} h_i,$$

parametrizing $\hat{\mathcal{K}}$ are determined by (3.1.24) with

$$L_\pm = \sum_{1\le j \le r-1} \varphi_{\pm j}(z_\pm)X_{\pm\pi_j} + \varphi_{\pm r}(z_\pm)X_{\pm(\pi_{r-1}+\pi_r)}$$

$$+ \varphi_{\pm(r+1)}(z_\pm)X_{\pm(\pi_{r-1}+2\pi_r)}$$

and by the formulas

$$A_i^{(\pm)} = \pm \sum_{1\le j \le r} (k_{ij}^{B_r})^{-1} \ln \varphi_{\pm j}, \ 1\le i \le r-2;$$

$$D_\pm \equiv \varphi_{\pm r}^2 - \varphi_{\pm(r-1)}\varphi_{\pm(r+1)};$$

$$A_{r-1}^{(\pm)} - A_{r-2}^{(\pm)} = \pm(1/2)\ln D_\pm, \tag{3.7}$$

$$A_{(\pm)}^\mp = \pm(1/2)\varphi_{\pm(r-1)}D_\pm^{-1/2};$$

$$A_{(\pm)}^\pm = \pm\varphi_{\pm(r-1)}^{-1}[\varphi_{\pm r} - D_\pm^{1/2}]$$

(up to the singularity curves in (3.7), i.e., for regular $g_0^\pm$).

In the one-dimensional case the functions ρ_j, ω depend only on one variable, say $\hat{r} \equiv z_+ + z_-$, and the solutions can be obtained from the solutions of two-dimensional equations under the substitution $\varphi_{\pm\alpha} = c_{\pm\alpha}\exp m_\alpha z_\pm$ with constants $c_{\pm\alpha}$ and m_α, where $m_r \equiv (1/2)(m_{r-1}+m_{r+1})$. Therefore, they are characterized by $2(r+1)$ arbitrary numerical parameters m_α; $d_\alpha \equiv c_\alpha c_{-\alpha}$ for $\alpha = \{\pi_i, \ 1 \le i \le r-1; \ \pi_{r-1} + 2\pi_r\}$ and $c_{\pm(\pi_{r-1}+\pi_r)}$ and can be represented in the form of finite polynomials of $\exp(n_a\hat{r})$, where n_a are some functions in m_α. (Recall that the solutions of the generalized Toda lattice are expressed only in terms of the diagonal elements $D_{ii}(\hat{\mathcal{K}}) = \xi^{\{i\}}(\hat{K})$ and in the one-dimensional problem are parametrized via m_i, d_i, $1 \le i \le r$.

2) $\mathfrak{g}_0^{(r-1)} = \mathfrak{gl}(1) \oplus B_{r-1}$. For this embedding we have

$$\mathfrak{g}_0^f = \{H, X_{\pm\pi_r}, X_{\pm(\pi_{r-1}+\pi_r)}, \ldots, X_{\pm(\pi_2+\ldots+\pi_r)}\},$$

$$\mathfrak{g}_{\pm 1} = \{X_{\pm\pi_1}, X_{\pm(\pi_1+\pi_2)}, \ldots, X_{\pm\pi}; X_{\pm(\pi+\pi_r)}, \ldots, X_{\pm(\pi+\pi_2+\ldots+\pi_r)}\}, \tag{3.8}$$

where $H = 2 \sum_{1\le j \le r-1} h_j + h_r$, $J_\pm = X_{\pm\pi}$. (This case coincides with the preceding one for $r = 2$.)

Setting

$$f_i \equiv x_i^{(1)}\exp\sigma, \qquad\qquad f_{2r-i} \equiv x_i^{(2)}\exp\sigma, \ 1\le i \le r-1,$$

$$f_r \equiv x_0\exp\sigma, \qquad x_0^2 + (\boldsymbol{x}^{(1)}\boldsymbol{x}^{(2)}) = 1,$$

we rewrite (3.2.6) in the form (cf. (3.2))

$$\sigma_{,z_+z_-} = 2[1 - (\boldsymbol{x}^{(1)}\boldsymbol{x}^{(2)})]^{1/2}\exp\sigma,$$

$$\{[1 - (\boldsymbol{x}^{(1)}\boldsymbol{x}^{(2)})]^{-1/2}\boldsymbol{x}_{,z_+}^{(\varepsilon)}\}_{,z_-} = 2\boldsymbol{x}^{(\varepsilon)}\exp\sigma. \tag{3.9}$$

In parametrization $x_i^{(\varepsilon)} = (-1)^{\varepsilon-1}\operatorname{sh}\delta\exp[(-1)^{\varepsilon-1}\theta_i]n_i$, $\boldsymbol{n}^2 = 1$, we reduce the system to the form

$$\sigma_{,z_+z_-} = 2\operatorname{ch}\delta\exp\sigma, \quad \delta_{,z_+z_-} = 2\operatorname{sh}\delta\exp\sigma - \operatorname{th}\delta\cdot\Delta,$$

$$n_i\hat{D}\theta_i + (\theta_{i,z_+}n_{i,z_-} + \theta_{i,z_-}n_{i,z_+}) = 0, \tag{3.10}$$

$$\hat{D}n_i + n_i\theta_{i,z_+}\theta_{i,z_-} - n_i\Delta = 0,$$

where

$$\hat{D} \equiv \partial^2/\partial z_+\partial z_- + 2\operatorname{sh}^{-1}(2\delta)\cdot\delta_{,z_-}\cdot\partial/\partial z_+ + \operatorname{cth}\delta\cdot\delta_{,z_+}\,\partial/\partial z_-,$$

$$\Delta \equiv \sum_j(\theta_{j,z_+}\cdot\theta_{j,z_-}\cdot n_j^2 - n_{j,z_+}n_{j,z_-}).$$

Expressing an $(r-1)$-dimensional unit vector $\boldsymbol{n}$ in generalized Euler angles it is easy to perform the further reduction in dimension and get a form symmetric in z_+ and z_-.

These are just the equations which describe the 2-dimensional minimal surfaces in N-dimensional (pseudo) Euclidean space, cf. [6, 116].

In conclusion, notice that from the systems (3.2.11) by a method similar to the above we can obtain the concrete realization of completely integrable systems generated by the local part of infinite-dimensional Lie algebras. In particular, we may thus obtain the Lund-Regge model and its multicomponent generalizations.

§ 4.4 The case of a generic Lie algebra*)

Consider in detail the application of the general construction of Chapter 3 to nonlinear systems associated with a generic Lie algebra. The most interesting is the case of a Lie algebra with an abelian invariance subalgebra $\mathfrak{g}_0$ which leads thanks to (3.1.4) in the canonical grading to the nonlinear systems of the form

$$\rho_{a,z_+z_-} = \Phi_a(\rho), \quad \rho_a \equiv \rho_a(z_+, z_-). \tag{4.1}$$

Then in accordance with the commutation relations (1.2.7) the structure of this system is the same as (1.2.6), namely, it contains as a subsystem equations of the type (3.1.9) corresponding to the semisimple part $\mathfrak{g}_{(s)}$ in the

---

*) [69, 72]

Levi decomposition $\mathfrak{g} = \mathfrak{w} \dotplus \mathfrak{g}_{(s)}$. There is no criterion yet how to describe all the $\mathbb{Z}$-graded Lie algebras for which $\mathfrak{g}_0$ is an abelian subalgebra.

Now take for $\mathfrak{g}$ the semidirect sum of a simple Lie algebra $\mathfrak{g}$ of rank r and the space of its adjoint representation $\mathfrak{w}^{(1)} = \{H_j, \mathcal{Y}_{\pm j}, 1 \le j \le r\}$ with the trivial bracket on $\mathfrak{w}^{(1)}$. Consider the canonical grading on $\mathfrak{g}$ and $\mathfrak{w}^{(1)}$. Thanks to (1.2.7) the elements of $\mathfrak{w}^{(1)}$ satisfy

$$[H_i, X_{\pm j}] = \pm k_{ji} \mathcal{Y}_{\pm j}, [H_i, h_j] = 0, [h_i, \mathcal{Y}_{\pm j}] = \pm k_{ji} \mathcal{Y}_{\pm j}, [X_{+i}, \mathcal{Y}_{-j}] = \delta_{ij} H_j. \tag{4.2}$$

Then substituting, cf. (3.1.2),

$$A_{\pm} = \sum_{1 \le j \le r} (u_{\pm j} h_j + f_{\pm j} X_{\pm j} + U_{\pm j} H_j + F_{\pm j} \mathcal{Y}_{\pm j}) \tag{4.3}$$

into (3.1.1) we get equations (3.1.8) for $u_{\pm j}$, $f_{\pm j}$ and

$$(F_{-j}/f_{-j})_{,z_+} = (kU_+)_j, \quad (F_{+j}/f_{+j})_{,z_-} = -(kU_-)_j,$$
$$U_{+j,z_-} - U_{-j,z_+} = F_{-j} f_{+j} + F_{+j} f_{-j}. \tag{4.4}$$

This is a particular case of the general system (3.1.4) for $\mathfrak{g} = \mathfrak{w}^{(1)} \dotplus \mathfrak{g}_{(s)}$. Setting $\rho_j \equiv \ln f_{+j} f_{-j}$ and $\sigma_j \equiv F_{-j}/f_{-j} + F_{+j}/f_{+j}$ we get (cf. (3.1.9))

$$\rho_{j,z_+z_-} = \sum_i k_{ji} \exp \rho_i, \quad \sigma_{j,z_+z_-} = \sum_i k_{ji} \sigma_i \exp \rho_i. \tag{4.5}$$

For finite-dimensional simple Lie subalgebras $\mathfrak{g}_{(s)}$ in $\mathfrak{g}$ the system (4.5) is completely integrable in quadratures via the general scheme since $\mathfrak{g} = \mathfrak{w}^{(1)} \dotplus \mathfrak{g}_{(s)}$ is the result of the Inönü-Wigner contraction applied to the direct sum of two simple Lie algebras. Therefore, the explicit solutions of (4.5) can be obtained from the known solutions (4.1.11) for the two-dimensional generalized Toda lattice corresponding to $\mathfrak{g}_{(s)} \oplus \mathfrak{g}_{(s)}$. We will skip them.

The contractions of self-symmetric simple Lie algebras enables us to get a number of other nonlinear integrable systems making use of symmetry properties of root systems of these algebras.

In particular, the contraction of D_r with respect to the symmetry $\pi_{r-1} \rightleftarrows \pi_r$ in the root system gives the semidirect sum of B_{r-1} with the space of its identity (standard) representation. Under the transformation

$$x_{r-1} \to x_r \equiv x'_{r-1}, \ x_{r-1} - x_r \equiv x'_r; \ x_j \equiv x'_j, \ 1 \le j \le r - 2$$

the system (3.1.10) corresponding to D_r becomes

$$x'_{j,z_+z_-} = \exp(k'x')_j, \ 1 \le j \le r - 1; \ x'_{r,z_+z_-} = 2x'_r \exp(-x'_{r-2} + 2x'_{r-1}), \tag{4.6}$$

where k' is the Cartan matrix of B_{r-1}. From the physical point of view (cf. 4.1.1) equation (4.6) describes the r-th "particle" in the external field created by the solutions of the first subsystem.

Similarly, the contraction of B_3 leading to the semidirect product of G_2 with its 7-dimensional module corresponds to the passage to the limit with respect to solutions x_1 and x_3 of (3.1.10) for B_3 and leads to the integrable equations

$$x'_{1,z_+z_-} = \exp(2x'_1 - x'_2),\ x'_{2,z_+,z_-} = \exp(-3x'_1 + 2x'_2);$$
$$x'_{3,z_+z_-} = 2x'_3 \exp(2x'_1 - x'_2),$$

(4.7)

the first two of which constitute the system (3.1.10) for G_2.

In conclusion, consider the system (3.1.14) associated with $A_1 \niplus \mathfrak{w}^{(l)}$. The representation (3.1.1) leads to equations (3.1.8) for $u_\pm$ and $f_\pm$, and

$$(f^1_+/f_+)_{,z_-} = [l(l+1)]^{1/2} f^0_-,\ (f^1_-/f_-)_{,z_+} = [l(l+1)]^{1/2} f^0_+,$$
$$f^0_{+,z_-} - f^0_{-,z_+} = [l(l+1)]^{1/2}(f_+ f^1_- - f_- f^1_+),$$

(4.8)

which implies (3.1.14). It is not difficult to see that the general solutions are

$$x = (1/2) \ln[\psi_{+,z_+} \psi_{-,z_-} (\psi_+ + \psi_-)^{-2}],$$
$$y = (\psi_+ + \psi_-)^{l+1}\{(d/d\psi_-)^l[\varphi_-(\psi_-)(\psi_+ + \psi_-)^{-(l+1)}] \qquad (4.9)$$
$$+ (d/d\psi_+)^l[\varphi_+(\psi_+)(\psi_+ + \psi_-)^{-(l+1)}]\}$$

for integer l, where $\varphi_\pm(\psi_\pm)$ and $\psi_\pm(z_\pm)$ are arbitrary functions. (The expression (4.9) is an equivalent form of (4.1.15) in which y takes the simplest form.

§ 4.5 Supersymmetric equations*)

Representation (3.1.1) for nonlinear systems on supermanifolds is realized by operators $A_\pm$, elements of an appropriate Lie superalgebra, $\mathfrak{g} = \mathfrak{g}_{\bar{0}} \oplus \mathfrak{g}_{\bar{1}}$. Here the elements e_0, $e^{(2a)}_\pm$ and $e^{(2a-1)}_\pm$, $a \geq 1$ from the subspaces $\mathfrak{g}_0$, $\mathfrak{g}_{\pm 2a}$ and $\mathfrak{g}_{\pm(2a-1)}$ in expansions $\mathfrak{g}_{\bar{0}} = \oplus \mathfrak{g}_{2a}$, $\mathfrak{g}_{\bar{1}} = \oplus \mathfrak{g}_{2a+1}$, which are $\mathbb{Z}$-graded consistently with the parity of $\mathfrak{g}$, in the products (3.1.2) are multiplied by even functions $u_\pm(z_+, z_-)$, $f^{(2a)}_\pm(z_+, z_-) \equiv f^a_\pm(z_+, z_-)$ (resp. by odd functions $f^{(2a-1)}_\pm(z_+, z_-) \equiv f'^a_\pm(z_+, z_-)$) which take values in an auxiliary supercommutative superalgebra C,

$$A_\pm = (e_0 u_\pm) + \sum_{1 \leq a \leq m_\pm} (e^{(2a)}_\pm f^a_\pm) + \sum_{1 \leq a \leq m'_\pm} (e^{(2a-1)}_\pm f'^a_\pm). \qquad (5.1)$$

In particular, for the simple Lie superalgebras in a principal grading (see Appendix) with $m_\pm = m'_\pm = 1$, the formula (3.1.7) is written as

$$A_\pm = (h u_\pm) + (X_\pm f_\pm) + (Y_\pm f'_\pm),\ \varepsilon_X \equiv \bar{0},\ \varepsilon_Y \equiv \bar{1}. \qquad (5.2)$$

---

*) [71, 150]

In condition (3.1.17) on the vector (A_+, A_-), the element g belongs to the complex hull of the adjoint Lie supergroup G of the Lie superalgebra $\mathfrak{g}$.

The regular elements g can be represented by the Gauss expansion (1.5.3), where $\mathcal{M}_\pm$, $\mathcal{N}_\pm$ are elements of complex hull of the nilpotent subsupergroups of G spanned by $X_\pm$, $Y_\pm$, while $\hat{H}_+$, $\hat{H}_-$ belong to the Cartan subgroup of G. The main stages of the general integration scheme developed in §3.1 for two-dimensional nonlinear systems associated with Lie algebras are also valid for Lie superalgebras. At the same time, there are some specific features pertaining only to Lie superalgebras. In particular, to solve nonlinear equations, for example, (3.1.9), we have to know the element $\exp\hat{H}(\equiv \exp(\hat{H}_+ - \hat{H}_-))$ in (4.1.5) whose parameters are expressed by (4.1.8) in terms of the highest matrix elements of the fundamental representations, which depend on the element $\mathcal{M}_+^{-1}\mathcal{M}_-$. However, this is not sufficient for the description of complete solutions of the supersymmetric generalizations of equations (3.1.9) or (3.1.8) obtained by the substitution (5.2) in (3.1.1).

We confine ourselves to the simplest case — the supersymmetric Liouville equation (3.1.16) related to the superalgebra $osp(1/2)$ clearly illustrates the characteristic features of supersymmetric systems. By substituting the operators

$$A_\pm = hu_\pm + X_\pm f_\pm + Y_\pm f'_\pm \qquad (5.3)$$

in representation (3.1.1) and using the commutation relations from 3.1.2.6 we get the system

$$f_{-,z_+} = 2u_+ f_-, \qquad f_{+,z_-} = -2u_- f_+,$$

$$f'_{-,z_+} - u_+ f'_- = f_- f'_+, \qquad f'_{+,z_-} + u_- f'_+ = f_+ f'_-, \qquad (5.4)$$

$$u_{-,z_+} - u_{+,z_-} + f_+ f_- + f'_+ f'_- = 0,$$

which, after the change of variables $f_+ f_- \equiv \exp 2x$, $f'_\pm = f_\pm^{1/2}\omega^\pm$ reduces to the equation (3.1.16$_1$) which realizes the component (with respect to the Taylor series expansion in odd variables θ_+, θ_-) form of equation (3.1.16$_2$).

According to the general construction of §3.1, formulas (5.3) and (3.1.17) in the case under consideration imply that the elements $\mathcal{M}_\pm$ can be represented as

$$\mathcal{M}_\pm = \exp(v^\pm X_\pm + \varepsilon^\pm Y_\pm), \qquad (5.5)$$

where $v^+(z_+)$, $v^-(z_-)$ (resp. $\varepsilon^+(z_+)$, $\varepsilon^-(z_-)$) are even (resp. odd) functions of their arguments with values in an auxiliarly supercommutative superalgebra C. (Expression (5.5) is a direct consequence of (3.1.25) taking into account that $[X_+, Y_+] = [X_-, Y_-] = 0$.) In what follows, we take for simplicity $H_- = 0$ ($\hat{H} = H_+$), where $u_+ = 0$, $f_{-,z_+} = 0$. The identity (3.2.17) makes it possible to define the group parameters of the elements $\mathcal{N}_\pm \equiv \exp(\tilde{v}^\pm X_\pm + \tilde{\varepsilon}^\pm Y_\pm)$ and

$\exp \hat{H} \equiv \exp h\tau$ through arbitrary functions $v^{\pm}$ and $\varepsilon^{\pm}$ which parametrize $\mathcal{M}_{\pm}$, namely

$$\exp(-\tau) = 1 - v^+ v^- - \varepsilon^+ \varepsilon^-, \quad \tilde{\varepsilon}^{\pm} = (\varepsilon^{\pm} + v^{\pm}\varepsilon^{\mp})\exp \tau,$$
$$\tilde{v}^{\pm} = v^{\pm}\exp \tau. \tag{5.6}$$

Substituting the Gauss decomposition of the elements $\mathcal{M}_{\pm}$, $\mathcal{N}_{\pm}$ and $\exp \hat{H}$, which are known from (5.5) and (5.6), in (3.1.17) and comparing with (5.3) we come to the final formula for the functions $u_{\pm}$, $f_{\pm}$ and $f'_{\pm}$:

$$f_+ = (v^+_{,z+} + \varepsilon^+ \varepsilon^+_{,z+})\exp 2\tau, \quad f_- = v^-_{,z-} + \varepsilon^- \varepsilon^-_{,z-},$$
$$u_- = -(\varepsilon^+ \varepsilon^-_{,z-} + v^+ v^-_{,z-})\exp \tau, \quad \lambda \equiv 1 - v^+ v^-,$$
$$f'_+ = \lambda^{-1}\varepsilon^+_{,z+} + \lambda^{-2}v^+_{,z+}(\varepsilon^- + \varepsilon^+ v^-), \tag{5.7}$$
$$f'_- = \varepsilon^-_{,z-}(1 - \lambda^{-1}\varepsilon^+ \varepsilon^-) + \lambda^{-1}v^-_{,z-}(\varepsilon^+ + \varepsilon^- v^+).$$

Thus, general solution of the supersymmetric Liouville equation, i.e., of system $(3.1.16_1)$, is of the form:

$$\exp 2x = (v^+_{,z+} + \varepsilon^+ \varepsilon^+_{,z+})(v^-_{,z-} + \varepsilon^- \varepsilon^-_{,z-})(1 - v^+ v^- - \varepsilon^+ \varepsilon^-)^{-2},$$
$$\omega^{\pm} = (v^{\pm}_{,z\pm} + \varepsilon^{\pm}\varepsilon^{\pm}_{,z\pm})^{-1/2}[(1 - v^+ v^-)^{-1}\varepsilon^{\pm}_{,z\pm} \tag{5.8}$$
$$+ (1 - v^+ v^-)^{-2}v^{\pm}_{,z\pm}(\varepsilon^{\mp} + \varepsilon^{\pm} v^{\mp})][1 - v^+ v^- - \varepsilon^+ \varepsilon^-].$$

This integration procedure for the supersymmetric equation $(3.1.16_2)$ is generalized naturally for the systems related to arbitrary $\mathbb{Z}$-graded Lie superalgebras. Here the main part of the problem lies in the construction of the elements $\mathcal{N}_{\pm}$ and $\exp \hat{H}$ in terms of the known $\mathcal{M}_{\pm}$ which satisfy equations (3.1.23), as in the case of Lie algebras. We emphasize that even in the canonical grading with $m_{\pm} = m'_{\pm} = 1$ it is not sufficient to know the highest matrix elements of fundamental representations (of the element $\mathcal{M}_+^{-1}\mathcal{M}_-$) of the simple supergroup G to find general solutions. This can be seen even in the simplest example considered above where the functions $\exp(2x)$ and $\omega^{\pm}$ cannot be reconstructed only from the highest matrix element $\xi(\mathcal{M}_+^{-1}\mathcal{M}_-) = 1 - v^+ v^- - \varepsilon^+ \varepsilon^-$.

§ 4.6 The formulation of the one-dimensional system (3.2.13) based on the notion of functional algebra*)

Here we will apply the approach developed in 4.1.5 (formulas (1.44)–(1.47)) for the one-dimensional Toda lattice to a considerably more general system

---

*) [108]

(3.2.13). For this consider the functional algebra $\mathfrak{g}^f$ associated with a non-compact real Lie group G with generators $F_{\mu_0}^R$, $Z_{\underline{\alpha}}^L$ and $Z_{-\underline{\alpha}}^R$ (see formula (2.1.17$_2$)) and satisfying (cf. (1.46))

$$\{F_{\mu_0}^R, F_{\nu_0}^R\} = B_{\mu_0, \nu_0}^{\sigma_0} F_{\sigma_0}^R, \quad \{F_{\mu_0}^R, Z_{\underline{\alpha}}^L\} = \{F_{\mu_0}^R, Z_{-\underline{\alpha}}^R\} = \{Z_{\underline{\alpha}}^L, Z_{-\underline{\beta}}^R\} = 0,$$

$$\{Z_{\underline{\alpha}}^L, Z_{\underline{\beta}}^L\} = \mathcal{N}_{\underline{\alpha}, \underline{\beta}} Z_{\underline{\alpha}+\underline{\beta}}^L, \quad \{Z_{-\underline{\alpha}}^R, Z_{-\underline{\beta}}^R\} = \mathcal{N}_{-\underline{\alpha}, -\underline{\beta}} Z_{-(\underline{\alpha}+\underline{\beta})}^R. \tag{6.1}$$

Here $B_{\mu_0 \nu_0}^{\sigma_0}$ and $\mathcal{N}_{\underline{\alpha}, \underline{\beta}}$ are structure constants and $\{Z_{\underline{\alpha}}^L, g_0\} = \{Z_{-\underline{\alpha}}^R, g_0\} = 0$. Recall that thanks to arguments from 3.2.3 the $F_{\mu_0}^R$ play the role of generalized momenta and the corresponding group parameters g_0 are generalized coordinates: $\{F_{\mu_0}^R, g_0\} = -g_0 \mathcal{X}_{\mu_0}$.

Further, consider the quadratic Casimir operator $C^f = C(\mathfrak{g}^f)$ given by (2.2.5$_4$) as the Hamiltonian of the corresponding classical system. It gives rise to the dynamics:

$$\dot{Q} = \{C^R, Q\}$$

where Q is either $Z_{-\underline{\alpha}}^R$ or $Z_{\underline{\alpha}}^L$. With (6.1) we get (as in (1.46)) the equations:

$$\dot{Z}_{\underline{\alpha}}^L = \sum_{\underline{\beta}, \underline{\gamma}} (\mathcal{X}^{\underline{\beta}}, g_0 \mathcal{X}^{-\underline{\gamma}} g_0^{-1}) \mathcal{N}_{-\underline{\beta}, \underline{\alpha}} Z_{-\underline{\gamma}}^R Z_{\underline{\alpha}+\underline{\beta}}^L,$$

$$\dot{Z}_{-\underline{\alpha}}^R = \sum_{\underline{\beta}, \underline{\gamma}} (\mathcal{X}^{\underline{\beta}}, g_0 \mathcal{X}^{-\underline{\gamma}} g_0^{-1}) \mathcal{N}_{-\underline{\gamma}, -\underline{\alpha}} Z_{\underline{\beta}}^L Z_{-(\underline{\alpha}+\underline{\gamma})}^R. \tag{6.2}$$

Thanks to (6.2) we can express the $\dot{Z}_{\underline{\alpha}}^L$ for $\underline{\alpha} \in R_+$ in terms of the $Z_{\underline{\alpha}+\underline{\beta}}^L$, which thanks to (1.2.2) correspond to elements of $\mathfrak{g}$ of greater degree (with respect to the principal grading). Setting the initial conditions $Z_{\underline{\alpha}}^L \big|_{t=0} = 0$ for $\underline{\alpha} \in R_+ \setminus R_+^1$ (where R_+^i are elements of degree i) we get $Z_{\underline{\alpha}}^L(t) = 0$ for $\underline{\alpha} \in R_+ \setminus R_+^1$ and $Z_\alpha^L(t) = c(\alpha)$ for $\alpha \in R_+^1$ for some constant $c(\alpha)$.

Similarly, $Z_{-\underline{\alpha}}^R(t) = 0$ for $\underline{\alpha} \in R_+ \setminus R_+^1$ and $Z_{-\alpha}^R(t) = c(-\alpha)$ for $\alpha \in R_+^1$.

Define $c(\pm\alpha)$ from equations $Z_\alpha^L = v_\alpha c(\alpha)$, $Z_{-\alpha}^R = v_\alpha c(-\alpha)$, where $v_\alpha = (X_\alpha, X_{-\alpha})$ and $c(\pm\alpha)$ coincide with coefficients of decomposition of generators $J_\pm = \sum_{\alpha \in R_+^1} c(\pm\alpha) X_{\pm\alpha}$ of $A_1 \subset \mathfrak{g}$ with the grading of $\mathfrak{g}$ being determined by $H = [J_+, J_-]$.

Then the remaining equations on $Q = F_{\mu_0}^R$ and $Q = g_0$ lead due to (6.1) to the desired system (3.2.13). Note that to ensure (3.2.13$_2$) we must require $F_{\mu_0}^R \equiv 0$ for the generators of $\mathfrak{g}_0^0$. Clearly, in this case (2.2.5$_4$) reduces identically to (3.2.14) which is the Hamiltonian of (3.2.13). Thus, an embedding of A_1 into $\mathfrak{g}$ is directly connected with the initial conditions (6.2).

Chapter 5
Internal symmetries of integrable dynamical systems

In the preceding two chapters we have illustrated the effectiveness of the method for integration of nonlinear dynamical systems based on the "spectral" properties of the operators entering representation (3.1.1) and taking values in a $\mathbb{Z}$-graded Lie algebra. However, as we have mentioned it is impossible to describe all such systems since there is no uniform parametrization of structure constants of a generic Lie algebra and nor is there a general description of all their gradings. Therefore, though we can describe explicitly via the general construction the group element uniquely determining the corresponding solutions we cannot describe in terms of a Lie group or its Lie algebra a compact form of the equations themselves.

Therefore, it is desirable and even necessary to have criteria enabling us to judge the properties of the group of internal symmetries of a system by looking at its form. If the group is nontrivial then by assigning to these equations operators (3.1.2) with prescribed spectra we guarantee the integrability of the equations. (In what follows we will assume that any groups of internal symmetries are nontrivial.)

In other words we have the following problem: given a system of equations, what is its group of internal symmetries? (Recall that in Chapters 3 and 4 the starting point was a $\mathbb{Z}$-graded Lie algebra to which, via formulas (3.1.1) and (3.1.2), we have associated a system whose concrete form has been determined by a choice of gauge conditions. The exact integrability of this system was known beforehand; it followed from the spectral properties of the operators $A_{\pm}$.)

Therefore, a complete solution of this problem would have enabled us to answer which exactly integrable two-dimensional systems are contained among those considered in the preceding chapter.

These problems are associated with the criteria for the existence of the group of internal symmetries of dynamical systems and the construction of exact solutions using the group. This chapter is devoted to the solution of these problems.

In what follows we will concentrate on equations of the form $u_{a,z_+z_-} = f_\alpha(u)$, whose theory is more or less developed, cf. [75, 126, 128].

Also, in the last sections we establish an explicit form of the Bäcklund transformation connecting solutions of a nonlinear and the corresponding linearized systems with the help of and in terms of the perturbation theory known from classical mechanics and field theory. This will enable us to get explicit expressions for solutions of exactly integrable systems from Chapter 4 in one more way. The condition for exact integrability of systems with exponential diagonal nonlinearities (cf. 5.1) is that the matrix k from (4.1.1) is equivalent (under the symmetry of the equation) to the Cartan matrix of a simple Lie algebra.

§ 5.1 Lie-Bäcklund transformations. The characteristic algebra and defining equations of exponential systems*)

First notice one evident circumstance which is essential in the formulation of criteria for the existence of a nontrivial group of internal symmetries of nonlinear systems. Consider a compatible nonlinear system of equations in a space of any dimension

$$\mathcal{F}_\alpha(u_i^b) = 0, \tag{1.1}$$

where the functional $\mathcal{F}$ depends on the unknown functions u_b and their derivatives $(u_b)_s \equiv \partial^s u_b/\partial x_j \ldots \partial x_l$ $(1 \leq s \leq n)$ up to a certain finite order. Let this system be integrable in the class of functions which depend on a (finite or infinite) number of parameters. Then the linear *system of equations* for the functions v_a, which in what follows will be called the *defining* one,

$$\sum_{b,r} (v_b)_r \partial \mathcal{F}_\alpha/\partial(u_b)_r = 0, \tag{1.2}$$

is exactly integrable in the same class of functions as the initial system (1.1). (Here $(v_b)_r$ is the r-th derivative of v_b with respect to variables $x_1, \ldots, x_l$ over which the function u^b in (1.1) is differentiated.)

Let us elucidate the above with some examples. The Liouville equation $u_{,z_+z_-} = \exp 2u$ is exactly integrable, i.e., there exists an explicit algebraic expression for its solution (4.1.15) depending on two arbitrary functions $\varphi_+(z_+)$ and $\varphi_-(z_-)$. The corresponding linear equations $v_{,z_+z_-} = 2v \exp 2u$ also has explicit solution depending on two (additional) arbitrary functions $\psi_+(z_+)$ and $\psi_-(z_-)$ with given $\varphi_+(z_+)$ and $\varphi_-(z_-)$ defining a solution of Liouville's equation.

---

*) [37, 38, 75, 126, 128, 131]

An explicit form of solutions of this type is rather easy to find. Let u_b be an n-parameter solution of (1.1) characterized by arbitrary parameters a_α, $1 \leq \alpha \leq n$. Then $v_b = \sum_{1 \leq c \leq n} B_c \partial u_b / \partial a_c$ with parameters B_c is a solution of (1.2). In particular, if a solution of (1.1) depending on n arbitrary functions φ_α and their derivatives with respect to the spatial variables is known then a solution of (1.2) can be expressed in the form

$$v_a = \sum_{b,r} (\psi_b)_r \partial u_a / \partial (\varphi_b)_r. \tag{1.3}$$

Here the ψ_b are arbitrary functions expressing the solutions of (1.2) for given φ_α.

Let us apply these arguments to get a solution of the linearized equation associated with the Liouville one. The general solution of the latter, $u = (1/2) \ln \varphi_{+,z_+} + (1/2) \ln \varphi_{-,z_-} - \ln(\varphi_+ + \varphi_-)$, enables us to get the known general solution of the corresponding linear equation

$$v = \psi_{+,z_+} \partial u / \partial \varphi_{+,z_+} + \psi_{-,z_-} \partial u / \partial \varphi_{-,z_-} + \psi_+ \partial u / \partial \varphi_+ + \psi_- \partial u / \partial \varphi_-$$

$$= (1/2)(\psi_{+,z_+} / \varphi_{+,z_+} + \psi_{-,z_-} / \varphi_{-,z_-}) - (\psi_+ + \psi_-)/(\varphi_+ + \varphi_-).$$

So far the procedure for finding explicit solutions of (1.2) from the known solutions of (1.1) has been formulated in such generality that it involves neither the dimension of the space nor the complete set of parameters defining the solutions of the initial nonlinear equations (1.1).

To answer the question of what such a complete set is we have to study the group of internal symmetries of (1.1), i.e., the group of Lie-Bäcklund transformations of the given system. These are the one-parameter transformations (with parameter τ) given by the relation

$$\partial u^a / \partial \tau \equiv u_\tau^a = f^a(u_r^b). \tag{1.4}$$

The invariance of (1.1) with respect to the transformation (1.4) yields

$$d\mathcal{F}_\alpha / d\tau = \sum_{b,r} (u_\tau^b)_r \partial \mathcal{F}_\alpha / \partial u_r^b = \sum_{b,r} f_r^b \partial \mathcal{F}_\alpha / \partial u_r^b = 0. \tag{1.5}$$

In other words, this means that the solutions of the linear system (1.2) can be constructed as some functions $v^b = f^b(u_r^a)$ in the solutions of the initial systems (1.1) and their derivatives up to a certain order.

Listing all the possible f^a's we define the Lie-Bäcklund algebra, the properties of the group of internal symmetries of (1.1) and as a corollary the information on a complete set of its parametric solutions. Notice that (1.4) always admits a set of solutions of the form $f^a = \partial u^a / \partial x_i$, which in what follows will be assumed to be obvious ones.

Hereafter in this chapter we will confine ourselves to the systems of the form

$$u_{\alpha, z_+ z_-} = \mathcal{F}_\alpha(u_\beta), \quad 1 \leq \alpha, \ \beta \leq n, \tag{1.6}$$

in the two-dimensional space. By definition the functions f_α from (1.4) depend only on $u_s^\alpha \equiv \partial^s u_\alpha / \partial z_+^s$ and $u_{\underline{s}}^\alpha \equiv \partial^{\underline{s}} u_\alpha / \partial z_-^{\underline{s}}$ since all the mixed derivatives can be excluded with the help of (1.5). Therefore, $f_\alpha \equiv f_\alpha(u_\beta, u_s^\beta, u_{\underline{s}}^\beta)$, $1 \leq s, \underline{s} \leq n$. The derivatives realized on these functions with such a set of independent variables are of the form

$$\partial / \partial z_+ = \sum_{\beta, s} u_{s+1}^\beta \partial / \partial u_s^\beta + \sum_{\beta, \underline{s}} (\mathcal{F}_\beta)_{\underline{s}-1} \partial / \partial u_{\underline{s}}^\beta, \quad (\mathcal{F}_\beta)_{\underline{s}} \equiv \partial^{\underline{s}} \mathcal{F}_\beta / \partial z_-^{\underline{s}},$$

$$\partial / \partial z_- = \sum_{\beta, s} (\mathcal{F}_\beta)_{s-1} \partial / \partial u_s^\beta + \sum_{\beta, \underline{s}} u_{\underline{s}+1}^\beta \partial / \partial u_{\underline{s}}^\beta, \quad (\mathcal{F}_\beta)_s \equiv \partial^s \mathcal{F}_\beta / \partial z_+^s. \tag{1.7}$$

Substituting them in the defining equation for f_α we get

$$f_{\alpha, z_+ z_-} = \sum_\beta f_\beta \partial \mathcal{F}_\alpha / \partial u_\beta, \tag{1.8}$$

and therefore we see that the requirement for f_α to depend on a finite number of derivatives of u_α with respect to z_+ and z_- leads to the constraint

$$f_\alpha = f_{+\alpha}(u_s^\beta) + f_{-\alpha}(u_{\underline{s}}^\beta).$$

Thus, the Lie-Bäcklund transformations of (1.6) do not depend on u_α and split along two characteristic directions.

Let us confine ourselves to considering $f_\alpha \equiv f_{-\alpha}(u_{\underline{s}}^\beta)$ and denote $\partial / \partial z_-$ by $\mathcal{D}$. Then on this class of functions the equations (1.8) take the form

$$\sum_{\beta, s} (\mathcal{F}_\beta)_{s-1} \partial / \partial u_s^\beta \mathcal{D} f_\alpha = \sum_\beta f_\beta \partial \mathcal{F}_\alpha / \partial u_\beta, \tag{1.9}$$

where the solutions u_α of (1.6) are exterior parameters and with respect to them (1.9) should be identity. Therefore, if the derivatives of $\mathcal{F}_\alpha$ with respect to u_β are not linearly expressed in derivatives of lower orders starting from a certain order then the only solution of (1.9) is the identity solution $f_\alpha = u_1^\alpha$ leading to the trivial Lie-Bäcklund algebra.

The simplest possibility for the existence of nontrivial solutions of (1.9) arises when all the first-order derivatives are linearly expressed in terms of the functions themselves, i.e., when

$$\partial \mathcal{F} / \partial u_\alpha = A^\alpha \mathcal{F}. \tag{1.10}$$

Here A^α is a set of n commuting $n \times n$ matrices.

Integrating (1.10) is trivial:

$$\mathcal{F} = \exp(uA) \cdot C \equiv \exp\left(\sum_\alpha u_\alpha A^\alpha\right) \cdot C, \tag{1.11}$$

where C is a constant column. If the system (1.6) has right-hand side of the form (1.11) it will be called the *exponential* one and if all the $\mathcal{A}^\alpha$ are diagonal $((\mathcal{A}^\alpha)_{ij} = \delta_{ij}k_{\alpha i})$ the *diagonally exponential* one, cf. (4.1.1).

Substituting (1.11) into (1.9) and simplifying we get

$$C\exp(u\mathcal{A}^T)\left\{\exp(-(u\mathcal{A}^T))\sum_s(\exp(u\mathcal{A}^T))_{s-1}\partial/\partial u_s\right\}\mathcal{D}f_\alpha$$

$$= C\exp(u\mathcal{A}^T)\cdot(f\mathcal{A}^T)_\alpha,$$

where $(f\mathcal{A}^T) \equiv \sum_\alpha f_\alpha \mathcal{A}^{\alpha T}$ and C and $\partial/\partial u_s$ are n-vectors. Taking into account that the f_α do not depend on the u_β we get finally

$$\hat{X}_\nu \mathcal{D}f_\alpha \equiv \left\{\exp(-(u\mathcal{A}^T))\sum_s(\exp(u\mathcal{A}^T))_{s-1}\partial/\partial u_s\right\}_\nu \mathcal{D}f_\alpha = (f\mathcal{A}^T)_{\nu\alpha}.$$

(1.12)

The form of the defining equations (1.12) does not already depend on u_α.

In order to find the explicit solutions of (1.12) and to establish the conditions for $\mathcal{A}^\alpha$ which guarantee the existence of such nontrivial solutions we will study the operators $\hat{X}_\nu$ and their commutation relations with each other and with $\mathcal{D}$.

By consecutive bracketing with new elements:

$$\hat{X}_{\alpha\beta} \equiv [\hat{X}_\alpha, \hat{X}_\beta], \quad \hat{X}_{\alpha\beta\gamma} = [\hat{X}_\alpha[\hat{X}_\beta, \hat{X}_\gamma]], \text{ etc.}$$

we extend the set of operators $\hat{X}_\alpha$ to a Lie algebra (finite- or infinite-dimensional) which will be called the *characteristic Lie algebra* because it determines the Lie-Bäcklund transformations of the system we are interested in. The explicit expressions for the operators $\hat{X}_\alpha$ follow from their definition (1.12)

$$\hat{X}_\nu = \partial/\partial u_\nu^1 + (\tilde{A}_1\partial/\partial u_2)_\nu + ([\tilde{A}_2 + \tilde{A}_1^2]\partial/\partial u_3)_\nu$$

$$+ ([\tilde{A}_3 + 3\tilde{A}_1\tilde{A}_2 + \tilde{A}_1^3]\partial/\partial u_4)_\nu + \ldots,$$

where $\tilde{A}_j \equiv \sum_\mu u_j^\mu A^{T\mu}$ are $n \times n$ matrices, $\partial/\partial u_j$ is a n-vector with the components $\partial/\partial u_j^\alpha$.

The bracket of $\hat{X}_\alpha$ and $\mathcal{D}$ is presentable in the form

$$[\hat{X}, \mathcal{D}] = \mathcal{A}_1\hat{X} \quad ([\hat{X}_\nu, \mathcal{D}] = \sum_\mu(\mathcal{A}_1)_{\nu\mu}\hat{X}_\mu),$$

(1.13)

whose validity is easiest to establish by a direct verification starting from the definition of the operators (1.7) and (1.12).

To calculate the bracket of $\mathcal{D}$ and $\hat{X}_{\alpha_1...\alpha_m}$ we need the obvious relations

$$[\hat{X}_\nu, (\tilde{A}_1)_{\alpha\beta}] = (\tilde{A}_\nu)_{\alpha,\beta}, \quad [\hat{X}_{\alpha_1...\alpha_m}, (\tilde{A}_1)_{\nu\mu}] = 0,$$

(1.14)

which are corollaries of (1.12), of the definition of $\tilde{A}_1$ and also of the fact that $\hat{X}_{\alpha_1 \dots \alpha_m}$, $m \geq 2$, does not contain derivatives with respect to u_μ^1. Then making use of the Jacobi identity and relations (1.14) and (1.13) we get

$$
\begin{aligned}
[\hat{X}_{\alpha_1 \dots \alpha_m}, \mathcal{D}] = & \sum_{1 \leq i \leq m} \sum_{1 \leq j \leq n} (\tilde{A}_1)_{\alpha_i j} \hat{X}_{\alpha_1 \dots \alpha_{i-1} j \alpha_{i+1} \dots \alpha_m} \\
& + \sum_{1 \leq i \leq m-1} \sum_{1 \leq j \leq n} \sum_{i+1 \leq k \leq m} (\tilde{A}_{\alpha_i})_{\alpha_k j} \hat{X}^{\tilde{\alpha}_i}_{\alpha_1 \dots \alpha_{k-1} j \alpha_{k+1} \dots \alpha_m} \\
& - \sum_{1 \leq j \leq n} (\tilde{A}_{\alpha_m})_{\alpha_{m-1} j} \hat{X}^{\tilde{\alpha}_m}_{\alpha_1 \dots \alpha_{m-2} j},
\end{aligned}
\tag{1.15}
$$

where $\hat{X}^{\tilde{\alpha}_i}_{\alpha_1 \dots \alpha_m}$ denotes the $(m-1)$-multiple bracket without index α_i. The simplest way to establish this formula is by the direct calculation of the 2nd and 3rd order brackets and the subsequent reduction.

In what follows we need one obvious statement. Consider a set of elements $\mathcal{Y}_i$ linear in the first derivatives $\partial / \partial u_i^\alpha$ with coefficients which are functions in independent variables u_i^α. Now, if we require that $\mathcal{Y}_i$ do not contain the first derivatives in u_1^α then the operator equation

$$
[\mathcal{Y}_i, \mathcal{D}] = \sum_j c_{ij} \mathcal{Y}_j,
\tag{1.16}
$$

where c_{ij} are some functions in u_s^α, has only trivial solution. Indeed, from $[\mathcal{Y}_i, u_1^\alpha] = 0$, the bracket of (1.16) with u_1^α, Jacobi identity, the definition of $\mathcal{D}$ and $[\mathcal{D}, u_1^\alpha] = u_2^\alpha$ we get $[\mathcal{Y}_i, u_2^\alpha] = 0$. Continuing this procedure we get $[\mathcal{Y}_i, u_s^\alpha] = 0$ for an arbitrary s, which proves the above statement.

Now we are set to study the characteristic algebra. For this introduce $2n + 1$ generators: $\mathcal{D}$ and $\mathcal{Z}_\alpha$, h_α, $1 \leq \alpha \leq n$ of an abstract algebra with relations

$$
[\mathcal{D}, \mathcal{Z}_\alpha] = h_\alpha, \quad [h_\alpha, \mathcal{Z}_\beta] = \sum_{1 \leq \gamma \leq n} A^\alpha_{\beta \gamma} \mathcal{Z}_\gamma, \quad [h_\alpha, h_\beta] = 0,
\tag{1.17}
$$

where the A_α are commuting matrices. In particular, we may assign (1.17) to the elements of the local part of the graded Lie algebras $\mathfrak{g}$ whose subalgebra $\mathfrak{g}_0$ is abelian. Namely, let us identify $\mathcal{Z}_\alpha$ with elements from $\mathfrak{g}_{+1}$ and assume that $\mathcal{D} = \sum_\alpha X_{-\alpha}$ is a linear combination of elements from $\mathfrak{g}_{-1}$. Then making use of formulas (1.2.8) we see that the relations (1.17) are automatically satisfied. The elements from the subspace $\mathfrak{g}_{+m}$ are m-th order brackets of elements $\mathcal{Z}_\alpha$. Clearly, the considered algebra is finite-dimensional if for some m_0 all the subspaces $\mathfrak{g}_{+a}$, $a \geq m_0$, are zero. Recall that an element from $\mathfrak{g}_{+a}$, $a \geq 1$, is considered to be zero if its bracket with all the elements from $\mathfrak{g}_{-1}$ is zero. Now, calculate $\mathcal{Z}_{\alpha_1 \dots \alpha_m} \equiv [\mathcal{Z}_{\alpha_1}[\dots [\mathcal{Z}_{\alpha_{m-1}}, \mathcal{Z}_{\alpha_m}] \dots]]$ with the help of

(1.17). Then by induction we get

$$
[\mathcal{Z}_{\alpha_1\ldots\alpha_m}, \mathcal{D}] = \sum_{1\leq j\leq n} \sum_{1\leq i\leq m-1} \sum_{i+1\leq k\leq m} (\mathcal{A}_{\alpha_i})_{\alpha_k j} \mathcal{Z}^{\tilde{\alpha}_i}_{\alpha_1\ldots\alpha_{k-1}j\alpha_{k+1}\ldots\alpha_m}
$$
$$
- \sum_{1\leq j\leq n} (\mathcal{A}_{\alpha_m})_{\alpha_{m-1}j} \mathcal{Z}^{\tilde{\alpha}_m}_{\alpha_1\ldots\alpha_{m-2}j}. \tag{1.18}
$$

If the algebra is finite-dimensional (i.e. $\mathcal{Z}_{\alpha_1\ldots\alpha_{m_0}} = 0$ beginning from some m_0) then the right-hand side of (1.18) vanishes identically for $m \geq m_0$ which is a direct corollary of the properties of the matrices $\mathcal{A}^\alpha$ from (1.17) which determine the algebra.

Now return to relation (1.15), where some of the terms in the right-hand side coincide with the right-hand side of (1.18). Consecutively, for those matrices $\mathcal{A}^\alpha$ for which (1.18) vanishes so do the corresponding terms in (1.15). Therefore, (1.15) leads to the relation

$$
[\hat{X}_{\alpha_1\ldots\alpha_m}, \mathcal{D}] = \sum_{1\leq j\leq n} \sum_{1\leq i\leq m} (\mathcal{A}_1)_{\alpha_i j} \hat{X}_{\alpha_1\ldots\alpha_{i-1}j\alpha_{i+1}\ldots\alpha_m}, \tag{1.19}
$$

which, when considered as an equation in $\hat{X}_{\alpha_{p_1}\ldots\alpha_{p_r}}$, has only zero solution since for $p \geq 2$ the latter do not contain the first derivatives with respect to u_1^α (cf. Remark after (1.16)). The equation $\hat{X}_{\alpha_1\ldots\alpha_m} = 0$ with $m \geq m_0$ means that the characteristic algebra of (1.6) with $\mathcal{F} = \exp(u\mathcal{A}) \cdot C$ is finite-dimensional under the above assumptions on $\mathcal{A}$.

Concluding this section recall once more the domain of application of the above results. We have considered only systems of a certain form, namely, $u_{\alpha, z_+ z_-} = \mathcal{F}_\alpha(u)$, which automatically narrows the search for the functions $\mathcal{F}_\alpha$ and splits the Lie-Bäcklund transformation into two families associated with the two characteristic directions of the two-dimensional space. Even in the framework of such a restriction only the simplest case is studied, when the number of generators of the Lie-Bäcklund algebra is equal to the rank of the initial system (1.1), i.e., when the derivatives of the right-hand sides of (1.1) are linearly expressed in terms of the functions themselves.

The case of a general Lie-Bäcklund algebra with a finite number of generators corresponds to a linear dependence between the derivatives $\partial^n \mathcal{F}_\alpha / \partial u_{\alpha_1} \ldots \partial u_{\alpha_n}$, $n \leq s$. The complete investigation of this problem would have given the answer in the form of the right-hand side $(\mathcal{F}_\alpha)$ for which the characteristic algebra of the corresponding system is finitely generated.

§ 5.2 Systems of type (3.2.8), their characteristic algebra and local integrals*)

The exponential systems considered in the preceding sections are directly related to the systems described in 3.2.1. To see that let us reformulate the former in terms of the group element $g_0 = \exp(hu) \equiv \exp \sum_\alpha h_\alpha u_\alpha$, where h_α are abelian generators of an algebra (1.17). Using the commutation relations between $\mathcal{D}$ and $\mathcal{Z}_\alpha$, h_α being taken into account, we rewrite the exponential system in the form (3.2.8):

$$\partial(g_0^{-1}\partial g_0/\partial z_-)/\partial z_+ = \left[\mathcal{D}, g_0^{-1}\left(\sum_\alpha \mathcal{Z}_\alpha\right)g_0\right] = [J_-, g_0^{-1}J_+g_0], \qquad (2.1)$$

with the obvious identification $\mathcal{D} = J_-$, $\sum_\alpha \mathcal{Z}_\alpha = J_+$. Therefore, from the algebraic point of view the exponential systems are associated with the local parts of Z-graded Lie algebras $\mathfrak{g}$ whose subalgebras $\mathfrak{g}_0$ are abelian. Therefore, in the subsequent sections of this chapter we will study the Lie algebraic and group properties and their corollaries for the equations of type (2.1) bearing in mind that the exponential systems constitute a particular case of the general one from 5.1.

To make use of the methods of perturbation theory for the explicit construction of exact solutions of (2.1) we will need their Hamiltonian structure. In general, we can not find a Lagrangian (or, directly, a Hamiltonian) corresponding to these equations in terms of g_0 (except for the exponential diagonal systems of the same type as the generalized Toda lattice).

If such a Lagrangian existed it would enable us to introduce all the paraphernalia of the canonical formalism, in particular, simultaneous Poisson brackets between various dynamical variables or the corresponding commutation relations in the quantum domain. However, the system (2.1) supplemented by its defining equation enables us to perform this.

Indeed, the linear system of defining equations for an element f taking values in $\mathfrak{g}_0^*$ (dual to $\mathfrak{g}_0$) is obtained from (2.1) by formal differentiation of g_0 with respect to a parameter τ with the subsequent identification $\dot{g}_0 g_0^{-1} \equiv f$.

This gives rise to the following two equivalent forms of the defining equations

$$\partial(g_0^{-1}f_{,z_-}g_0)/\partial z_+ = [J_-, g_0^{-1}[J_+, f]g_0], \qquad (2.2_1)$$

$$\partial(g_0\bar{f}_{,z_+}g_0^{-1})/\partial z_- = [J_+, g_0[\bar{f}, J_-]g_0^{-1}], \quad \bar{f} \equiv g_0^{-1}fg_0. \qquad (2.2_2)$$

---

*) [37, 38, 126, 128]

A direct verification shows that the scalar function

$$\mathcal{L} = \mathrm{tr}\{(\bar{f}_{,z_+}\, g_0^{-1} g_{0,z_-}) + (\bar{f}[J_-, g_0^{-1} J_+ g_0])\} \tag{2.3}$$

leads (under the independent variation of g_0 and $\bar{f}$) both to (2.1) and to the definition of the equations (2.2). In other words, the function $\mathcal{L}$ is the Lagrange function for the systems (2.1) and (2.2).

Now, pass to the study of the structure of the group of internal symmetries of (2.1). Set

$$\Delta^s \equiv \partial^s(g_0^{-1} g_{0,z_-})/\partial z_-^s, \quad \Delta^0 \equiv \Delta \equiv g_0^{-1} g_{0,z_-};$$

$$\tilde{J}_+ \equiv g_0^{-1} J_+ g_0 = \bar{\Delta}_{,z_+} \quad ([\bar{\Delta}, J_-] = -\Delta).$$

The representation of $\tilde{J}_+$ in the form of a derivative with respect to z_+ follows directly from (2.1).

Let us rewrite the defining equation (2.2_2) in the form

$$\bar{f}_{,z_+ z_-} + [\Delta, \bar{f}_{,z_+}] + [[\bar{f}, J_-], \tilde{J}_+] = 0 \tag{2.4}$$

and find its solutions in the class of solutions of (2.1). Repeating the corresponding arguments which justify the passage from (1.15) to (1.18) we see that $\bar{f}$ can depend only on independent variables $\Delta_\alpha^s \equiv (\tilde{\alpha}, \Delta^s)$. Here $\{\tilde{\alpha}\}$ is a set of linearly independent vectors orthogonal to a basis $\{\alpha\}$ of $\mathfrak{g}_0$. In order to formulate (2.4) in terms of the elements of the characteristic algebra, it is necessary to decompose the operator $\partial/\partial z_+$ on the class of functions depending on Δ_α^s with respect to the system of linearly independent functions $\tilde{J}_{\beta_1}^+ \equiv (\tilde{J}_+, \beta_1)$, where $\{\beta_1\}$ is a basis of $\mathfrak{g}_{-1}$. It follows directly from the definition of Δ_α^s and (2.1) that

$$\partial/\partial z_+ = \sum_{\alpha,s} \partial \Delta_\alpha^s/\partial z_+ \partial/\partial \Delta_\alpha^s = \sum_{\alpha,s} \partial^s(\tilde{\alpha}, [J_-, \tilde{J}_+])/\partial z_-^s \partial/\partial \Delta_\alpha^s$$

$$= \sum_{\alpha,s} (\tilde{J}_+, \omega_\alpha^s)\partial/\partial \Delta_\alpha^s = \sum_{\beta_1} \tilde{J}_{\beta_1}^+ \sum_{s,\alpha} (\beta_1, \omega_\alpha^s)\partial/\partial \Delta_\alpha^s \tag{2.5}$$

$$\equiv \sum_{\beta_1} \tilde{J}_{\beta_1}^+ \hat{X}_{\beta_1}.$$

Here the operator-valued function ω_α^s is determined due to (2.5) and (1.2.11) by the recursive relations

$$\omega_\alpha^0 \equiv [\tilde{\alpha}, J_-], \quad (\tilde{J}_+, \omega_\alpha^{s+1}) = \partial(\tilde{J}_+, \omega_\alpha^s)/\partial z_-$$

$$= (\tilde{J}_+, \partial \omega_\alpha^s/\partial z_- + [\Delta, \omega_\alpha^s]),$$

hence $\omega_\alpha^{s+\alpha} = \partial \omega_\alpha^s/\partial z_- + [\Delta, \omega_\alpha^s]$.

Thus, (2.5) determines the generators $\hat{X}_{\beta_1}$ of the characteristic algebra of the system (2.1) in terms of which we rewrite (2.4) in the following form:

$$\hat{X}_{\beta_1} \mathcal{D}\bar{f} + [\Delta, \hat{X}_{\beta_1}\bar{f}] + [[\bar{f}, J_-], \hat{X}_{\beta_1}] = 0.$$

(As in 5.1 we have set $\mathcal{D} = \partial/\partial z_-$ and $\mathcal{D}\bar{f} = \sum\limits_{\alpha,s} \Delta_\alpha^{s+1} \partial/\partial \Delta_\alpha^s \bar{f}$.) The brackets between $\hat{X}_{\beta_1}$ and $\mathcal{D}$ are easiest to find from the identity $[\partial/\partial z_+, \partial/\partial z_-] \equiv 0$ applied to a function depending on Δ_α^s:

$$[\partial/\partial z_+, \partial/\partial z_-] = \sum_{\beta_1}\{\partial \tilde{J}_{\beta_1}^+/\partial z_- \hat{X}_{\beta_1} + \tilde{J}_{\beta_1}^+[\hat{X}_{\beta_1}, \mathcal{D}]\}$$

$$= \sum_{\beta_1}\{-[\Delta, \tilde{J}_+]_{\beta_1} \hat{X}_{\beta_1} + \tilde{J}_{\beta_1}^+[\hat{X}_{\beta_1}, \mathcal{D}]\} \equiv 0.$$

Therefore,

$$[\hat{X}_{\beta_1}, \mathcal{D}] = \sum_{\gamma_1}(\Delta, [\beta_1, \tilde{\gamma}_1])\hat{X}_{\gamma_1}, \tag{2.6}$$

where $\{\tilde{\gamma}_1\}$ is a set of vectors from $\mathfrak{g}_{-1}^*$ (dual to $\mathfrak{g}_{-1}$). This formula enables us to compute the bracket of $\mathcal{D}$ with the m-th order brackets of the elements of the characteristic algebra by making use of the Jacobi identity and the relation

$$\hat{X}_{\beta_1}\Delta_\alpha = (\beta_1, \omega_\alpha^0) = (J_-, [\beta_1, \tilde{\alpha}]).$$

The remaining stages of the study of the properties of the Lie-Bäcklund algebra of (2.1) coincide with those given in 5.1 for exponential systems.

The final deduction can be formulated as follows. The Lie-Bäcklund algebra of (2.1) is finite-dimensional if this system is generated in the framework of the scheme of 3.2.1 by the local part of a finite-dimensional $\mathfrak{g}$.

As in Chapter 3, for a finite-dimensional graded Lie algebra $\mathfrak{g}$ the corresponding system of type (3.2.8) is exactly integrable. The comparison of these facts enables us to fomulate a criterion for exact integrability starting directly from the properties of the Lie-Bäcklund algebra of the system. Namely, *equations (2.1) are exactly integrable if the number of functionally independent solutions of the system*

$$\hat{X}_{\beta_1} W = 0, \quad 1 \le \beta_1 \le d_1 \equiv \dim \mathfrak{g}_1, \tag{2.7}$$

equals $d_0 = \dim \mathfrak{g}_0$.

The system (2.7) can be rewritten in an equivalent form. For this recall that $\partial/\partial z_+ = \sum\limits_{\beta_1} \tilde{J}_{\beta_1}^+ \hat{X}_{\beta_1}$ on the class of functions depending on Δ_α^s and therefore

$$\partial/\partial z_+ W(\Delta_\alpha^s) = 0. \tag{2.8}$$

In what follows the system (2.7) or (2.8) will be called the *characteristic equation* and its functionally independent solutions the *local integrals*.

Remark: In the modern terms, these integrals are just the elements of so-called W-algebras (for the Toda systems) and their generalizations.

Therefore, a criterion for the exact integrability of (2.1) or, equivalently, of the finite dimensionality of its Lie-Bäcklund algebra is the existence of d_0 functionally independent local integrals.

Now pass to the explicit construction of local integrals of the system (2.1), which will be indexed by an *order p* defining the maximal value of the index s of functions Δ_α^s entering the integrals. Taking the projection of (2.1) onto an arbitrary vector $\tilde{\alpha} \in \mathfrak{g}_0$, we get

$$\partial \Delta_\alpha / \partial z_+ = (\tilde{J}_+, [\tilde{\alpha}, J_-]).$$

This implies that if $\mathfrak{g}_0$ contains d_0^f linearly independent elements $\tilde{\alpha}^q$ invariant with respect to A_1, i.e., $[J_\pm, \tilde{\alpha}^q] = 0$, then (2.1) possesses integrals of 0-th order: $W_q^{(0)} = \Delta_{\alpha^q}$.

Differentiating (2.1) with respect to z_- and simplifying we get

$$\partial / \partial z_+ \{\Delta^1 + (1/2)[[\bar{\Delta}, \Delta], J_-]\} = -(1/2)[[\bar{\Delta}, \tilde{J}_+], J_-], J_-], \qquad (2.9)$$

where we have made use of the identity

$$[[\tilde{J}_+, \Delta], J_-] = (1/2)\partial / \partial z_+ [[\bar{\Delta}, \Delta], J_-] + (1/2)[[\bar{\Delta}, \tilde{J}_+], J_-], J_-].$$

In what follows $\mathcal{A}_{(s)}$ denotes the s-th order bracket of $\mathcal{A}$ with J_-. Taking the projection of (2.9) onto an arbitrary vector $\tilde{\alpha}^t \in \mathfrak{g}_0$, we get

$$\partial / \partial z_+ \{(\Delta^1, \tilde{\alpha}^t) + (1/2)([\bar{\Delta}, \Delta], [J_-, \tilde{\alpha}^t])\} = -(1/2)([\bar{\Delta}, \tilde{J}_+], (\tilde{\alpha}^t)_{(2)}).$$

Therefore, if the algebra of (2.1) decomposed into irreducible representations of A_1 contains representations with the unit value of the orbital momentum l, i.e., if $[[\tilde{\alpha}^t, J_-], J_-] = 0$, then the system has $d_1 (1 \le t \le d_1)$ of the 1st order integrals. Notice that this system always possesses at least one integral with $p = 1$, since the elements of $A_1 = \{H, J_\pm\}$ constitute the multiplet with $l = 1$ ($[H, J_-] = -2J_-$, $[[H, J_-]J_-] = 0$).

Continuing this procedure let us differentiate (2.9) with respect to z_-. We get

$$\partial / \partial z_+ \{\Delta^2 + [\dot{\bar{\Delta}}, \Delta]_{(1)} + (1/3)[\bar{\Delta}[\bar{\Delta}, \Delta]]_{(2)}\} = -(1/3)[\bar{\Delta}[\bar{\Delta}, \tilde{J}_+]]_{(3)}, \qquad (2.10)$$

which, thanks to the arguments similar to the ones above, implies that to every multiplet with $l = 2$, i.e., such that $[[\alpha^f, J_-]J_-]J_-] = 0$, there corresponds a 2nd order local integral of the form

$$W_f^{(2)} = (\tilde{\alpha}^f, \{\Delta^2 + [\dot{\bar{\Delta}}, \Delta]_{(1)} + (1/3)[\bar{\Delta}[\bar{\Delta}, \Delta]]_{(2)}\}); \quad \dot{\bar{\Delta}} \equiv \partial \bar{\Delta} / \partial z_-.$$

Local integrals of order 3 are obtained by differentiating (2.10), and their explicit expressions are more complicated than $W^{(p)}$, $p \le 2$. Therefore, we have skipped them. Notice though, that with every multiplet corresponding to an embedding A_1 into $\mathfrak{g}$ is associated a local integral $W^{(p)}$ whose order coincides with the momentum l of the multiplet. The relations of type (2.9) and (2.10) are, generally, of the form

$$\partial/\partial z_+ W^{(p)} \sim \underbrace{[\bar{A}[\bar{A}[\ldots[\bar{A},\tilde{J}_+]\ldots]]]}_{p}{}_{(p+1)},$$

where $W^{(p)}$ takes values in $\mathfrak{g}_0$. The total number of functionally independent integrals $W^{(p)}$ coincides with the number of independent multiplets with respect to A_1 or with dim $\mathfrak{g}_0$. (Since the system of characteristic equations is linear, an arbitrary function in functionally independent integrals and their derivatives of an arbitrary order is also a local integral.)

Let us give one more method for finding local integrals, illustrating it with the example of a diagonal exponential system described by equations

$$x_{\alpha,z_+z_-} = \exp(kx)_\alpha$$

with an arbitrary $n \times n$-matrix k. As a byproduct we will establish the restrictions on the form of k starting from the requirement of the existence of 1st and 2nd order integrals of this system. In the considered case, $\Delta_\alpha^s = \partial^s x_\alpha/\partial z_-^s$. Let the system possess a 1st order integral. Its most general form is

$$W^{(2)} = \sum_\alpha \hat{A}_\alpha \ddot{x}_\alpha + (1/2) \sum_{\alpha,\beta} \hat{B}_{\alpha\beta} \dot{x}_\alpha \dot{x}_\beta, \quad 1 \le \alpha, \beta \le n;$$

$$\hat{B}_{\alpha\beta} = \hat{B}_{\beta\alpha}; \quad \dot{x} \equiv x_{,z_-} .$$

The main equation (2.8) yields

$$\sum_\alpha \hat{A}_\alpha \partial/\partial z_- \exp \rho_\alpha + \sum_{\alpha,\beta} \hat{B}_{\alpha\beta} \dot{x}_\beta \exp \rho_\alpha = 0, \quad \rho_\alpha \equiv (kx)_\alpha.$$

Collecting the terms at n linearly independent functions $\exp \rho_\alpha$, we get

$$\hat{A}_\alpha (k\dot{x})_\alpha + \sum_\beta \hat{B}_{\alpha\beta} \dot{x}_\beta = 0, \quad \text{or} \quad \hat{B}_{\alpha\beta} = k_{\alpha\beta} \hat{A}_\alpha.$$

Therefore, the restriction on the form of k under which a 1st order integral exists is that it should be symmetrizable with the help of a diagonal matrix $\hat{A}$, i.e. $\hat{A}_\alpha k_{\alpha\beta} = \hat{A}_\beta k_{\beta\alpha}$ (since $\hat{B}$ is symmetric by definition).

A general expression for a second order integral is of the form

$$W^{(2)} = \sum_\alpha \tilde{A}_\alpha \dddot{x}_\alpha + \sum_{\alpha,\beta} \tilde{A}_{\alpha\beta} \dot{x}_\alpha \ddot{x}_\beta + \tilde{B}(\dot{x}),$$

where $\tilde{B}$ is a function homogeneous in $\dot{x}$ of degree 3, i.e., $\tilde{B}(\lambda\dot{x}) = \lambda^3 \tilde{B}(\dot{x})$. In accordance with (2.8) we have

$$\partial W^{(2)}/\partial z_+ = \sum_\alpha \tilde{A}_\alpha (\ddot{\rho}_\alpha + \dot{\rho}_\alpha^2) \exp \rho_\alpha + \sum_{\alpha,\beta} \tilde{A}_{\alpha\beta} (\ddot{x}_\beta \exp \rho_\alpha$$

$$+ \dot{x}_\alpha \dot{\rho}_\beta \exp \rho_\beta) + \sum_\alpha \partial\tilde{B}/\partial\dot{x}_\alpha \exp \rho_\alpha = 0.$$

As above, equating to zero the coefficients of the linearly independent functions $\exp \rho_\alpha$ and comparing the coefficients of the second derivatives $\ddot{x}_\alpha$ we get $\tilde{A}_{\alpha\beta} = -\tilde{A}_\alpha k_{\alpha\beta}$. The terms corresponding to the first derivatives lead to a system defining $\tilde{B}$

$$\tilde{A}_\alpha \dot{\rho}_\alpha^2 - \sum_\beta \tilde{A}_\beta \dot{x}_\beta \dot{\rho}_\alpha k_{\beta\alpha} + \partial \tilde{B}/\partial \dot{x}_\alpha = 0.$$

The compatibility condition for this system $(\partial/\partial \dot{x}_\alpha \partial/\partial \dot{x}_\beta \tilde{B} = \partial/\partial \dot{x}_\beta \partial/\partial \dot{x}_\alpha \tilde{B})$ yields

$$(\tilde{A}_\alpha k_{\alpha\beta} - \tilde{A}_\beta k_{\beta\alpha})(k_{\alpha\gamma} + k_{\beta\gamma}) + (\tilde{A}_\alpha k_{\alpha\gamma} - \tilde{A}_\gamma k_{\gamma\alpha}) k_{\alpha\beta}$$
$$+ (\tilde{A}_\gamma k_{\gamma\beta} - \tilde{A}_\beta k_{\beta\gamma}) k_{\beta\alpha} = 0. \tag{2.11$_1$}$$

There is an obvious solution $\tilde{A}_\alpha k_{\alpha\beta} = \tilde{A}_\beta k_{\beta\alpha}$, of (2.11_1) which is equivalent to the existence of a 1st order integral. Then a 2nd order integral arises as a derivative of the 1st order integral. Therefore, such a solution is trivial and we will not consider it.

The system (2.11_1) splits into subsystems with independent indices 1, 2, 3. Set $y_i = \sum_{j,l} \varepsilon_{ijl} \tilde{A}_j k_{jl}$ $(1 \le i, j, l \le 3)$, where ε_{ijl} is a totally antisymmetric tensor. Then (2.11_1) takes the form

$$\begin{cases} (k_{13} + k_{23})y_3 - k_{12}y_2 - k_{21}y_1 = 0, \\ -k_{13}y_3 + (k_{32} + k_{12})y_2 - k_{31}y_1 = 0, \\ -k_{23}y_3 - k_{32}y_2 + (k_{21} + k_{31})y_1 = 0; \end{cases} \tag{2.11$_2$}$$

$$\begin{cases} y_3(2k_{12} + k_{22}) = 0, \\ y_3(2k_{21} + k_{11}) = 0, \end{cases} \begin{cases} y_2(2k_{13} + k_{11}) = 0, \\ y_2(2k_{31} + k_{33}) = 0, \end{cases} \begin{cases} y_1(2k_{32} + k_{22}) = 0, \\ y_1(2k_{23} + k_{33}) = 0; \end{cases} \tag{2.11$_3$}$$

which has a trivial solution $y_j = 0$, $1 \le j \le 3$, whose meaning has been clarified above.

There exists a generating function for the integrals of the exactly integrable system (3.1.3). Indeed, in the gauge $E_-^0 = 0$ rewrite (3.1.17) in the form

$$g_{,z_+} = g \left(E_+^0 + \sum_{1 \le a \le m_+} E_+^a \right), \quad g_{,z_-} = g \sum_{1 \le a \le m_-} E_-^a \tag{2.12}$$

and consider the matrix elements $\langle \alpha | g | \beta \rangle \equiv \Psi_{\alpha\beta}^{\{l\}}$ of the l-th fundamental representation of $\mathfrak{g}$, the states $\langle \alpha |$ and $| \beta \rangle$ of which satisfy $\langle \alpha | \mathfrak{g}_{+a} = 0$ and $\mathfrak{g}_{-a} | \beta \rangle = 0$ for $a \ge 1$. Thanks to (2.12) we get $\partial \Psi_{\alpha\beta}^{\{l\}}/\partial z_- = 0$. Let us calculate the derivatives of $\Psi_{\alpha\beta}^{\{l\}}$ with respect to z_+ up to order N_l, where N_l is the dimension of the l-th representation of $\mathfrak{g}$. We get

$$\partial\Psi_{\alpha\beta}^{\{l\}}/\partial z_+ = \langle\alpha|g\left(E_+^0 + \sum_{1\leq a\leq m_+} E_+^a\right)|\beta\rangle = \sum_{\gamma}\langle\alpha|g|\gamma\rangle f_{\gamma\beta}^1(E_+^a, 0\leq a\leq m_+)$$

and similarly

$$\partial^n\Psi_{\alpha\beta}^{\{l\}}/\partial z_+^n = \sum_{\gamma}\langle\alpha|g|\gamma\rangle f_{\gamma\beta}^n.$$

Excluding the expressions for the derivatives of N_l matrix elements $\langle\alpha|g|\gamma\rangle$ from the $(N_l + 1)$-th relation we get one equation of order N_l.

$$\sum_{0\leq n\leq N_l} W_n\partial^n\Psi_{\alpha\beta}^{\{l\}}/\partial z_+^n = 0, \tag{2.13}$$

whose solution does not depend on z_-. Therefore, the coefficient functions (2.13) do not depend on z_- either, i.e., $\partial/\partial z_-(W_n/W_m) = 0$ and therefore W_n/W_m are integrals of the characteristic equation of (3.1.3).

§ 5.3 A complete description of Lie-Bäcklund algebras for the diagonal exponential systems of rank 2*)

Here we will consider the simplest case, i.e., the case of the system of two equations of exponential type:

$$x_{j,z_+z_-} = \exp\rho_j, \quad \rho_j = k_{j1}x_1 + k_{j2}x_2; \quad j = 1,2. \tag{3.1}$$

This example will illustrate the method of direct investigation of solvability of the characteristics and defining equations which does not make use of the classification of simple Lie algebras.

For $k_{12}k_{21} = 0$ the system (3.1) splits and takes the form

$$u_{,z_+z_-} = \exp u + \exp\lambda u; \quad v_{,z_+z_-} = \exp u. \tag{3.2}$$

Clearly, the integration of this system reduces to that of the scalar equation $u_{,z_+z_-} = \exp u + \exp\lambda u$. The solvability conditions for the characteristic and defining equations for the system (3.2) coincide with the corresponding solvability conditions ($\lambda = -1, -2$) of the scalar equation and in what follows we will assume that

$$k_{12}k_{21} \neq 0, \quad \rho_1 \neq \rho_2. \tag{3.3}$$

In the case considered the characteristic equation takes the form (cf. 5.2)

$$\hat{X}_1W = \hat{X}_2W = 0, \tag{3.4}$$

___
*) [75]

which is a compatible system and for any k_{ij} possesses a 2nd order integral:

$$W^{(2)} = k_{21}x_1 + k_{12}x_2 - (1/2)k_{11}k_{21}(x_1^1)^2 - k_{12}k_{21}x_1^1x_2^1$$
$$- (1/2)k_{22}k_{12}(x_2^1)^2; \quad x_j^i \equiv \partial x_j/\partial z_i, \ z_1 \equiv z_+, \ z_2 \equiv z_-. \tag{3.5}$$

Suppose there exists an integral $W = W(x^1, \ldots, x^m)$ of order m independent on $W^{(2)}$. Making use of the known solutions $W^{(2)}$, $\mathcal{D}W^{(2)}$, etc. of (3.4) this additional solution can be presented in the form

$$W = x_1^m + \alpha_1 x_1^{m-1} + \alpha_2 x_1^{m-2} + \ldots, \tag{3.6}$$

where α_j, $j \le m$, are generalized-homogeneous polynomials in $x^1, x^2, \ldots$ $(\deg(x_n^j)^\alpha = j\alpha)$. In other words, the function W is linear with respect to the highest order variables and the coefficients $\alpha_j = \partial W/\partial x^{m-j}$ satisfy

$$\hat{X}_{s_1}\hat{X}_{s_2}\ldots\hat{X}_{s_n}\alpha_j = 0, \quad n > j,$$

for any set $s_1, s_2, \ldots, s_n = 1, 2$. The first necessary conditions for the existence of an additional solution take the form

$$2k_{21}/k_{11}, \ 2k_{12}/k_{22} = -1, -2, -3, \ldots; \tag{3.7}$$

$$(k_{11} + 2k_{21})(k_{22} + 2k_{12}) = 0. \tag{3.8}$$

To verify the first of these conditions it suffices to notice that the main formula (1.15) implies

$$\hat{X}_{11\ldots12}x_1^m = (2k_{21}+k_{11})\ldots(2k_{21}+(m-1)k_{11})[-k_{12}\partial/\partial x_1^m + k_{21}\partial/\partial x_2^m + \ldots],$$

$$\hat{X}_{22\ldots21}x_1^m = (2k_{12}+k_{22})\ldots(2k_{12}+(m-1)k_{22})[k_{12}\partial/\partial x_1^m - k_{21}\partial/\partial x_2^m + \ldots], \tag{3.9}$$

where the left-hand side contains brackets of order m. The second necessary condition is obtained if we compare the coefficients of x^{m-3} in expressions

$$\hat{X}_{112}W = \hat{X}_{112}x_1^m + (\hat{X}_{112}\alpha_3)x_1^{m-3} + \ldots = 0,$$

$$\hat{X}_{212}W = \hat{X}_{212}x_1^m + (\hat{X}_{212}\alpha_3)x_1^{m-3} + \ldots = 0.$$

Indeed, the same formula (1.15) implies also that

$$\hat{X}_{112}x_1^m = -k_{12}(k_{11} + 2k_{21})\left\{\left[\binom{m-1}{2} + 1\right]\rho_1^{m-3} + (m-2)\rho_2^{m-3} + \ldots\right\},$$

$$\hat{X}_{212}x_1^m = -k_{12}(k_{22} + 2k_{12})\left\{\binom{m-1}{2}\rho_1^{m-3} + (m-1)\rho_2^{m-3} + \ldots\right\}.$$

On the other hand,

$$\hat{X}_{112}\alpha_3 = (k_{11} + 2k_{21})(-k_{12}\partial\alpha_3/\partial x_1^3 + k_{21}\partial\alpha_3/\partial x_2^3),$$

$$\hat{X}_{212}\alpha_3 = (k_{22} + 2k_{12})(-k_{12}\partial\alpha_3/\partial x_1^3 + k_{21}\partial\alpha_3/\partial x_2^3).$$

Therefore, if (3.8) fails we see that $k_{22} - k_{12} = 0$ contradicting the first necessary condition (3.7).

We similarly verify that if $k_{22} + 2k_{12} = 0$ then for the existence of an additional integral it is necessary that

$$(k_{11} + 2k_{21})(2k_{11} + 2k_{21})(3k_{11} + 2k_{21}) = 0. \tag{3.10}$$

The condition (3.10) concludes the list of necessary conditions for $r(\equiv \text{rank } \mathfrak{g})$ $= 2$ and is obtained by comparison of the coefficients of x_1^{m-5} in the relations

$$\hat{X}_{11112}W = \hat{X}_{11112}x_1^m + (\hat{X}_{11112}\alpha_5)x_1^{m-5} + \ldots = 0,$$

$$\hat{X}_{21112}W = \hat{X}_{21112}x_1^m + (\hat{X}_{21112}\alpha_5)x_1^{m-5} + \ldots = 0.$$

Formulas (1.15) enable us to calculate the dimension of the characteristic algebra and verify that the conditions (3.10) are also sufficient. The minimal m for which an additional solution exists is equal to 3, 4 or 6 and we get, respectively, $k_{11} + 2k_{21} = 0$, $k_{11} + k_{21} = 0$ and $3k_{11} + 2k_{21} = 0$. In what follows these additional solutions of the characteristic equation are denoted by $W^{(m)}$, $m = 3, 4, 6$.

Now, let us show that in the general case the Lie-Bäcklund algebra of the equations (3.1) is completely described by the formulas

$$k_{11}f_1 + k_{12}f_2 = (\mathcal{D} + \rho_1^1)\varphi(W^{(2)}, \mathcal{D}W^{(2)}, \ldots),$$

$$k_{21}f_1 + k_{22}f_2 = (\mathcal{D} + \rho_2^1)\varphi(W^{(2)}, \mathcal{D}W^{(2)}, \ldots) \tag{3.11}$$

(here φ is an arbitrary function in a finite number of variables) and let us list the cases when this algebra can be enlarged. To this end pass in the defining equations

$$\hat{X}_1\mathcal{D}f_1 = k_{11}f_1 + k_{12}f_2, \quad \hat{X}_2\mathcal{D}f_2 = k_{21}f_1 + k_{22}f_2,$$

$$\hat{X}_2\mathcal{D}f_1 = \hat{X}_1\mathcal{D}f_2 = 0 \tag{3.12}$$

to the functions

$$g_1 = \hat{X}_1f_1, \quad g_2 = \hat{X}_2f_2, \quad g = g_1 - g_2. \tag{3.13}$$

It follows from (3.12) that

$$k_{j1}f_1 + k_{j2}f_2 = (\mathcal{D} + \rho_j^1)g_j, \quad \hat{X}_1g_1 = \hat{X}_2g_2 = 0, \tag{3.14}$$

and

$$(\mathcal{D} + \rho_1^1 + \rho_2^1)\hat{X}_1g + k_{21}g = (\mathcal{D} + \rho_1^1 + \rho_2^1)\hat{X}_2g + k_{12}g = 0. \tag{3.15}$$

We will call (3.15) *modified defining equations*.

It is not difficult to see that any solution g of the modified defining equations generates (cf. (3.13), (3.14)) a solution f of (3.12). In particular, to the trivial solution $g = 0$ of (3.14) a solution f of defining equations given by (3.11) corresponds. Now the problem reduces to the listing of the cases when (3.15) has a nontrivial solution.

The first necessary condition for the existence of solutions of (3.15) coincides with the conditions (3.7) of solvability of the characteristic equations. This is proved by reduction in the number of independent variables. Suppose that (3.15) has a solution g of order m, i.e., $g = g(x^1, \ldots, x^m)$, $\partial g / \partial x^m \neq 0$. Set

$$g^1 \equiv \hat{X}_1 g, \quad g^{11} \equiv \hat{X}_1^2 g, \quad g^{21} \equiv \hat{X}_2 \hat{X}_1 g, \ldots$$

From (3.15) we get

$$(\mathcal{D} + v_1^1 + v_2^1)g^1 + k_{21}g = 0, \tag{3.16}$$

and therefore, g^1 does not depend on x^m. Replace g by g^1 in (3.15) using relation (3.16). Then we see after dividing by $\mathcal{D} + v_1^1 + v_2^1$ that the following modification of the defining equations for g^1:

$$(\mathcal{D} + 2v_1^1 + v_2^1)\hat{X}_1 g + (k_{11} + 2k_{21})g^1$$
$$= (\mathcal{D} + v_1^1 + 2v_2^1)\hat{X}_2 g^1 + (k_{22} + 2k_{12})g^1 = 0, \tag{3.17}$$

has a solution of order $m - 1$. If (3.7) fails then the corresponding modified defining equation for g^α possesses a 0-th order solution, i.e., $g^\alpha = \text{const}$. This is possible only for $g^\alpha = 0$, that means that there are not any nontrivial solutions $g \neq 0$ of (3.15).

If $k_{11} + 2k_{21} = k_{22} + 2k_{12} = 0$, then from (3.15) we get

$$\hat{X}_1 g^1 = \hat{X}_2 g^1 = 0 \rightleftharpoons g^1 = \varphi(W^{(2)}, W^{(3)}, \mathcal{D}W^{(2)}, \mathcal{D}W^{(3)}, \ldots),$$

where $W^{(3)}$ is an additional 3rd order solution of the characteristic equations.

Returning to the solution f of the initial defining equations (3.12) we see that the additional solution $W^{(3)}$ determines via (3.4) a series of solutions of (3.12) containing an arbitrary function of a large number of independent variables $W^{(2)}$, $W^{(3)}$, $\mathcal{D}W^{(2)}$, $\mathcal{D}W^{(3)}$, $\ldots$

A similar extension of the Lie-Bäcklund algebra takes place also when the characteristic equation has solutions of order 4 or 6.

Now consider the case when the characteristic equation has no additional solutions, i.e. either (3.8) or (3.9) fails. Our problem is to derive additonal necessary conditions for the solvability of the modified defining equation (3.15). We will confine ourselves to the case where (3.15) possesses a solution of however large order m, i.e., to the case where the corresponding Lie-Bäcklund algebra is infinite-dimensional.

It is convenient to set $k_{11} = k_{22} = 2$, $k_{12} = \lambda_2$, $k_{21} = \lambda_1$, which does not cause the loss of generality since thanks to (3.3) and (3.7) we have $k_{11}k_{22} \neq 0$.

Suppose that there exists a solution $g = g(x^1, \ldots, x^m)$ of (3.15) of a sufficiently large order m. Then

$$g = \alpha x_1^m + \beta x_2^m + \alpha_1 x_1^{m-1} + \beta_1 x_2^{m-1} + \ldots, \tag{3.18}$$

where the derivatives $\alpha_j = \partial g / \partial x_1^{m-j}$, $\beta_j = \partial g / \partial x_2^{m-j}$ satisfy (cf. (3.6))

$$\hat{X}_{s_1} \hat{X}_{s_2} \ldots \hat{X}_{s_n} \alpha_j (\beta_j) = 0 \quad \text{for} \quad n > j.$$

Substitution of (3.18) into (3.15) gives us a series of equations to define α_j, β_j consecutively and, since $\hat{X}_1 \alpha = \hat{X}_2 \alpha = \hat{X}_1 \beta = \hat{X}_2 \beta = 0$, the first coefficients depend only on $W^{(2)}$ and its derivatives. The further arguments only slightly differ of the deduction of the necessary conditions (3.8) and (3.10) for solvability of the characteristic equations. From a system of equations for α_3 we get

$$(\hat{X}_{112}\alpha_3)/2(1 + \lambda_1) = (\hat{X}_{212}\alpha_3)/2(1 + \lambda_2) \to (2 - \lambda_2)\alpha + (2 - \lambda_1)\beta, \tag{3.19}$$

for $(1 + \lambda_1)(1 + \lambda_2) \neq 0$. Making use of this relation between α and β and comparing the expressions $\hat{X}_{1112}\alpha_4$, $\hat{X}_{2212}\alpha_4$, $\hat{X}_{1212}\alpha_4$ obtained from the equations for α_4 we see that for the solvability of (3.15) for $(1+\lambda_1)(1+\lambda_2) \neq 0$ it is necessary that $\lambda_1 = \lambda_2 = -2$. Since k is not invertible, the system reduces to the form (cf. 3.1.12))

$$u_{,z_+z_-} = \exp 2u + \exp(-2u), \quad v_{,z_+z_-} = \exp 2u. \tag{3.20}$$

It remains only to consider the case when one of the numbers λ_1, λ_2 is equal to -1. For $\lambda_2 = -1$, $\lambda_1 = \lambda$ it is convenient to pass to the system (3.17) for $g^1 = g^1(x^1, \ldots, x^m)$

$$[\mathcal{D} + (4 + \lambda)x_1^1]\hat{X}_1 g^1 + 2(1 + \lambda)g^1 = 0, \quad \hat{X}_2 g^1 = 0. \tag{3.21}$$

Directly from this system we get $\hat{X}_{11112}\partial g^1 / \partial x^{m-5}$ and $\hat{X}_{21112}\partial g^1 / \partial x^{m-5}$ and comparing them we get

$$(\lambda + 1)(\lambda + 2)(\lambda + 3)(\lambda + 4) = 0, \tag{3.22}$$

which is necessary for solvability of (3.21). Comparing (3.22) with (3.10) we see that $\lambda = -4$.

The corresponding system (3.1) reduces to the form (cf. (3.1.12))

$$u_{,z_+z_-} = \exp 2u + \exp(-4u); \quad v_{,z_+z_-} = \exp 2u. \tag{3.23}$$

Description of infinite-dimensional Lie-Bäcklund algebras for systems (3.20) and (3.22) reduces to the description of the corresponding algebras for the scalar equations.

Notice that by now we have obtained the complete classification of all exponential diagonal systems. The result can be formulated as the following theorem.

If an exponential diagonal system is exactly integrable, i.e., if its Lie-Bäcklund algebra is finite-dimensional, then the matrix k determining the system is equivalent to the Cartan matrix of a finite-dimensional semisimple Lie algebra.

A necessary condition for the existence of an infinite set of solutions of the defining equations (existence of a wide spectrum of soliton equations) is the equivalence of k to the Cartan matrix of a direct sum of infinite-dimensional Lie algebras of finite growth.

§ 5.4 The Lax-type representation of systems (3.2.8) and explicit solution of the corresponding initial value (Cauchy) problem*)

For the systems of type (3.2.8) an explicit (formal) solution of the Cauchy problem can be obtained, i.e., we can recover g_0 on the whole two-dimensional space knowing its values and the values of its first derivatives on a contour. For this let us rewrite the representation (3.1.1) with the operators $A_\pm$ taking values in the local part of $\mathfrak{g}$ in the gauge $E_+^0 = 0$ (then $E_-^1 = J_-$) (cf. 3.3):

$$[\partial/\partial z_+ + g_0^{-1}J_+g_0, \; \partial/\partial z_- + g_0^{-1}g_{0,z_-} + J_-] = 0. \tag{4.1}$$

As a result we get the system (3.2.8) and the gradient relations (3.1.17)

$$g_0^{-1}g_{0z_-} + J_- = g^{-1}g_{,z_-}, \; g_0^{-1}J_+g_0 = g^{-1}g_{,z_+}, \tag{4.2}$$

implying

$$g_0 J_- g_0^{-1} = g_0 g^{-1} g_{,z_-} g_0^{-1} - g_{0,z_-} g_0^{-1}$$
$$= (gg_0^{-1})^{-1}(gg_0^{-1})_{,z_-} \equiv -(g_0 g^{-1})_{,z_-} (g_0 g^{-1})^{-1}.$$

Consider the set of highest states $|s\rangle$ annihilated by $\mathfrak{g}_{+1}$, i.e., $X_{+\alpha}|s\rangle = 0$. Since $g_0^{-1}X_{+\alpha}g_0$ takes values in $\mathfrak{g}_{+1}$, these elements constitute a representation of G_0 and

$$X_{+\alpha}g_0|s\rangle = g_0(g_0^{-1}X_{+\alpha}g_0)|s\rangle = 0.$$

The definition of the highest states and (4.2) imply that $g(z_+^1, z_-^2)|s\rangle$ does not depend on z_+^1. Indeed,

$$\partial/\partial z_+^1 g(z_+^1, z_-^2)|s\rangle \equiv g(g^{-1}g_{,z_+^1})|s\rangle = g(g_0^{-1}J_+g_0)|s\rangle = 0.$$

Similarly we verify that $\langle s|g_0(z_+^2, z_-^1)g^{-1}(z_+^2, z_-^1)$ does not depend on z_-^1, i.e.,

$$\partial/\partial z_-^1 \langle s|g_0 g^{-1} = -\langle s|(g_0 J_- g_0^{-1})g_0 g^{-1} = 0.$$

---

*) [126]

From these two statements we conclude that

$$\langle s|g_0(z_+^2, z_-^1)g^{-1}(z_+^2, z_-^1)g(z_+^1, z_-^2)|s'\rangle$$

does not depend on z_+^1 and z_-^1. Setting $z_-^1 = z_-^2$, $z_+^1 = z_+^2$, we get

$$\langle s|g_0(z_+^2, z_-^1)g^{-1}(z_+^2, z_-^1)g(z_+^1, z_-^2)|s'\rangle = \langle s|g_0(z_+^2, z_-^2)|s'\rangle. \quad (4.3)$$

where $\langle s|$, $|s'\rangle$ are arbitrary highest states of the algebra in the above sense. The relations (4.3) connect the matrix elements of the group structures at three different points $P(z_+^2, z_-^2)$, $A(z_+^2, z_-^1)$ and $B(z_+^1, z_-^2)$ on the plane (z_+, z_-), the last two of which are situated on the characteristics $z_- = \mathrm{const}$ and $z_+ = \mathrm{const}$ intersecting at P. Then (4.3) takes the form

$$\langle s|g_0(P)|s'\rangle = \langle s|g_0(A)g^{-1}(A)g(B)|s'\rangle. \quad (4.4)$$

Let us connect A and B by a path γ, along which thanks to (4.2) we get

$$g^{-1}\partial g/\partial\gamma d\gamma = (g_0^{-1}g_{0,z_-} + J_-)dz_-^{(\gamma)} + g_0^{-1}J_+g_0 dz_+^{(\gamma)} \equiv L_\gamma dl_\gamma. \quad (4.5)$$

The last relation enables us to recover $g^{-1}(A)g(B)$ as the solution of an S-matrix type (3.1.23) equation with Lagrangian given by the right-hand side of (4.5):

$$g^{-1}(A)g(B) = \hat{\gamma}\exp\int_A^B L_\gamma dl_\gamma, \quad (4.6)$$

where $\hat{\gamma}$ is the ordering along the path γ.

Therefore, (4.4) connects a relation of (3.2.8) at an arbitrary point P with the initial values on an arbitrary path through it. Then the following schemes correspond to solutions of the Cauchy and Goursát problems (Fig. 4.7.C and 4.7.G respectively).

The solution of the Cauchy and Goursát problems for equations (3.1.3) generated by the operators A_+ and A_- of the form (3.1.2) which take values in the subspaces $\bigoplus_{0\leq a\leq m_+} \mathfrak{g}_{+a}$ and $\bigoplus_{0\leq a\leq m_-} \mathfrak{g}_{-a}$, respectively, is quite similar. Here one should choose an appropriate gauge condition in order to match the number of equations to the number of unknown functions which is necessary for the Cauchy and Goursát problems for such systems to be well-defined. (Note that the described approach has been suggested in [88] for a particular case of the Toda lattice with fixed end-points.)

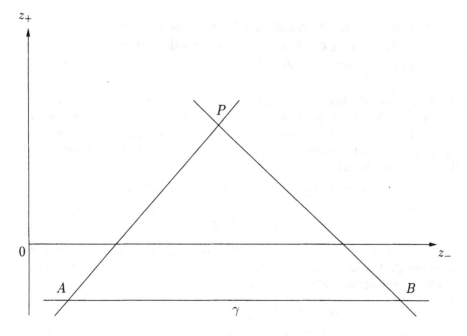

Figure 4.7C.

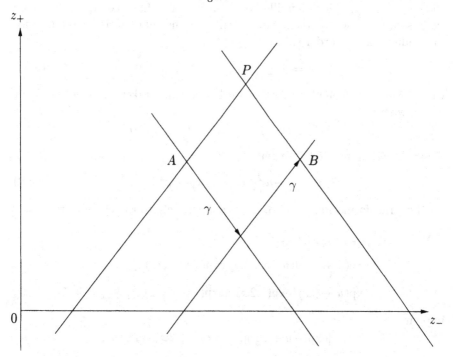

Figure 4.7G.

§ 5.5 The Bäcklund transformation of the exactly integrable systems as a corollary of a contraction of the algebra of their internal symmetry*)

In this section we describe a general scheme for construction of explicit solutions of exactly integrable systems when a complete set of solutions of the corresponding characteristic equation exists, i.e., when the considered system is completely integrable due to the properties of the corresponding Lie-Bäcklund algebra.

Let us start with the Liouville equation with an explicitly introduced coupling constant λ

$$u_{,z_+z_-} = \lambda \exp 2u. \tag{5.1}$$

As has been mentioned in Chapter 3 this equation for $\lambda = 1$ arises from A_1 with the grading operator H. For an arbitrary λ the algebra generating (5.1) is also A_1 under the Inönü-Wigner contraction such that

$$[H, X_\pm] = \pm 2X_\pm, \quad [X_+, X_-] = \lambda H.$$

For $\lambda = 0$ A_1 turns into the algebra of the group of motions of the pseudoeuclidean plane and the Liouville equation into the Laplace equation whose exact solution is well known. Now, recall that the characteristic equation for (5.1) admits a first order solution

$$W_+ = u_{,z_+z_+} - u_{,z_+}^2 ; \quad W_{+,z_-} = 0, \tag{5.2_1}$$

which does not depend on λ. Obviously the same takes place in the limit $\lambda = 0$, where

$$u_0 = \varphi_+(z_+) + \varphi_-(z_-), \ u_{0,z_+z_-} = 0.$$

Substituting u_0 into (5.2) we get

$$u_{,z_+z_+} - u_{,z_+}^2 = \varphi_{+,z_+z_+} - \varphi_{+,z_+}^2. \tag{5.3}$$

Further transformations of (5.3) lead consecutively to the chain of relations

$$(u - \varphi_+)_{,z_+z_-} - (u + \varphi_+)_{,z_+} (u - \varphi_+)_{,z_+}$$

$$\equiv \exp(u + \varphi_+)[(u - \varphi_+)_{,z_+} \exp(-u - \varphi_+)]_{,z_+}$$

$$\equiv \exp(u + \varphi_+)\{\exp(-2\varphi_+)[\exp(-u + \varphi_+)]_{,z_+} \}_{,z_+} = 0,$$

implying

$$[\exp(-u + \varphi_+)]_{,z_+} = c_2 \exp 2\varphi_+(z_+)$$

---

*) [126]

and therefore

$$\exp(-u) = \exp(-\varphi_+) \left\{ c_1 + c_2 \int^{z_+} dz'_+ \exp 2\varphi_+(z'_+) \right\}.$$

Here c_1 and c_2 are functions in z_- which can be determined by substituting the last expression into a solution of the characteristic equation

$$W_- = z_{,z_- z_-} - u^2_{,z_-} = \varphi_{-,z_- z_-} - \varphi^2_{-,z_-}; \; W_{-,z_+} = 0 \qquad (5.2_2)$$

which is symmetric with respect to the change $z_+ \to z_-$.

Therefore, knowing the explicit expressions for the integrals of the characteristic equation and the properties of the asymptotic solution which follow from the investigation of the "rectified" version of the initial algebra enabled us to get an explicit solution of the Liouville equation.

Notice that the Ricatti equation $u_{,tt} - u^2_{,t} = f(t)$ with an arbitrary right-hand side is not generally solvable in elementary functions. The arguments connected with the "rectifiability" of the algebra enable one to establish the form of the function $f(t)$ (see (5.3)) when such a solution is possible.

The system

$$u_{,z_\pm z_\pm} - u^2_{,z_\pm} = \varphi_{\pm,z_\pm z_\pm} - \varphi^2_{\pm,z_\pm}$$

can be considered as one more pair of equations (now of order 2) which connect solutions of the Liouville and Laplace equations. In other words, it corresponds to the Bäcklund transformation which in the above formulation connects solutions of the Liouville equation (5.1) with their asymptotic values described by the Laplace equation. (The linear Bäcklund transformations for this problem are representable in the form (4.2.9) with $N_1 N_3 = 1$.)

In general, the Bäcklund transformation for the systems of type (3.2.8) is obtained if we equate the complete set of the integrals of motion to their values at $\lambda = 0$

$$\begin{aligned}
W_+^{(n)}(g_0, g_{0,z_+}, \ldots) &= W_+^{(n)}(g_0^{(0)}, g_{0,z_+}^{(0)}, \ldots), \\
W_-^{(n)}(g_0, g_{0,z_-}, \ldots) &= W_-^{(n)}(g_0^{(0)}, g_{0,z_-}^{(0)}, \ldots).
\end{aligned} \qquad (5.4)$$

In other words, this transformation is a system of equations describing a connection of solutions of nonlinear equations of the considered type with the solutions of the same system at $\lambda = 0$.

At present there is no method for constructing solutions of (5.4) though we feel that one can be obtained along the same lines of the perturbation theory which were discussed in 5.6. First, let us integrate (5.4) in one of the simplest cases, the generalized Toda lattice associated with A_2 for which A_2 is the algebra of internal symmetries. The equations describing this lattice are expressed thanks to the general formula (3.1.10) in the form

$$x_{1,z_+ z_-} = \exp(2x_1 - x_2) = \exp \rho_1, \; x_{2,z_+ z_-} = \exp(-x_1 + 2x_2) = \exp \rho_2. \quad (5.5)$$

In agreement with the result from 5.2 there are two independent integrals of orders 2 and 3 of the characteristic equation:

$$W^{(2)} = x_{1,zz} + x_{2,zz} - x_{1,z}^2 + x_{1,z}x_{2,z} - x_{2,z}^2,$$

$$W^{(3)} = x_{1,zzz} - x_{2,zzz} + (x_{1,z} + 2x_{2,z})x_{2,zz} \qquad (5.6)$$

$$- (x_{2,z} + 2x_{1,z})x_{1,zz} + 2x_{1,z}x_{2,z}(x_{1,z} - x_{2,z}), \quad (z = z_\pm).$$

The simplest way to verify this is to verify that $W_{\pm,z_\mp}^{(2)} = W_{\pm,z_\mp}^{(3)} = 0$ using (5.5).

The system $W^{(2)} = f_2(z)$, $W^{(3)} = f_3(z)$ is the simplest generalization of the Ricatti equation which does not generally have an exact solution for arbitrary functions f_2 and f_3. Using (5.4) to equate these integrals to their limit values determined by the functions $\varphi_j(z)$, $1 \le j \le 2$, taking their difference and adding the derivative of $W^{(2)}$ with respect to z to $W^{(3)}$ we get

$$x_{1,zzz} - x_{1,z}\rho_{1,zz} + x_{1,z}x_{2,z}(x_{1,z} - x_{2,z})$$

$$= \varphi_{1,zzz} - \varphi_{1,z}\sigma_{1,zz} + \varphi_{1,z}\varphi_{2,z}(\varphi_{1,z} - \varphi_{2,z}), \qquad (5.7)$$

$$x_{2,zzz} - x_{2,z}\rho_{2,zz} - x_{1,z}x_{2,z}(x_{1,z} - x_{2,z})$$

$$= \varphi_{2,zzz} - \varphi_{2,z}\sigma_{2,zz} - \varphi_{1,z}\varphi_{2,z}(\varphi_{1,z} - \varphi_{2,z}),$$

where $\sigma_1 \equiv 2\varphi_1 - \varphi_2$, $\sigma_2 \equiv 2\varphi_2 - \varphi_1$. The system (5.7) (and also (5.3)) is symmetric with respect to the transformation $x_1 \rightleftharpoons \varphi_1$, $x_2 \rightleftharpoons \varphi_2$; and the second equation is obtained from the first one under the change of indices $1 \rightleftharpoons 2$. Substituting

$$W^{(2)}(\varphi_1, \varphi_2) - x_{1,zz} + x_{1,z}^2 - x_{1,z}x_{2,z} + x_{2,z}^2 = x_{2,zz},$$

instead of $x_{2,zz}$ which enters $\rho_{1,zz}$ in the first of the equations (5.7) (this is possible thanks to (5.6)) we get

$$x_{1,zzz} - 3x_{1,z}x_{1,zz} + x_{1,z}^3 - x_{1,z}(\varphi_{1,zz} + \varphi_{2,zz} - \varphi_{1,z}^2$$

$$+ \varphi_{1,z}\varphi_{2,z} - \varphi_{2,z}^2) = \varphi_{1,zzz} - \varphi_{1,z}\sigma_{1,zz} + \varphi_{1,z}\varphi_{2,z}(\varphi_{1,z} - \varphi_{2,z}).$$

This equation can be rewritten in the form

$$[\partial/\partial z \exp(-2\sigma_2)[\partial/\partial z \exp(-2\sigma_1)[\partial/\partial z \exp(-x_1 + \varphi_1)]]] = 0 \qquad (5.8_1)$$

or equivalently as

$$[\partial/\partial z \exp(-2\rho_2)[\partial/\partial z \exp(-2\rho_1)[\partial/\partial z \exp(\varphi_1 - x_1)]]] = 0,$$

which reflects the symmetry of (5.7) with respect to the change $x_1 \rightleftharpoons \varphi_1$, $x_2 \rightleftharpoons \varphi_2$.

Similarly, we reduce the second equation in (5.7) to the form

$$[\partial/\partial z \exp(-2\sigma_1)[\partial/\partial z \exp(-2\sigma_2)[\partial/\partial x \exp(-x_2 + \varphi_2)]]] = 0. \qquad (5.8_2)$$

It is easy to integrate (5.8):

$$\exp(x_1) = \exp(-\varphi_1) \left\{ c_1 + c_2 \int^{z_+} dz'_+ \exp 2\sigma_1(z'_+) \right.$$

$$\left. + c_3 \int^{z_+} dz'_+ \int^{z'_+} dz''_+ \exp 2\sigma_1(z'_+) \exp 2\sigma_2(z''_+) \right\},$$

$$\exp(-x_2) = \exp[-x_1(\varphi_1 \rightleftharpoons \varphi_2; \; c_\alpha \to d_\alpha)].$$

The dependence of c_α and d_α, $1 \le \alpha \le 3$, on z_- is determined when we solve the second pair of Bäcklund equations obtained from (5.7) under the change $z \to z_-$, $\varphi_{+j} \to \varphi_{-j}$, $1 \le j \le 2$.

The final answer exactly reproduces the formula (4.1.16) obtained earlier for the solution of a generalized Toda lattice corresponding to A_2.

Now consider the procedure for constructing general solutions of the Toda lattice with fixed end-points (3.1.10) starting from (5.4) and inserting λ into the right-hand side of (3.1.10). For this let us rewrite (2.12) for the given system in the form

$$\text{(a) } g_{,z_+} = g \left(x_{,z_+} + \sum_j k_j^{1/2} X_{+j} \right), \quad \text{(b) } g_{,z_-} = g \sum_j k_j^{1/2} \exp(kx)_j X_{-j}$$
$$(5.9)$$

where $(x \equiv \sum_j h_j x_j)$. The desired solutions $\exp x_i = \langle i|g|i \rangle = \Psi_{ii}\{i\}$ satisfy (2.13) in which the coefficient functions do not depend on z_- thanks to (5.9b). (Here for $|i\rangle$ we take the lowest weight vector of the i-th fundamental representation of $\mathfrak{g}$.) Therefore, they can be chosen as in the simplest case of the Liouville equation also for $\lambda = 0$: replace x in (5.9a) by an arbitrary operator-valued function $\psi(z_+) = \sum_j h_j \psi_j(z_+)$. Taking this into account we conclude from (5.9a) that $g \exp(-\psi)$ satisfies (3.1.23) with $L_+(\psi) = \sum_j k_j^{1/2} \exp(k\psi)_j X_{+j}$ and is represented by a $\tilde{Z}_+$-ordered exponential

$$g \exp(-\psi) = g_0(z_-) \tilde{Z}_+ \exp \int^{z_+} dz'_+ L_+(\psi(z'_+)), \qquad (5.10)$$

where $g_0(z_-)$ is so far an arbitrary element from G_0. Identifying $\exp[(k\psi)_j + 1/2 \ln k_j]$ with $\varphi_{+j}(z_+)$ and determining $g_0(z_-)$ from (5.9b) we get (4.1.11) expressing the desired solutions of the system (3.1.10).

§ 5.6 Application of the methods of perturbation theory in the search for explicit solutions of exactly integrable systems (the canonical formalism)*)

In this section we apply the methods of perturbation theory for constructing explicit solutions of exactly integrable dynamical systems. We emphasize that we speak not about some approximate results but about exact expressions which are the results of summation of the perturbation theory series; in the considered cases this summation can be performed.

Therefore, the Bäcklund transformation connecting the nonlinear and the corresponding linearized systems can be described explicitly. Namely, it is a canonical transformation relating the solutions of a nonlinear dynamical system with the solutions of the system obtained from the initial one when the coupling constant vanishes. (In the simplest cases the roles of the nonlinear and the linearized systems are played by the Liouville and Laplace equations, respectively.)

In what follows we will need some notions from the perturbation theory of dynamical systems of classical mechanics and field theory which we will recapitulate here briefly referring for the details to [9, 10].

The Hamilton-Jacobi equation completely determining a conservative system is of the form

$$\partial S/\partial t + \mathfrak{h}(\partial S/\partial q_\alpha; q_\alpha) = 0 \qquad (6.1)$$

where $\mathfrak{h}(p; q)$ is the Hamiltonian of the system, i.e., a function of generalized coordinates q_α and momenta p_α, $1 \leq \alpha \leq n$. The Poisson brackets of dynamical variables $\varphi(p; q)$ and $\psi(p; q)$ are defined as usual, cf. 1.1.3:

$$\{\varphi, \psi\} = \sum_{1 \leq \alpha \leq n} (\partial\varphi/\partial p_\alpha \partial\psi/\partial q_\alpha - \partial\varphi/\partial q_\alpha \partial\psi/\partial p_\alpha).$$

In particular,

$$\{p_\alpha, q_\beta\} = \delta_{\alpha\beta}; \quad \{F(\varphi), G(\psi)\} = F_{,\varphi} G_{,\psi} \{\varphi, \psi\}.$$

To integrate a dynamical system we have to find a total integral (6.1) $S = S(t; q_\alpha; a_\alpha)$ depending on n arbitrary parameters a_α. The equations of motion follow from n relations $\partial S/\partial a_\alpha = b_\alpha$, where b_α is a set of n arbitrary parameters. When the Hamiltonian of the system can be represented in the form $\mathfrak{h} = \mathfrak{h}_0 + \lambda\mathfrak{h}_I$, where λ is a "small" parameter. We assume that the motion of the system does not differ "much" from its behaviour for $\lambda = 0$ and all the dynamical variables are expressed as series in the powers of λ. There are several equivalent formulations of the perturbation theory. To understand

---

*) [106, 108]

the subsequent arguments we will need the following explicit expression for a dynamical variable $\varphi(p;q)$ in order m, i.e., the coefficient of λ^m in the above-mentioned series:

$$[\varphi(p;q)]_m = \int_{-\infty}^{\infty} \cdots \int \prod_{1 \leq \alpha \leq m} dt_\alpha \theta(t_{\alpha-1} - t_\alpha)\{\mathfrak{h}_m\{\mathfrak{h}_{m-1} \cdots \{\mathfrak{h}_1, \varphi_t\} \cdots \}\},$$

(6.2)

where $\theta(\tau) \equiv \begin{Bmatrix} 1, & \tau \geq 0 \\ 0, & \tau < 0 \end{Bmatrix}$ is the Heaviside step-function; $t_0 \equiv t$, $\mathfrak{h}_\alpha \equiv$ $\mathfrak{h}_I(p(t_\alpha); q(t_\alpha))$ and $\varphi_t \equiv \varphi(p(t), q(t))$. Here $\mathfrak{h}_I$ is the Hamiltonian of the interaction with the generalized coordinates and momenta replaced by the corresponding dynamical variables depending on time t_α and the trajectories described by the free Hamiltonian $\mathfrak{h}_0$. The arguments φ_t correspond to the variables of $\mathfrak{h}_0$ at time t.

Therefore, with the help of perturbation theory we establish explicit formulas connecting the dynamical quantities described by the Hamiltonians $\mathfrak{h}_0$ and $\mathfrak{h}_0 + \lambda\mathfrak{h}_I$.

On the quantum level such a description of dynamical systems corresponds to the interaction representation.

The most essential part of this is that for the exactly integrable systems the series $\sum_m \lambda^m[\varphi(p;q)]_m$ are just polynomials and therefore, solve the considered problem exactly. For details see 7.5.

§ 5.7 Perturbation theory in the Yang-Feldmann formalism*)

The perturbation theory machinery applied in the above section to the systems of exponential type is essentially based on the canonical formalism which cannot be formulated for all integrable systems of type (3.2.8). Therefore, it is convenient to refer directly to the integration of the corresponding equations

$$\partial[g_0^{-1}\partial g_0/\partial z_-]/\partial z_+ = \lambda[J_-, g_0^{-1}J_+g_0] \tag{7.1}$$

by expanding them into series with respect to powers of the coupling constant λ introduced explicitly into (3.2.8) with consecutive calculation of all the orders of perturbation theory. In the physical application such a formalism which is essentially just rewriting a differential equation in an integrable form with an automatic use of the boundary condition, is known as the Yang-Feldmann method.

---

*) [108, 126]

Apply this method to (7.1). Let us expand the desired group element g_0 into the series with respect to λ considered as a parameter of "smallness":

$$g_0 = g_0^{(0)} \sum_{m \geq 0} \lambda^m u_m \equiv g_0^{(0)} p(u), \ u_0 \equiv 1;$$

$$g_0^{-1} = p^{-1}(u)[g_0^{(0)}]^{-1} = [1 - \lambda u_1 + \lambda^2(u_1^2 - u_2) + \ldots][g_0^{(0)}]^{-1}.$$

(7.2)

Obviously, the zero order term is of the form $g_0^{(0)} = h_+(z_+)h_-(z_-)$, where $h_\pm(z_\pm) \in G_0 = \exp \mathfrak{g}_0$ are arbitrary differentiable functions. Substituting (7.2) into (7.1) we get

$$\partial[p^{-1}h_-^{-1}\partial(h_-p)/\partial z_-]/\partial z_+ = \lambda[J_-, p^{-1}h_-^{-1}h_+^{-1}J_+h_+h_-p],$$

or, simplifying,

$$\partial[p^{-1}(\tilde{u})\partial p(\tilde{u})/\partial z_-]/\partial z_+ = \lambda[h_-J_-h_-^{-1}, p^{-1}(\tilde{u})h_+^{-1}J_+h_+p(\tilde{u})]$$

$$\equiv \lambda[J_-(h_-), p^{-1}(\tilde{u})J_+(h_+)p(\tilde{u})],$$

(7.3)

where

$$\tilde{u}_m \equiv h_-u_mh_-^{-1}, \ J_+(h_+) \equiv h_+^{-1}J_+h_+, \ J_-(h_-) \equiv h_-J_-h_-^{-1}.$$

Equating powers of λ in both sides of equation (2.3) we get

$$\partial \left\{ \sum_{0 \leq s \leq m-1} p_s^{-1}\partial p_{m-s}/\partial z_- \right\} / \partial z_+$$

$$= \left[J_-(h_-), \sum_{0 \leq s \leq m-1} p_s^{-1}J_+(h_+)p_{m-s-1} \right],$$

(7.4)

where p_s and p_s^{-1} are the coefficients of λ^s in the decompositions of $p(\tilde{u})$ and $p^{-1}(\tilde{u})$, respectively. In particular, for $m = 1, 2$ we get from (7.4)

$$\tilde{u}_{1,z_+z_-} = [J_-(h_-), \ J_+(h_+)],$$

$$\tilde{u}_{2,z_+z_-} = [J_-(h_-), \ [J_+(h_+), \tilde{u}_1]] + (\tilde{u}_1\tilde{u}_{1,z_-})_{,z_+},$$

(7.5)

and integrating we get

$$\tilde{u}_1 = \left[\int^{z_-} dz'_- J_-(h_-(z'_-)), \ \int^{z_+} dz'_+ J_+(h_+(z'_+)) \right],$$

$$\tilde{u}_2 = \left[\int^{z_-} dz'_- J_-(h_-(z'_-)) \right] \left[\int^{z_+} dz'_+ J_+(h_+(z'_+)), \right.$$

$$\left[\int^{z'_-} dz''_- J_-(h_-(z''_-)), \int^{z'_+} dz''_+ J_+(h_+(z''_+))\right]\Bigg]\Bigg] \tag{7.6}$$

$$+ \left[\int^{z_-} dz'_- J_-(h_-(z'_-)), \int^{z_+} dz'_+ J_+(h_+(z'_+))\right]$$

$$\times \left[\int^{z_-} dz''_- J_-(h_-(z''_-)), \int^{z_+} dz''_+ J_+(h_+(z''_+))\right].$$

Finally, we are interested in concrete matrix elements of g_0 whereas the above formulas for the terms of the series of perturbation theory contain the corresponding values for arbitrary states of the initial Lie algebra $\mathfrak{g}$.

In turn, the desired element g_0 is obtained after exponentiating the subalgebra $\mathfrak{g}_0 \subset \mathfrak{g}$. The repeated brackets of $J_-(h_-)$ and $J_+(h_+)$ belonging to $\mathfrak{g}_{-1}$ and $\mathfrak{g}_{+1}$, respectively, entering formulas of type (7.6) take values in $\mathfrak{g}_0^f \subset \mathfrak{g}_0$. Indeed, consider the states annihilated by J_+, i.e., $J_+|\rangle = 0$, and $\langle|J_- = 0$, respectively. Then the expansion (7.2) of the solutions of (7.1) (in terms of these matrix elements) can be represented in the form

$$\langle g_0\rangle = \langle h_+(z_+) \Bigg\{ 1 - \lambda \int^{z_+} dz'_+ J_+(h_+(z'_+)) \int^{z_-} dz'_- J_-(h_-(z'_-))$$

$$+ \lambda^2 \int^{z_+} dz'_+ J_+(h_+(z'_+)) \int^{z_+} dz''_+ J_+(h_+(z''_+)) \int^{z_-} dz'_- J_-(h_-(z'_-)) \int^{z'_-} dz''_- J_-(h_-(z''_-))$$

$$+ \dots \Bigg\} h_-(z_-)\rangle \equiv \langle h_+(z_+) \Bigg\{ \sum_{m\geq 0} (-\lambda)^m \int \dots \int \prod_{1\leq i\leq m} dz_+^{(i)} dz_-^{(i)}$$

$$\times \theta(z_+^{(i-1)} - z_+^{(i)})\theta(z_-^{(i-1)} - z_-^{(i)}) J_+(h_+(z_+^{(1)})) \dots J_+(h_+(z_+^{(m)}))$$

$$\times J_-(h_-(z_-^{(m)})) \dots J_-(h_-(z_-^{(1)})) \Bigg\} h_-(z_-)\rangle,$$

$$(z_\pm^{(0)} \equiv z_\pm),$$

where the rule for constructing multiple integrals is evident.

Finite-dimensional Lie algebras possess finite-dimensional representations for which a repeated application of lowering operators from the left, or raising operators from the right, annihilates the basis states. Therefore in this case the series of perturbation theory are finite and (7.7) is an equivalent form of expression of solutions for the exactly integrable dynamical systems of type (3.2.8) constructed in Chapters 3 and 4.

Skipping the reduction procedure which justifies (7.7) let us only indicate how to construct $\tilde{u}_m$ explicitly. For this let us consecutively move all the elements $J_+(J_-)$ onto the rightmost (leftmost) position when the corresponding matrix elements vanish.

§ 5.8   Methods of perturbation theory in the one-dimensional problem*)

Here we consider the exactly integrable dynamical systems that arise from two-dimensional ones of type (3.2.8) under some constraints on the dependence of unknown function on their arguments, say, $g_0 = g_0(z_+ + z_- = t)$ and are described by a system of ordinary differential equations (3.2.13). Their solutions in the classical domain can be obtained, as we have already mentioned, from the general solutions of the corresponding two-dimensional systems by an appropriate choice of asymptotic functions which in the final expression leads to the right dependence on one (time) variable.

This is exactly how the explicit formulas for the solutions of one-dimensional generalized Toda lattice (4.1.48) have been obtained. (In the quantum domain the situation is essentially different since the commutation relations differ the in one- and two-dimensional cases.) At the same time, while it is always possible, in principle, to find the solutions, it is not always trivial. Therefore, it is necessary to have an apparatus which enables us to construct solutions of the corresponding one-dimensional system explicitly. However paradoxical this may sound, the calculational part of the question in the one-dimensional case is not in the least simpler than in the two-dimensional one.

Unlike the two-dimensional case, the considered system

$$d(g_0^{-1}dg_0/dt)/dt = \lambda[J_-, g_0^{-1}J_+g_0] \tag{8.1}$$

possesses a Hamiltonian (3.2.14) expressible in terms of $g_0 \in G_0$ and shift operators on G_0 with the factor λ at the potential term. Thanks to 3.2.3 this leads to the equation (8.1).

Therefore, we may make use of the canonical formalism and construct the solutions of (8.1) starting from formula (5.6.2) with the help of formulas from 5.7. However, we will make use of the perturbation theory machinery in the Yang-Feldmann formalism since the technique in this case is closest to the two-dimensional case and a number of results obtained for the latter can be used directly in the former.

---

*) [108]

As in the preceding section, let us expand $g_0(t)$ into power series in λ as in (7.2), considering λ as a small parameter. The zero order term satisfies

$$d[(g_0^{(0)})^{-1}d(g_0^{(0)})/dt]/dt = 0 \ (dg_0^{(0)}/dt = g_0^{(0)}\hat{p}_-)$$

and the solution is

$$g_0^{(0)} = h\exp(\hat{p}_-t) \equiv (h\exp(\hat{p}_-t))h^{-1}h \equiv \exp(\hat{p}_+t)h,$$

where $\hat{p}_+$ and $\hat{p}_-$ take values in $\mathbf{g}_0^f$ and do not depend on t. In this notation (8.1) takes the form

$$d[(g_0^{(0)})^{-1}p^{-1}(u)dp(u)/dtg_0^{(0)}]/dt = \lambda[J_-, (g_0^{(0)})^{-1}p^{-1}(u)J_+p(u)g_0^{(0)}]. \quad (8.2)$$

Having expanded $p(u)$ into power series in λ and simplified (8.2) we get a system of equations for consecutive approximations for calculating $\tilde{u}_m$. The first terms of the perturbation give

$$d[(g_0^{(0)})^{-1}d\tilde{u}_1/dtg_0^{(0)}]/dt = [J_-, (g_0^{(0)})^{-1}J_+g_0^{(0)}],$$

$$d[(g_0^{(0)})^{-1}(d\tilde{u}_2/dt - \tilde{u}_1 d\tilde{u}_1/dt)g_0^{(0)}] = [J_-, (g_0^{(0)})^{-1}[J_+, \tilde{u}_1]g_0^{(0)}].$$

Set

$$J_+^{\tau_+} \equiv \exp(\hat{p}_+\tau_+)J_+\exp(-\hat{p}_+\tau_+),$$

$$J_-^{\tau_-} \equiv \exp(-\hat{p}_-\tau_-)J_-\exp(\hat{p}_-\tau_-).$$

This makes it possible to formulate the result in a more compact form

$$u_1 = \int\limits^0 d\tau_- \int\limits^0 d\tau_+ [g_0^{(0)}J_-^{\tau_-}(g_0^{(0)})^{-1}, J_+^{\tau_+}],$$

$$u_2 = \int\limits^0 d\tau_{-1} \int\limits^{\tau_{-1}} d\tau_{-2} \left\{ \int\limits^0 d\tau_{+1} \int\limits^0 d\tau_{+2} [J_+^{\tau_{+1}}, g_0^{(0)}J_-^{\tau_{-1}}(g_0^{(0)})^{-1}] \right.$$

$$\times [J_+^{\tau_{+2}}, g_0^{(0)}J_-^{\tau_{-2}}(g_0^{(0)})^{-1}] \quad (8.3)$$

$$+ \int\limits^0 d\tau_{+1} \int\limits^{\tau_{+1}} d\tau_{+2} [J_+^{\tau_{+1}}, [[J_+^{\tau_{+2}}, g_0^{(0)}J_-^{\tau_{-2}}(g_0^{(0)})^{-1}],$$

$$\left. g_0^{(0)}J_-^{\tau_{-1}}(g_0^{(0)})^{-1}]] \right\}.$$

Comparing (8.3) with (7.6) we see that the algebraic structures of these formulas are absolutely equivalent to each other under an obvious change of the ordering of the integration with respect to z_+, z_- to that with respect to τ_+, τ_- and the corresponding change of the limits of integration. Therefore, the value of g_0 between the states annihilated by the left action of J_+ and the right action of J_-, i.e., $J_+|\rangle = 0$, $\langle|J_- = 0$, is given by the expansion

$$\langle g_0 \rangle = \sum_{m \geq 0} (-\lambda)^m \int \ldots \int \prod_{1 \leq i \leq m} \theta(\tau_{+(i-1)} - \tau_{+i}) \theta(\tau_{-(i-1)} - \tau_{-i})$$

$$\times \langle J_+^{\tau+1} \ldots J_+^{\tau+m} g_0^{(0)}(t) J_-^{\tau-m} \ldots J_-^{\tau-1} \rangle \equiv \langle \hat{M}^+ g_0^{(0)}(t) \hat{M}^- \rangle, \tag{8.4}$$

$$\hat{M}^+ \equiv \hat{\tau}_+ \exp \lambda \int_0^0 d\tau_+ J_+^\tau, \; \hat{M}^- \equiv \hat{\tau}_- \exp \int_0^0 d\tau_- J_-^\tau.$$

The $\hat{\tau}_\pm$-ordered exponentials $\hat{M}^\pm$ which enter (8.4) do not depend on time. Let us emphasize that in deducing the latter equality we have made use of the result known from 3.2.4 for solutions of a system of type (3.2.8) in the two-dimensional case which states that $\langle g \rangle = \langle|(g_0^+)^{-1} \mathcal{M}_+^{-1} \mathcal{M}_- g_0^-|\rangle$. The structure of the series of the perturbation theory for the one-dimensional problem leads to exactly such a form for the matrix element of g_0 we are interested in.

To perform the integration in (8.4) we have to express $J_+^{\tau+}$, $J_-^{\tau-}$ taking into account (1.2.4) and $J_\pm = \sum_\alpha C_{\pm\alpha} X_\alpha^\pm$, $\hat{p}_- \equiv (X^0 p) \equiv \sum_\alpha X_\alpha^0 p_\alpha$, $\hat{p}_+ \equiv (X^0 \tilde{p})$. Set $f_{\alpha\gamma}^+ \equiv \sum_\beta C_{\beta\alpha}^{+\gamma} \tilde{p}_\beta$, $f_{\alpha\gamma}^- = -\sum_\beta C_{\beta\alpha}^{-\gamma} p_\beta$. Then $[\hat{p}_+, J_+] = \sum_\alpha C_{+\alpha}(f^+ X^+)_\alpha$ and $[\hat{p}_-, J_-] = \sum_\alpha C_{-\alpha}(f^- X^-)_\alpha$, implying $J_\pm^{\tau\pm} = \sum_\alpha C_{\pm\alpha}(\exp(f^\pm \tau_\pm) . X^\pm)_\alpha$. To calculate the integrals we have to diagonalize the matrices $f^\pm = S^\pm d^\pm (S^\pm)^{-1}$ with $d_{\alpha\beta}^\pm \equiv \delta_{\alpha\beta} d_\alpha^\pm$ which are eigenvalues of the operators

$$\tilde{X}_\alpha^\pm \equiv ((S^\pm)^{-1} X^\pm)_\alpha, \; [\hat{p}_+, \tilde{X}_\alpha^+] = d_\alpha^+ \tilde{X}_\alpha^+,$$

$$[\hat{p}_-, \tilde{X}_\alpha^-] = -d_\alpha^- \tilde{X}_\alpha^-, \; [\hat{q}, \tilde{X}_\alpha^-] = -e_\alpha^- \tilde{X}_\alpha^-,$$

where $\hat{q} \equiv (X^0 q)$, $h \equiv \exp \hat{q}$. The final result takes the form

$$g_0(t) = \sum_{m \geq 0} (-\lambda)^m \sum_{\{\alpha_\pm\}} G_{\{\alpha_\pm\}}^{(m)} \langle \tilde{X}_{\alpha+1}^+ \ldots \tilde{X}_{\alpha+m}^+ g_0^{(0)}(t) \tilde{X}_{\alpha-m}^- \ldots \tilde{X}_{\alpha-1}^- \rangle$$

$$\equiv \sum_{m \geq 0} \lambda^m \sum_{\{\alpha_\pm\}} G_{\{\alpha_\pm\}}^{(m)} \langle \tilde{X}_{\alpha+1}^+ \ldots \tilde{X}_{\alpha+m}^+ \tilde{X}_{\alpha-m}^- \ldots \tilde{X}_{\alpha-1}^- \rangle$$

$$\times \exp(X^0 q) \exp[(X^0 p)t] \exp \left[-\sum_{1 \leq s \leq m} (d_{\alpha-s}^- t + e_{\alpha-s}^-) \right], \tag{8.5}$$

$$G_{\{\alpha_\pm\}}^{(m)} \equiv \tilde{c}_{\{\alpha\}} \left[\prod_{1 \leq i \leq m} \prod_{\varepsilon=\pm} \left(\sum_{m-i+1 \leq l_\varepsilon \leq m} d_{\alpha_\varepsilon l_\varepsilon} \right) \right]^{-1},$$

$$\tilde{c}_{\{\alpha\}} \equiv \prod_{1 \leq j \leq m} \prod_{\varepsilon=\pm} \sum_{\{\beta_\pm\}} S_{\alpha_{\varepsilon j} \beta_{\varepsilon j}} c_{\varepsilon \beta_{\varepsilon j}}.$$

In the particular case of a generalized (finite, nonperiodic) Toda lattice, i.e., when

$$J_\pm = \sum_{1 \le j \le r} k_j^{1/2} X_{\pm j}, \quad f_{ij}^\pm \equiv \sum_l k_{il} \delta_{ij} p_l \equiv \delta_{ij}(kp)_j,$$

the expression (8.5) is considerably simplified. In the above form of this system $\ddot{x}_i = \lambda \exp(kx)_i$ the solutions are related with $g_0(t)$ by the formula

$$g_0(t) = \exp \left[-\sum_i h_i(x_i - \sum_j k_{ij}^{-1} \log k_j) \right],$$

implying

$$\exp(-x_i) = \prod_j k_j^{-(k^{-1})_{ij}} \langle i | g_0 | i \rangle$$

where $|i\rangle$ is the highest state of the i-th fundamental representation of $\mathfrak{g}$ satisfying (1.4.19). Thanks to (1.4.22) we express $\langle i | X_{+j_1} \ldots X_{+j_m} X_{-i_m} \cdots X_{-i_1} | i \rangle$ directly in terms of the elements of k getting an equivalent expression for solutions (4.1.48) which is a polynomial of λ.

§ 5.9 Integration of nonlinear systems associated with infinite-dimensional Lie algebras*)

The methods for integrating nonlinear systems developed in this chapter and based on the perturbation theory with a "small" parameter λ are also applicable to equations associated with infinite-dimensional Lie algebras. However, unlike the finite-dimensional case when the solutions obtained are polynomials in λ the series for infinite-dimensional Lie algebras are generally infinite ones. They converge absolutely if the algebra is of finite growth.

Therefore, the integration procedure of the nonlinear systems based on the above approach (see 3.2.4 and 4.1.2) should also be modified: we consider solutions of (3.2.16) as a series of consecutive approximations introducing a small parameter λ into $L_\pm$ or, equivalently, into $\varphi_{+\alpha}\varphi_{-\alpha}$. Then the main bulk of the integration procedure for the systems of type (3.2.11) associated with infinite-dimensional Lie algebras is connected with the investigations of the conditions for convergence of the series of the perturbation theory.

In this section we will illustrate this problem with the periodic Toda lattice, i.e., with the equations (3.1.10) where k is Cartan matrix of an affine Lie algebra.

---

*) [72, 73]

In accordance with general scheme substitute the expressions for $\mathcal{M}_\pm$ into K represented in the form (3.2.17) up to order m and define $\mathcal{N}_\pm$ and $\exp H$ with the same accuracy from the relation

$$K \approx \mathcal{N}_- \cdot \exp \hat{H} \cdot \mathcal{N}_+^{-1}. \qquad (9.1)$$

Hereafter $\approx$ denotes congruence up to order m with respect to powers of $\varphi_{+j_1} \cdots \varphi_{+j_m}$ and $\varphi_{-j_1} \cdots \varphi_{-j_m}$.

From here, repeating the arguments from 4.1.2 literally, we deduce that $\hat{H} \equiv \sum\limits_{1 \le j \le r} H_j h_j$ satisfies the system

$$\hat{H}_{,z_+z_-} \approx [L_-, \exp(-\hat{H})L_+ \exp \hat{H}] \quad \text{or} \quad H_{j,z_+z_-} \approx -\varphi_{+j}\varphi_{-j} \exp[-(k\hat{H})_j].$$
$$(9.2)$$

In order to perform the passage from (9.2) to (3.1.10) we have to take into account that φ_{+j}, φ_{-j} are related by

$$\sum_j \lambda_j^{(s)} \ln \varphi_{+j} = 0, \quad \sum_j \lambda_j^{(s)} \ln \varphi_{-j} = 0,$$

where $\lambda_j^{(s)}$ is a set of (left) eigenvectors of k with zero eigenvalues, i.e., $(\lambda^{(s)}k)_j = 0$. (Recall that for invertible Cartan matrices, i.e., for finite-dimensional simple Lie algebras, all the functions $\varphi_{\pm j}$ are functionally independent.)

As in the finite-dimensional case, to get the closed explicit expressions for the solutions $\exp(-x_j)$ of (3.1.10) we may make use of the "generators of highest vectors" $R_j^\pm$ satisfying (1.6.8) and (1.6.9). With their help we derive from (9.1), as we derived (4.1.12) and (4.1.38), that

$$\exp H_i = (\mathcal{M}_+ R_i^- \mathcal{M}_+^{-1}, \quad \mathcal{M}_- R_i^+ \mathcal{M}_-^{-1})$$
$$= \sum_{m=0}^\infty (-1)^m (p_m \ldots p_1)_+ (q_1 \ldots q_m)_- \cdot \langle i | p_m \ldots p_1, q_1 \ldots q_m | i \rangle, \qquad (9.3)$$

where $(p_1 \ldots p_m)_\pm$ is determined by (4.1.31) and the scalar product of

$$q_1 \ldots q_m | i \rangle \equiv \delta_{q_m i}[X_{-q_1 \ldots q_m}, R_i^+] \text{ and}$$

$$\langle i | p_m \ldots p_1 \equiv \delta_{p_m i}[X_{+p_1 \ldots p_m}, R_i^-]$$

is calculated with the help of the Jacobi identity and the relations (1.2.8), (1.6.8) and (1.6.9) (cf. (1.4.22)).

For solutions of (3.1.10) in the considered case, with the help of (9.1)–(9.3), we get the following formal expression and then investigate the domain of its convergence

$$\exp(-x_i) = \exp\left(-\sum_{1\leq j\leq m} \mu_i^{(j)} \ln(f_{+j}f_{-j})\right) \exp\left(-\sum_{m+1\leq s\leq r} \lambda_i^{(s)} \ln f_{+s}f_{-s}\right)$$

$$\times (\mathcal{M}_+ R_i^- \mathcal{M}_+^{-1}, \mathcal{M}_- R_i^+ \mathcal{M}_-^{-1}),$$

(9.4)

where $(k\mu^{(j)})_i = 0$, $1 \leq j \leq m$; $(k\lambda^{(s)})_i = E_s\lambda_i^{(s)}$, $m+1 \leq s \leq r$, and $E_s \neq 0$.

The functions $\varphi_{\pm i}$ entering (3.2.16) are related with $f_{\pm j}$ via

$$\varphi_{+i}\varphi_{-i} = \prod_{m+1\leq s\leq r} (f_{+s}f_{-s})^{\lambda_i^{(s)}}.$$

(9.5)

It is possible to derive from (9.4) the formal solution of the Goursát problem for the system (3.1.10). Indeed, let $\mathcal{M}_+ = 1$ for $z_+ = a_+$ and $\mathcal{M}_- = 1$ for $z_- = a_-$ be the initial conditions for the solution (3.2.16). Then thanks to (9.3) and (9.4) we have

$$\exp(-x_i)|_{z_-=a_-} = \text{const} \cdot \exp\left[-\sum_{1\leq j\leq m} \mu_i^{(j)} \ln f_{+j}(z_+)\right.$$

$$\left. -\sum_{m+1\leq s\leq r} \lambda_i^{(s)} \ln f_{+s}(z_+)\right],$$

i.e. the functions f_{+j}, $1 \leq j \leq m$, are determined by the value of $\exp(-x_i)$ for a fixed $z_- = a_-$. Similarly the functions f_{-j}, $1 \leq j \leq m$, are recovered from the values of $\exp(-x_i)$ at $z_+ = a_+$.

For the case when k is Cartan's matrix of an affine Kac-Moody algebra we can estimate the terms of the m-th approximation in (9.3) and show that for arbitrary functions $\varphi_{\pm i}$ bounded on the segment $a_+ \leq z_+ \leq b_+$, $a_- \leq z_- \leq b_-$ the series (9.3) is absolutely convergent.

To prove this recall that the Cartan matrix of an affine Kac-Moody algebra is positively semi-definite and has exactly one zero eigenvalue.

The scalar products of the basis vectors of the fundamental representations containing the same number of raising (lowering) operators

$$R^+_{j_1 j_1 j_2 \ldots j_m} \equiv [X_{-j_m \ldots j_1}, R^+_{j_1}], \quad R^-_{j_1 j_1 j_2 \ldots j_m} \equiv [X_{+j_m \ldots j_1}, R^-_{j_1}]$$

preserve the sign: $(-1)^m (R^-_{j_1 j_1 j_2 \ldots j_m}, R^+_{j_1 j_1 j_2 \ldots j_m}) > 0$ is positive for even m and negative otherwise. Denote by μ_j^0 the unique eigenvector of k with eigenvalue zero, i.e., $(k\mu^0)_j = 0$. For all affine Kac-Moody algebras $\mu_j^0 > 0$ and we will norm this eigenvector setting $\min \mu_j^0 = 1$.

Obviously, (9.2) has a solution of the form $H_j = \mu_j^0 H_0$ provided $-\mu_j^0 (H_0)_{,z+z-} = \varphi_{+j}\varphi_{-j}$ for all j, i.e., $\varphi_{+j}\varphi_{-j} \equiv \mu_j^0 \varphi_+ \varphi_-$ or $\varphi_{+j} = \mu_j^0 \varphi_+$, $\varphi_{-j} = \varphi_-$. Then $H_0 = -\int \varphi_+ dz_+ \int \varphi_- dz_-$ and $L_+ = \sum_j \mu_j^0 X_{+j} \varphi_+$, $L_- = \sum_j X_{-j}\varphi_-$. Then for $\mathcal{M}_\pm$ we get finite expressions

$$\mathcal{M}_+ = \exp \sum_j \mu_j^0 X_{+j} \int \varphi_+ dz_+ \equiv \exp \hat{\mathcal{X}}_+(\varphi_+),$$

$$\mathcal{M}_- = \exp \sum_j X_{-j} \int \varphi_- dz_- \equiv \exp \hat{\mathcal{X}}_-(\varphi_-). \tag{9.6}$$

From the definition of $\hat{\mathcal{X}}_\pm$ we get

$$[\hat{\mathcal{X}}_+, \hat{\mathcal{X}}_-] = \sum_j \mu_j^0 h_j \equiv \hat{h}, \quad [\hat{h}, X_{\pm i}] = \sum_j \mu_j^0 k_{ij} X_{\pm i}.$$

Substituting (9.6) into (9.3) and making use of a property of the multiple integral

$$\int \varphi dz_1 \int^{z_1} \varphi dz_2 \ldots \int^{z_{m-1}} \varphi dz_m = \left(\int \varphi dz \right)^m / m!$$

we get

$$\exp H_i = (\mathcal{M}_+ R_i^- \mathcal{M}_+^{-1}, \mathcal{M}_- R_i^+ \mathcal{M}_-^{-1}) = \exp \left(-\mu_i^0 \int \varphi_+ dz_+ \int \varphi_- dz_- \right)$$

$$= \sum_m (-1)^m \left[\left(\int \varphi_+ dz_+ \right)^m / m! \right] \left[\left(\int \varphi_- dz_- \right)^m / m! \right]$$

$$\sum \mu_i^0 \mu_{i_2}^0 \ldots \mu_{i_s}^0 (R_{ii_2 \ldots i_s}^-, R_{ij_2 \ldots j_s}^+),$$

i.e., $\sum \mu_i^0 \mu_{i_2}^0 \ldots \mu_{i_s}^0 (R_{ii_2 \ldots i_s}^-, R_{ij_2 \ldots j_s}^+) \equiv (\mu_i^0)^m m!$.

To estimate the terms in the m-th approximation (cf. (9.3)) consider the functions φ_{+i} and φ_{-i} bounded on $[a_+, z_+]$ and $[a_-, z_-]$, respectively, so that $|\varphi_{-i}| \leq A_-$, $|\varphi_{+i}| \leq A_+ \leq \mu_i^0 A_+$ (where the last inequality is a corollary of the accepted norming for the vector μ: $\mu_i^0 \geq 1$). For the m-th repeated integrals from (9.3) we obviously have

$$|(ip_{m-1} \ldots p_1)_+| \leq [A_+^m (z_+ - a_+)^m / m!] \mu_i^0 \mu_{p_1}^0 \ldots \mu_{p_{m-1}}^0,$$

$$|(q_1 \ldots q_{m-1} i)_-| \leq A_-^m (z_- - a_-)^m / m!. \tag{9.7}$$

Substituting (9.7) into (9.3) and making use of the above summation rule and the above-mentioned positive definiteness of scalar products of basis vectors we get the following inequality for the m-th term $\mathcal{S}_m$ of the series:

$$|\mathcal{S}_m| \leq [A_+ A_-(z_+ - a_+)(z_- - a_-)]^m / m!.$$

Therefore, the series which determines the solutions of (3.1.10) when k is equivalent to Cartan matrix of an affine Kac-Moody algebra is absolutely convergent and is a solution of the Goursát problem for this system which depends on the right number of arbitrary functions.

Notice that *the formal series* (9.3) *gives a formal solution of* (3.1.10) *for arbitrary k.*

Among the equations (3.1.10) corresponding to Kac-Moody algebras the best known ones are equations (3.1.12). The solutions of these equations depending on two arbitrary functions are direct corollaries of the general formulas of this chapter:

$$\exp x = \varphi_{+1}^{1/2}(z_+)\varphi_{-1}^{1/2}(z_-)X_2^{(1)}(X_1^{(1)})^{-1} \qquad (\varepsilon = 0); \qquad (9.8)$$

$$\exp x = \varphi_{+1}(z_+)\varphi_{-1}(z_-)X_2^{(2)}(X_1^{(2)})^{-2} \qquad (\varepsilon = 1); \qquad (9.9)$$

$$X_j^{(1)} \equiv \bar{X}_j\big|_{\varphi_{\pm 2}=\varphi_{\pm 1}^{-1}}, \ X_j^{(2)} \equiv \bar{X}_j\big|_{\varphi_{\pm 2}=\varphi_{\pm 1}^{-2}}; \qquad j = 1,2; \qquad (9.10)$$

$$\bar{X}_j \equiv (\mathcal{M}_+ R_j^- \mathcal{M}_+^{-1}, \ \mathcal{M}_- R_j^+ \mathcal{M}_-^{-1}),$$

$$\mathcal{M}_\pm(z_\pm) = \tilde{Z}_\pm \exp \int^{z_\pm} dz'_\pm[\varphi_{\pm 1}(z'_\pm)X_{\pm 1} + \varphi_{\pm 2}(z'_\pm)X_{\pm 2}].$$

The infinite series which arise when we decompose the expressions for $\bar{X}_j$ in this case are absolutely convergent, as we have just shown for the general case of affine Lie algebras. Let us illustrate the situation for the first eight terms of the expansion of $\bar{X}_1$ with respect to the powers of $\varphi_{+i}\varphi_{-i}$ (clearly, $\bar{X}_2$ is obtained from $\bar{X}_1$ after the change $\varphi_{\pm 1} \rightleftharpoons \varphi_{\pm 2}$):

$$\bar{X}_1 = \sum \left[(-1)^j \sum_i c_i |X_i^j|^2\right]$$

$$= 1 - |X_1^1|^2 + 2|X_1^2|^2 - 4|X_1^3|^2 - 2|X_2^3|^2 + 4|X_1^4|^2 + 8|X_2^4|^2$$

$$- 8|X_1^5|^2 - 8|X_2^5|^2 - 16|X_3^5|^2 + 8|X_1^6|^2 + 16|X_2^6|^2$$

$$+ 16|X_3^6|^2 + 32|X_4^6|^2 - 16|X_1^7|^2 - 16|X_2^7|^2 - 16|X_3^7|^2$$

$$- 32|X_4^7|^2 - 64|X_5^7|^2 + 32|X_1^8|^2 + 32|X_2^8|^2 + 64|X_3^8|^2$$

$$+ 64|X_4^8|^2 + 64|X_5^8|^2 + 128|X_6^8|^2 + \dots .$$

Here $|X_i^j|^2 \equiv X_{+i}^j X_{-i}^j$; the superscript is the number of the approximation and the subscript the number of the term in the approximation:

$$X_{\pm 1}^1 = (1)_\pm, \ X_{\pm 1}^2 = (12)_\pm, \ X_{\pm 1}^3 = (122)_\pm, \ X_{\pm 2}^3 = (121)_\pm,$$

$$X_{\pm 1}^4 = (1212)_\pm + (1221)_\pm, \ X_{\pm 2}^4 = (1221)_\pm, \ X_{\pm 1}^5 = (12122)_\pm + (12212)_\pm,$$

$$X_{\pm 2}^5 = (12121)_\pm + 2(12211)_\pm, \ X_{\pm 3}^5 = (12211)_\pm,$$

$$X_{\pm 1}^6 = (122121)_\pm + (121221)_\pm, \ X_{\pm 2}^6 = (121211)_\pm + 3(122111)_\pm,$$

$$X_{\pm 3}^6 = 2(122112)_\pm + (121212)_\pm + (122121)_\pm + (121221)_\pm,$$

$$X_{\pm 4}^6 = (122112)_\pm.$$

Concluding the section, let us indicate one more approach to the solution of (3.1.10), which will be written in the form

$$x_{i,z+z-} = \varphi_{+i}\varphi_{-i}\exp(kx)_i, \ \hat{x}_{,z+z-} = [\exp \hat{x} L_+ \exp(-\hat{x}), L_-]. \tag{9.11}$$

where $\hat{x} \equiv \sum_j x_j h_j$. We will consider the functions $\varphi_{+i}\varphi_{-i}$ in (9.11) as small and perform the iteration procedure with respect to them. Expanding the solutions x_i of (9.11) with respect to these small terms we get

$$x_i = \sum_{m=1}^{\infty} x_i^{(m)}, \ \hat{x} = \sum_{m=1}^{\infty} \hat{x}_{(m)},$$

and for the terms of order m we get from (9.11)

$$\hat{x}_{(m),z+z-} = \sum_{l_1,\ldots,l_m} \left(\prod_{1\leq j\leq m} l_j! \right)^{-1} [[\hat{x}_{(m-1)}^{l_{m-1}}\ldots\hat{x}_{(1)}^{l_1}, L_+], L_-], \tag{9.12}$$

where $\sum_{1\leq j\leq m-1} jl_j = m-1$ and $[\hat{x}_{(m-1)}^{l_{m-1}}\ldots\hat{x}_{(1)}^{l_1}, L_+]$ is the sequence of multiple brackets of orders $l_1,\ldots,l_{m-1}$, where $\hat{x}_1,\ldots,\hat{x}_{(m-1)}$ commute. Then with the standard technique after rather cumbersome calculations we get the following formal series of the perturbation theory

$$x_i = \sum_m (-1)^m (ii_2\ldots i_m) + (ij_2\ldots j_m) - (R_i^- X_{+i_2}\ldots X_{+i_m}, X_{-j_m}\ldots X_{-j_2} R_i^+). \tag{9.13}$$

Passing from x_i to $\exp(-x_i)$ we get the expression for the solutions of (3.1.10) obtained earlier.

Notice that from the physical point of view, the last passage means that we have taken into account unconnected Feynman diagrams while calculating the vacuum expectation value of the S-matrix.

Chapter 6
Scalar Lax-pairs and soliton solutions of the generalized periodic Toda lattice

In this chapter we suggest a general scheme for constructing solitons of dynamical systems without referring to the matrix realization of the Lax-type representation. In the ISP method, which helps one find the soliton-type solutions are found, a lucky choice of an LA-pair facilitates the calculations considerably and, moreover, in principle enables one to perform them. The approach suggested in what follows is invariant with respect to a concrete representation of the algebra of internal symmetries and appeals directly to the properties of the algebra.

The LA-pair is replaced by a system of linear equations of higher dimensions in only one scalar function and the their compatibility condition is the equation of the initial dynamical system. To every irreducible representation $\{l\}$ of the algebra such a linear system can be assigned, and the number of equations that enter it is $N_l + 1$, where $N_l = \dim\{l\}$ and the maximal degree of the derivatives is N_l. The scalar function ψ_l thus associated with the representation $\{l\}$ will be called its *wave function* and the system of equations it satisfies its *scalar LA-pair*.

The general construction will be illustrated here with the example of the generalized periodic Toda lattice.

§ 6.1 A group-theoretical meaning of the spectral parameter and the equations for the scalar LA-pair*)

Recall that the equations describing the generalized periodic Toda lattice,

$$\rho_{I,z+z_-} = \sum_J \tilde{k}_{IJ} \exp \rho_J, \quad 1 \leq I, J \leq r + 1, \tag{1.1}$$

---

*) [124, 126]

arise from the general scheme of Chapter 3 applied to Kac-Moody affine algebras $\tilde{\mathfrak{g}}$ of rank $\tilde{r} \equiv r + 1$. Here $\tilde{k}$ is Cartan matrix of $\tilde{\mathfrak{g}}$; let t be its unique eigenvector with eigenvalue 0, i.e., $\sum_I t_I \tilde{k}_{IJ} = 0$. Therefore, thanks to (1.1), the unknown functions $\rho_J(z_+, z_-)$ satisfy $(\sum_j t_J \rho_J)_{,z_+ z_-} = 0$. In what follows we will assume that $\sum_J t_J \rho_J \equiv 0$, i.e., $\rho_{r+1} = - \sum_{1 \leq j \leq r} t_j \rho_j t_{r+1}^{-1} \equiv - \sum_j m_j \rho_j \equiv -(m \cdot \rho)$. Then the first r independent equations of (1.1) can be rewritten in the form

$$\rho_{i,z_+ z_-} = \sum_j k_{ij} \exp \rho_j + \tilde{k}_{ir+1} \exp(-m\rho), \quad 1 \leq i,\, j \leq r,$$

where $k_{ij} \equiv \tilde{k}_{ij}$ for $1 \leq i,\, j \leq r$ is the (invertible) Cartan matrix of the corresponding finite-dimensional simple Lie algebra $\mathfrak{g}$. Introducing the independent functions $x_i \equiv \sum_j k_{ij}^{-1} \rho_j \equiv (k^{-1} \rho)_i$ and taking the general properties of the construction on affine Kac-Moody algebras into account (see 1.2) we get

$$x_{i,z_+ z_-} = \exp(kx)_i - m_i v_i^{-1} \exp(-mkx), \tag{1.2_1}$$

where m_i are the coefficients in the decomposition of the maximal root of $\mathfrak{g}$ with respect to the simple roots. Hereafter when we speak about the equations of the periodic Toda lattice we mean the system (1.2_1). Its vector expression is

$$x_{,z_+ z_-} = \left[\sum_i X_{+i} + X_{-m}, \exp(-x) \left(\sum_j X_{-j} + X_{+m} \right) \exp x \right], \tag{1.2_2}$$

where $X_{\pm j}$ are root vectors corresponding to the simple roots of $\mathfrak{g}$, X_{+m} (X_{-m}) is the root vector corresponding to the maximal (minimal) root of $\mathfrak{g}$ and $x \equiv \sum_j h_j x_j$ (cf. 5.9). The equations (1.2) are a direct corollary of the Lax-type representation (see Chapter 3) in the form

$$\left[\partial/\partial z_+ + x_{,z_+} + \lambda \left(\sum_i X_{+i} + X_{-m} \right), \partial/\partial z_- + \lambda^{-1} \left(\sum_j X_{-j} \exp(kx)_j \right. \right.$$

$$\left. \left. + X_{+m} \exp(-mkx) \right) \right] = 0$$

$$\tag{1.3}$$

with spectral parameter λ. As applied to (1.3) the term "spectral" does not hold so far any meaning and is introduced by analogy with the terms of the ISP method. It should be considered as an arbitrary complex nonzero number; for any such λ the representation (1.3) implies (1.2) which do not depend on λ.

Here it is appropriate to investigate the problem of the relation of different forms of Lax-type representations (3.1.1), namely, between the realization of the operators $A_\pm$ of the form (3.1.7) taking values in $\tilde{\mathfrak{g}}$ and leading to (1.1) and the representation associated with the corresponding finite-dimensional simple Lie algebra $\mathfrak{g}$ and its finite-dimensional representations leading to (1.2), which differ from (1.1) by a conformal transformation.

To this end recall that any infinite-dimensional simple $\mathbb{Z}$-graded Lie algebra we consider is generated by its local part $\hat{\tilde{\mathfrak{g}}} \equiv \tilde{\mathfrak{g}}_{-1} \oplus \tilde{\mathfrak{g}}_0 \oplus \tilde{\mathfrak{g}}_{+1}$. The latter is generated by its $3\tilde{r}$ elements $X_{\pm J}$, h_J satisfying the commutation relations (1.3.1). If $\tilde{\mathfrak{g}} = \mathfrak{g}^{(1)}$, we can take for the simple roots of $\tilde{\mathfrak{g}}$ a set of simple roots of $\mathfrak{g}$ supplemented by the minimal root, i.e.,

$$\tilde{\mathfrak{g}}_{-1} = \{X_{-j}, \ 1 \le j \le r; \ X_{+m}\},$$

$$\tilde{\mathfrak{g}}_{+1} = \{X_{+j}, \ 1 \le j \le r; \ X_{-m}\},$$

$$\tilde{\mathfrak{g}}_0 = \{h_j, \ 1 \le j \le r; \}$$

$$\mathcal{H} \equiv [X_{+m}, X_{-m}].$$

Clearly, the commutation relations for the elements from $\hat{\tilde{\mathfrak{g}}}$ are automatically satisfied. Besides, in every subspace $\tilde{\mathfrak{g}}_a$ of $\tilde{\mathfrak{g}} = \bigoplus_{a \in \mathbb{Z}} \tilde{\mathfrak{g}}_a$ formed by $|a|$-multiple brackets of root vectors corresponding to simple roots there cannot be more elements than $\dim \mathfrak{g}$, see Chapter 1. (This was the way the infinite-dimensional Lie algebras of finite growth were introduced initially.)

A representation of $\tilde{\mathfrak{g}}$ constructed thus is very degenerate since the same element of $\mathfrak{g}$ labels an infinite number of elements from $\tilde{\mathfrak{g}}$. This degeneracy can be partly eliminated by introducing a complex (nonzero) parameter λ and taking the local part of $\tilde{\mathfrak{g}}$ in the form

$$\tilde{\mathfrak{g}}_{-1} = \lambda^{-1}\{X_{-j}; X_{+m}\}, \ \tilde{\mathfrak{g}}_0 = \{h_j, \mathcal{H}\}, \ \tilde{\mathfrak{g}}_{+1} = \lambda\{X_{+j}, X_{-m}\}.$$

Now every element from $\tilde{\mathfrak{g}}$ contains as a factor an appropriate power of λ. However, it is still impossible to distinguish in such a representation the element $\mathcal{H} \in \tilde{\mathfrak{g}}_0$ from the linear combination of Cartan generators of simple roots.

The Lax-type representation in the form (1.3) coincides with its realization by the operators (3.1.7), where in (1.3) we have selected a finite-dimensional degenerate representation of $\tilde{\mathfrak{g}}$ in which $\mathcal{H} \in \tilde{\mathfrak{g}}_0$ coincides with $\sum_j h_j m_j v_j^{-1}$.

This latter circumstance lowers the rank of (1.2_1) by one as compared with (1.1).

Therefore from the algebraic point of view it is advisable to define λ as a grading parameter of $\tilde{\mathfrak{g}}$. However, we will retain the traditional term "spectral parameter" whose meaning will be clarified in a short while.

The representation (1.3) means, as previously, that the Lax operators are gradient, i.e.,

$$\dot{g}g^{-1} = \dot{x} + \lambda\left(\sum_j X_{+j} + X_{-m}\right), \quad g'g^{-1} = \lambda^{-1}\left(\sum_j X_{-j}\exp(kx)_j\right.$$
$$\left. + X_{+m}\exp(-mkx)\right), \tag{1.4_1}$$

where $g \in \exp\mathfrak{g}$. Hereafter we denote for brevity the derivations with respect to z_+ and z_- by "." and "'" respectively. Having performed the transformation $g \to \exp(h\cdot c)g$ with constant parameters c and excluding therefore the dependence of the root vectors corresponding to simple roots on λ, let us rewrite (1.4_1) in the form

$$\dot{g} = \left(\dot{x} + \sum_j X_{+j} + \lambda^{M+1}X_{-m}\right)g,$$

$$g' = \left[\sum_j X_{-j}\exp(kx)_j + \lambda^{-(M+1)}X_{+m}\exp(-mkx)\right]g, \tag{1.4_2}$$

where M is the height of the maximal root of $\mathfrak{g}$. The expression (1.4) is the equation in g such that the corresponding current-like operators $\dot{g}g^{-1}$, $g'g^{-1}$ have the required spectra. They can be rewritten as a system of equations in parameters defining a given element g. Thus, if g is parametrized by the Gauss decomposition, $g = N_-\exp(h\tau)N_+$ (see 1.5), then substituting it into (1.4) we get a complicated nonlinear differential dependence between the parameters of $N_\pm$ (associated with the root vectors from $\mathfrak{g}_{\pm a}$, $a \geq 1$) and τ_j.

It is remarkable that the equations defining $\exp\tau_j$ are separated from the whole system and still remain essentially nonlinear. At the same time the functions $\exp\sum_j l_j\tau_j$ with arbitrary integers l_j (> 0) satisfy higher-order linear equations (which are extracted from the system in $\exp\tau_i$) whose coefficient functions depend on ρ_j and their derivatives. Their compatibility condition is just the initial system of equations for the periodic generalized Toda lattice. In what follows we will call $\exp(l\tau) \equiv \psi_l$ the *wave function* and the system of equations for them the *scalar LA-pair* of the representation $\{l\}$ of $\mathfrak{g}$.

Let us emphasize once more that the equations of the scalar pair itself is invariant with respect to the choice of a representation of the elements $X_{\pm j}$, h_j, $1 \leq j \leq r$, $X_{\pm m}$, in (1.4). The term "wave function of the representation $\{l\}$" means only that the parameter τ_j, or more precisely its exponential, is most simply determined as the matrix element of g between the highest vectors of the representation $\{l\}$, i.e.,

$$\Psi_l \equiv \exp \sum_j l_j \tau_j = \langle l|g|l\rangle.$$

For example, consider the classical Lie algebras A_2 and $C_2 \cong B_2$. (The simplest case of A_1 leading to the sine-Gordon equation will be considered from various points of view in the following section.)

The fundamental representations of A_2 are 3-dimensional. Their basis is constituted by

$$|1\rangle, X_{-1}|1\rangle, X_{-2}X_{-1}|1\rangle, \ (X_{+j}|1\rangle = 0, \ h_j|1\rangle = \delta_{j1}|1\rangle)$$

for the first fundamental representation and by

$$|2\rangle, X_{-2}|2\rangle; \ X_{-1}X_{-2}|2\rangle, \ (X_{+j}|2\rangle = 0, \ h_j|2\rangle = \delta_{j2}|2\rangle)$$

for the second one. Setting $\exp \tau_1 = \Psi_1 = \langle 1|g|1\rangle$ and making use of (1.4_2) we consecutively get

$$\dot\Psi_1 = \langle 1|(h_1\dot x_1 + h_2\dot x_2 + X_{+1} + X_{+2} + \lambda^3 X_{-12})g|1\rangle = \dot x_1\Psi_1 + \langle 1|X_{+1}g|1\rangle$$

or

$$\exp x_1(\Psi_1\exp(-x_1))^{\cdot} = \langle 1|X_{+1}g|1\rangle.$$

Similarly,

$$\exp x_1 \cdot (\Psi_1\exp(-x_1))^{\cdot} = \langle 1|X_{+1}g|1\rangle, \exp(-\rho_1 - \rho_2)\Psi_1' = \lambda^{-3}\langle 1|X_{+1}X_{+2}g|1\rangle;$$

$$\exp(x_2 - x_1)(\exp\rho_1(\Psi_1\exp(-x_1))^{\cdot})^{\cdot} = \langle 1|X_{+1}X_{+2}g|1\rangle,$$

$$\exp\rho_2(\Psi_1'\exp(-\rho_1 - \rho_2))' = \lambda^{-3}\langle 1|X_{+1}g|1\rangle;$$

$$\exp(-x_2)(\exp\rho_2(\exp\rho_1(\exp(-x_1)\cdot\Psi_1)^{\cdot})^{\cdot})^{\cdot} = \lambda^3\Psi_1,$$

$$(\exp\rho_2(\Psi_1'\exp(-\rho_1 - \rho_2))')' = \lambda^{-3}\Psi_1\exp(-\rho_1).$$

Excluding in (1.5) the matrix elements between the states of the basis we get the system

a) $(\exp\rho_2(\exp\rho_1(\Psi_1\exp(-x_1))^{\cdot})^{\cdot})^{\cdot} = \lambda^3\Psi_1\exp x_2,$

b) $(\exp\rho_1(\Psi_1\exp(-x_1))^{\cdot})^{\cdot} = \lambda^3\Psi_1'\exp(-2x_2),$

c) $(\Psi_1\exp(-x_1))^{\cdot} = \lambda^3\exp(2x_2 - 2x_1)(\Psi_1'\exp(-x_1 - x_2))',$

d) $\Psi_1 = \lambda^3\exp\rho_1(\exp\rho_2(\exp\rho_2(\Psi_1'\exp(-x_1 - x_2))')')',$

$$(1.6)$$

where $(\rho_1 = 2x_1 - x_2,\ \rho_2 = -x_1 + 2x_2)$, i.e., the scalar LA-pair for the first fundamental representation of A_2. As a corollary of (1.6) we get

$$\dot{\Psi}_1' - \dot{x}_1 \Psi_1' = \Psi_1 \exp(x_1 + x_2),$$

$$\dot{\Psi}_1' - \dot{x}_1 \Psi_1' - \dot{x}_1' \Psi_1 = \Psi_1 \exp(-\rho_1),$$

$$\Psi_1' = \exp 2x_2 (\exp x_2 (\Psi_1' \exp(-x_1 - x_2))^{\cdot})^{\cdot},$$

$$(\Psi_1 \exp(-x_1))^{\cdot} = \exp(2x_2 - x_1)\,(\exp(x_2 - x_1)\,(\exp \rho_1 (\Psi_1 \exp(-x_1))^{\cdot})^{\cdot})^{\cdot},$$

or

$$\ddot{\tilde{x}}_1' = \exp \tilde{\rho}_1 - \exp(-\tilde{x}_1 - \tilde{x}_2),\quad \ddot{\tilde{x}}_2' = \exp \tilde{\rho}_2 - \exp(-\tilde{x}_1 - \tilde{x}_2),$$

$$\tilde{x}_j \equiv -x_j,$$

i.e., the system of equations of the periodic Toda lattice for A_2. The equations of the scalar pair (1.6) elucidate the meaning of the term "spectral parameter for λ in the ISP method": it arises as the parameter of the corresponding Sturm-Liouville problem.

The second fundamental representation of A_2 is dual to the first one and the only difference in the scalar pair in this case manifests itself in the sign of λ^3 in (1.6).

Now consider C_2. The first fundamental representation is 4-dimensional and its basis vectors are

$$|1\rangle,\ X_{-1}|1\rangle,\ X_{-2}X_{-1}|1\rangle,\ X_{-1}X_{-2}X_{-1}|1\rangle,$$

while the second one is 5-dimensional and its basis vectors are

$$|2\rangle,\ X_{-2}|2\rangle,\ X_{-1}X_{-2}|2\rangle,\ X_{-1}X_{-1}X_{-2}|2\rangle,\ X_{-2}X_{-1}X_{-1}X_{-2}|2\rangle.$$

As usual, $X_{+j}|\rangle = 0$, $h_j|i\rangle = \delta_{ij}|i\rangle$. The system of equations connecting the multiple derivatives with the matrix elements of g for the first fundamental representations

$$\exp x_1 (\Psi_1 \exp(-x_1))^{\cdot} = \langle 1|X_{+1}g|1\rangle,\ \Psi_1' = -\lambda^{-4}\exp 2x_1 \langle 1|X_{+1}X_{+2}X_{+1}g|1\rangle,$$

$$\exp(-x_1 + x_2)(\exp \rho_1 (\Psi_1 \exp(-x_1))^{\cdot})^{\cdot} = \langle 1|X_{+1}X_{+2}g|1\rangle,$$

$$(\Psi_1' \exp(-2x_1))' = -\lambda^{-4}\exp(-\rho_1)\langle 1|X_{+1}X_{+2}g|1\rangle;$$

$$\exp(x_1 - x_2)(\exp \rho_2 (\exp \rho_1 (\Psi_1 \exp(-x_1))^{\cdot})^{\cdot})^{\cdot} = \langle 1|X_{+1}X_{+2}X_{+1}g|1\rangle,$$

$$(\exp \rho_1 (\Psi_1' \exp(-2x_1))')' = -\lambda^{-4}\exp(-\rho_2)\langle 1|X_{+1}g|1\rangle;$$

$$(\exp \rho_1 (\exp \rho_2 (\exp \rho_1 (\exp(-x_1)\cdot \Psi_1)^{\cdot})^{\cdot})^{\cdot})^{\cdot} = -\lambda^4 \Psi_1 \exp x_1,$$

$$(\exp \rho_2 (\exp \rho_1 (\Psi_1' \exp(-2x_1))')')' = -\lambda^{-4}\Psi_1 \exp(-\rho_1)$$

$$(1.7_1)$$

leads to five equations for the scalar LA-pair for this representation. Two of them coincide with the two last equations of (1.7_1), for the other three we get

$$(\exp \rho_1(\Psi_1' \exp(-2x_1))')' = -\lambda^4 \exp(3x_1 - 2x_2)(\Psi_1 \exp(-x_1))',$$

$$(\exp \rho_2(\exp \rho_1(\Psi_1 \exp(-x_1))')')' = -\lambda^4 \exp(x_2 - 3x_1)\Psi_1, \qquad (1.7_2)$$

$$(\Psi_1' \exp(-2x_1))' = -\lambda^{-4} \exp(-3x_1 + 2x_2)(\exp \rho_1(\exp(-x_1)\Psi_1)')'.$$

For the second fundamental representation the relations similar to (1.7_1) are

$$\exp x_2(\Psi_2 \exp(-x_2))' = \langle 2|X_{+2}g|2\rangle,$$

$$\Psi_2' = (1/2)\lambda^{-4}\langle 2|X_{+2}X_{+1}X_{+1}g|2\rangle \exp 2x_1,$$

$$\exp \rho_1(\exp \rho_2(\Psi_2 \exp(-x_2))')' = \langle 2|X_{+2}X_{+1}g|2\rangle,$$

$$(\Psi_2' \exp(-2x_1))' = \lambda^{-4}\langle 2|X_{+2}X_{+1}g|2\rangle;$$

$$(\exp \rho_1(\exp \rho_2(\Psi_2 \exp(-x_2))')')' = \langle 2|X_{+2}X_{+1}X_{+1}g|2\rangle,$$

$$(\exp \rho_1(\Psi_2' \exp(-2x_1))')' = 2\lambda^{-4} \exp(-\rho_1)\langle 2|X_{+2}g|2\rangle;$$

$$\exp(-\rho_1)(\exp \rho_1(\exp \rho_1(\exp \rho_2(\Psi_2 \exp(-x_2))')')')'$$
$$= 2\lambda^4\Psi_2 + \langle 2|X_{+2}X_{+1}X_{+1}X_{+2}g|2\rangle;$$

$$(\exp \rho_1(\exp \rho_1(\Psi_2' \exp(-2x_1))')')' = 2\lambda^{-4}\Psi_2 \exp(-\rho_2)$$
$$+\lambda^{-8} \exp 2x_1\langle 2|X_{+2}X_{+1}X_{+1}X_{+2}g|2\rangle,$$
$$(1.8_1)$$

and the system of equations for the scalar LA-pair they imply are

a) $\quad (\exp \rho_2(\exp \rho_1(\exp \rho_1(\exp \rho_2(\Psi_2 \exp(-x_2))')')')')' = 4\lambda^4 \exp x_2 \cdot \dot{\Psi}_2,$

b) $\quad (\exp(-2x_1))(\exp \rho_1(\exp \rho_1(\Psi_2' \exp(-2x_1))')')')'$
$$= 4\lambda^{-4} \exp(-x_2)(\Psi_2 \exp(-x_2))',$$

c) $\quad \lambda^{-4} \exp x_2(\exp \rho_1(\exp \rho_1(\exp \rho_2(\Psi_2 \exp(-x_2))')')')'$
$$+2[\Psi_2 \exp(-\rho_2) - \Psi_2 \exp 2x_1] = \lambda^4(\exp \rho_1(\exp \rho_1(\Psi_2' \exp(-2x_1))')')',$$

d) $\quad (\exp \rho_1(\exp \rho_2(\Psi_2 \exp(-x_2))')')' = 2\lambda^4(\Psi_2' \exp(-2x_1)),$

e) $\quad (\exp \rho_1(\Psi_2' \exp(-2x_1))')' = 2\lambda^{-4} \exp \rho_2(\Psi_2 \exp(-x_2))',$

f) $\quad (\Psi_2' \exp(-2x_1))' = \lambda^{-4}(\exp \rho_2(\Psi_2 \exp(-x_2))')'.$
$$(1.8_2)$$

We skip the deduction of the fact that the compatibility conditions for (1.7) and (1.8) lead to the equations for the periodic Toda lattice for C_2:

$$-\dot{x}'_1 = \exp(-\rho_1) - \exp 2x_1, \quad -\dot{x}'_2 = \exp(-\rho_2) - \exp 2x_1$$

$$(\rho_1 = 2x_1 - x_2, \; \rho_2 = -2x_1 + 2x_2).$$

For B_2 the first and second fundamental representations of C_2 just interchange since their Cartan matrices are transposes of each other.

The examples considered of rank 2 Lie algebras indicate the general scheme for constructing the scalar LA-pair for an arbitrary representation $\{l\}$ of $\mathfrak{g}$. The derivatives of arbitrary order of the wave function of this representation are linearly expressed via $N_l = \dim\{l\}$ matrix elements $\langle n|g|l\rangle$ between the basis vectors $\langle n|$ of this representation and its highest vector $|l\rangle$. The coefficients are homogeneous polynomials in unknowns ρ_j (with respect to the derivatives) which enter the expression for the multicomponent LA-pair (1.3). Thus, all the matrix elements $\langle n|g|l\rangle$ can be expressed as a linear combination of the wave function and its derivatives up to order $(N_l - 1)$ with respect to both z_+ and z_-. Equating the expressions obtained in this way we get the first $N_l - 1$ equations of the LA-pair. Excluding N_l matrix elements $\langle n|g|l\rangle$ from the derivatives (with respect to z_+ and z_-) of the wave function up to order N_l we get the two remaining equations.

Therefore, the scalar LA-pair of the representation $\{l\}$ of $\mathfrak{g}$ contains exactly $N_l + 1$ equations and the maximal degree of the derivative with respect to both z_+ and z_- equals N_l. The two of these equations which contain derivatives with respect to z_+ (or z_-) only will be called *spectral* ones.

A spectral equation of the representation $\{l\}$ is of the form

$$P_{N_l}(\mathcal{D})\Psi_l = \lambda^{M+1} P_{N_l - M - 1}(\mathcal{D})\Psi_l,$$

where $P_N(\mathcal{D})$ is the differential operator with highest degree of the derivative equal N. The derivative operation is not present in the right-hand side only if the dimension of the representation is greater by one than the height M of the maximal root. This is the case for representations of minimal dimension for A_r, C_r. The spectral equations of this type are also associated with $A_r^{(2)}$ whose simplest representative is the one-component Bulough-Dodd equation.

The solutions of equations of the periodic Toda lattice associated with $C_r^{(1)}$ and $A_r^{(2)}$ are obtained from the solutions for $A_r^{(1)}$ under additional constraints on their parameters compatible with the symmetry properties of the Dynkin diagram of $A_r^{(1)}$. The solutions corresponding to $A_r^{(1)}$ pass into those for $C_k^{(1)}$ if $r = 2k + 1$ or those for $A_r^{(2)}$ if $r = 2k$, cf. 4.1.4.1.

For the series B_r for $r \geq 3$ the height of the maximal root is $2r - 1$ and the dimension of minimal (vector) representation is $2r + 1$. Therefore the operator in the right-hand side of the spectral equations contains first degree derivatives. For D_r, where $r \geq 4$, the height of the maximal root is $2r - 3$ and the minimal dimension of the irreducible (standard in this case) representation is $2r$. Therefore the polynomial in the right-hand side of the spectral equation is quadratic with respect to derivatives.

Let us make some general comments on the coefficient functions of the differential operator in the left-hand side of the spectral equation in derivatives with respect to one of the arguments, say, z_+. To this end assume that the compatibility conditions (1.4) in the operator-valued functions $A_\pm$ contain no elements $X_{\pm m}$, i.e., we deal with the generalized Toda lattice with fixed end-points (cf. 4.1). Then the wave function $\Psi_l = \langle l | g | l \rangle$ satisfies $\Psi_{l,z_-} = 0$ ($\langle l | X_- \equiv 0$), and the spectral equation in derivatives with respect to z_+ coincides with the spectral equation of the corresponding periodic problem for the spectral parameter λ equal to zero. Since Ψ_l does not depend on z_-, neither do all the coefficient functions of the equation which defines it. They are homogeneous polynomials in derivatives with respect to z_+, i.e., $a \equiv a(\rho_{j,z_+}, \rho_{j,z_+z_+}, \ldots, \underbrace{\rho_{j,z_+\ldots z_+}}_{r})$ and $\partial a / \partial z_- = 0$. Therefore, a are the func-

tions in the solutions of the characteristic equations of the generalized Toda lattice with fixed end-points (see 5.2), i.e., functions of its local integrals.

From the group theory we can deduce certain facts about the quadratic combinations of different solutions of the equations of the scalar LA-pair of the representation $\{l\}$ of $\mathfrak{g}$. Quadratic combinations of any wave functions of these equations are the wave functions of the representation $\{2l\}$ of $\mathfrak{g}$. Indeed, by definition the latter ones equal $\Psi_{2l} = \exp 2 \sum_j l_j \tau_j$ and therefore the square of every wave function of the representation $\{l\}$ is a wave function for the representation $\{2l\}$. The superposition of any two different wave functions of the representation $\{l\}$ is again a wave function of the same representation and its square

$$\Psi^2 = c_1^2 \Psi_1^2 + c_2^2 \Psi_2^2 + 2c_1 c_2 \Psi_1 \Psi_2, \text{ where } c_1, c_2 = \text{const},$$

is a wave function of the representation $\{2l\}$. But thanks to linearity $\Psi_1 \cdot \Psi_2$ is a solution of the equations of the scalar LA-pair of the $\mathfrak{g}$-module $\{2l\}$ since Ψ_1^2 and Ψ_2^2 are the solutions. Therefore, the product of any two wave functions of the representation $\{l\}$ is a wave function of the representation $\{2l\}$. The total number of such quadratic combinations equals $N_l(N_l + 1)/2$, whereas the degree of the spectral equation of the LA-pair of the representation $\{2l\}$ equals N_{2l} and has N_{2l} different solutions. Therefore, between the quadratic combinations of the wave functions of the representation $\{l\}$ there should be $N_l(N_l + 1)/2 - N_{2l}$ linear relations.

Examples. The dimension of the vector representation of A_r is $r + 1$ and the $2l$-th representation is its 2nd rank symmetric tensor with $(r+1)(r+2)/2$ components. Therefore, from the above we get $(r+1)(r+2)/2 - (r+1)(r+$

$2)/2 \equiv 0$, i.e., all the quadratic combinations of different solutions of the scalar LA-pair of the vector representation are different.

The dimension of the vector representation of B_2 equals 5 and the $2l$-th representation is the traceless symmetric tensor of rank 2 and its dimension equals 14. Therefore, there are $(\frac{5 \cdot 6}{2} - 14 = 1)$ linear relations between the quadratic combinations of those 5 linearly independent solutions.

These general properties of the spectral equations of the scalar LA-pair are useful in concrete calculations of soliton solutions for the periodic Toda lattice.

§ 6.2 Soliton solutions of the sine-Gordon equation*)

The simplest well-studied representative of the generalized Toda lattice equations is the sine-Gordon equation with which we will illustrate all the details for finding soliton solutions with the help of the scalar LA-pair.

The sine-Gordon equation $\dot{x}' = \exp 2x - \exp(-2x)$ is a corollary of the compatibility conditions for the linear system

$$\dot{g}g^{-1} = h\dot{x} + \lambda(X_+ + X_-), \quad g'g^{-1} = \lambda^{-1}(X_+ \exp 2x + X_- \exp(-2x)), \quad (2.1)$$

where $X_\pm, h \in A_1$. Consider the Gauss decomposition $g = \exp(\alpha X_+)\exp(h\tau) \exp(\beta X_-)$ of an arbitrary element of $SL(2;\mathbb{C})$. Substituting it into (2.1) we get a system of equations connecting the parameters α, τ, β,

$$\dot{g}g^{-1} = h(\dot{\tau} + \alpha\dot{\beta}\exp(-2\tau))$$
$$+ X_+(\dot{\alpha} - 2\alpha\dot{\tau} - \alpha^2\dot{\beta}\exp(-2\tau)) + X_-\dot{\beta}\exp(-2\tau)$$

$$= h\dot{x} + \lambda(X_+ + X_-),$$

or

$$\dot{\beta}\exp(-2\tau) = \lambda, \qquad\qquad \alpha\lambda = \dot{x} - \dot{\tau},$$
$$\ddot{\psi} = (\lambda^2 - \ddot{x} + \dot{x}^2)\psi, \quad \psi \equiv \exp(-\tau). \qquad (2.2_1)$$

In complete analogy we get

$$g'g^{-1} = h(\tau' + \alpha\beta'\exp(-2\tau))$$
$$+ X_+(\alpha' - 2\alpha\tau' - \alpha^2\beta'\exp(-2\tau)) + X_-\beta'\exp(-2\tau)$$

$$= \lambda^{-1}(X_+ \exp 2x + X_- \exp(-2x));$$

$$\beta'\exp(-2\tau) = \lambda^{-1}\exp(-2x),$$

$$\alpha\lambda^{-1} = -\tau'\exp 2x,$$

$$(\psi\exp x)'' = (\lambda^{-2} + x'' + x'^2)(\psi\exp x). \qquad (2.2_2)$$

---

*) [124, 126]

As a corollary of (2.2), taking the compatibility conditions $((\dot\beta)' = (\beta')^{\cdot})$ into account we get the system

a)
$$\ddot\psi = (\lambda^2 - \ddot{x} + \dot{x}^2)\psi,$$

b)
$$(\psi \exp x)'' = (\lambda^{-2} + x'' + x'^{2})(\psi \exp x), \tag{2.3}$$

c)
$$\lambda^{-2}(\dot\psi + \dot{x}\psi) = \psi' \exp 2x.$$

The system (2.3) is a LA-pair of the only fundamental (spinor) representation (of dimension 2) of A_1. The above deduction is nothing but one of the methods to find the scalar LA-pair starting from the system of equations in parameters g which follows from (1.4). (We have mentioned this already in the preceding section.)

Notice that the parameters α, β of g can be calculated in quadratures in terms of x and τ with the help of (2.2). However, even the system (2.3) itself requires as its compatibility condition the fulfilment of the sine-Gordon equation. Indeed, equating the mixed derivatives $(\dot\psi)' = (\psi')^{\cdot}$ from (2.3c) and with (2.3a) and (2.3b) we get the sine-Gordon equation.

By the methods of 6.1 we can get (2.3) as follows. Set $\Psi = \langle 1|g|1\rangle$, where $|1\rangle$ is the lowest vector of the spinor representation, i.e., $X_{-1}|1\rangle = 0$, $h|1\rangle = -|1\rangle$. In the Gauss parametrization we have $\Psi = \exp(-\tau)$; furthermore

$$\dot\Psi = -\dot{x}\Psi + \lambda\langle 1|X_{-}g|1\rangle, \qquad \exp(-x)(\Psi \exp x)^{\cdot} = \lambda\langle |X_{-}g|1\rangle,$$

$$\Psi' = \lambda^{-1}\exp(-2x)\langle 1|X_{-}g|1\rangle, \qquad \exp 2x\,\Psi' = \lambda^{-1}\langle 1|X_{-}g|1\rangle.$$

Taking derivatives of the left-hand side equations with respect to z_+ and z_-, respectively, and making use of the properties of $|1\rangle$ we get (2.3). The scalar LA-pair (2.3) obtained somehow or other will be the point of reference for the further calculations.

Let us seek a solution of (2.3) in the form

$$\Psi = c\exp(\lambda z_+ + \lambda^{-1}z_-)\prod_{1\leq\alpha\leq n}(-\lambda + a_\alpha), \quad \exp(-x) = \prod_{1\leq\alpha\leq n}a_\alpha, \quad c = \text{const.}$$

Here a_α are the unknown functions of independent arguments z_+, z_- and we get the equations for them by substituting Ψ into (2.3) and comparing the factors of the same powers of λ. Consider first the first equation. We get

$$\dot\Psi = \left(\lambda + \sum_\alpha \frac{\dot{a}_\alpha}{-\lambda + a_\alpha}\right)\Psi,$$

$$\ddot\Psi = \left[\lambda^2 + 2\lambda\sum_\alpha \frac{\dot{a}_\alpha}{-\lambda + a_\alpha} + \sum_\alpha \frac{\ddot{a}_\alpha}{-\lambda + a_\alpha} + \sum_{\alpha\neq\beta} \frac{\dot{a}_\alpha \dot{a}_\beta}{(-\lambda + a_\alpha)(-\lambda + a_\beta)}\right]\Psi.$$

Substituting $\ddot{\Psi}$ into (2.3a) and equating the residues at the poles $(-\lambda + a_\alpha)^{-1}$ to zero we get

$$\ddot{a}_\alpha + 2a_\alpha \dot{a}_\alpha - 2\dot{a}_\alpha \sum_{\beta \neq \alpha} \frac{\dot{a}_\beta}{a_\alpha - a_\beta} = 0, \quad -2 \sum_\alpha \dot{a}_\alpha = -\ddot{x} + \dot{x}^2, \quad 1 \leq \alpha \leq n. \quad (2.4)$$

The $(n+1)$-th equation of (2.4) is obtained by equating the residue of (2.3a) at infinity to zero. It is a corollary of the first n equations (2.4) and the definition of x, i.e., $\exp(-x) = \prod_\alpha a_\alpha$.

There are n first integrals of (2.4). It is convenient to express them in the form resolved with respect to $\dot{a}_\alpha$:

$$\dot{a}_\alpha = -\frac{P_n(a_\alpha^2)}{\prod_{\beta \neq \alpha}(a_\alpha^2 - a_\beta^2)}, \quad P_n(a_\alpha^2) \equiv \prod_{1 \leq \beta \leq n}(a_\alpha^2 - \mu_\beta^2), \quad \dot{\mu}_\beta \equiv 0, \quad (2.5)$$

which is a direct consequence of (2.4) with n constants μ_β playing the role of the first n first integrals of (2.4).

It is possible to perform further integration of equation (2.5). To do this consider the symmetric function of n arguments a_α^2,

$$\Phi^s \equiv \sum_{1 \leq \alpha \leq n} \frac{a_\alpha^{2s}}{\prod_{\beta \neq \alpha}(a_\alpha^2 - a_\beta^2)} = \begin{cases} 0, & 1 \leq s \leq n-2, \\ 1, & s = n-1. \end{cases}$$

Indeed, reducing to the common denominator we express Φ^s as the ratio of two functions where the denominator (Vandermonde determinant) is antisymmetric with respect to any odd permutation of its arguments. Therefore so is the numerator, since Φ^s is symmetric. This is possible only for $s \geq n-1$ since otherwise the degree of the numerator would be less than that of the denominator. Hence, we derive from (2.5) that

$$\sum_\alpha \frac{\Pi(a_\alpha^2)\dot{a}_\alpha}{P_n(a_\alpha^2)} = -1,$$

where $\Pi(a_\alpha^2)$ is an arbitrary polynomial of degree $n-1$ with leading coefficient 1. Selecting it so that its $n-1$ roots coincide with the $n-1$ roots of the denominator we get

$$\sum_\alpha \frac{\dot{a}_\alpha}{a_\alpha^2 - \mu_\beta^2} = -1 \text{ or } \prod_\alpha \frac{a_\alpha - \mu_\beta}{a_\alpha + \mu_\beta} = \exp 2(-\mu_\beta z_+ + \nu_\beta). \quad (2.6)$$

Therefore, n relations (2.6) determine an implicit dependence of a solution of (2.3a) on $2n$ parameters μ_α, ν_α.

So far we have only integrated the first equation of the system (2.3). To find the dependence of "integrals" μ_α, ν_α on z_- we have to make use of the remaining two equations (2.3b, c). Since $\exp(-x) = \prod_\alpha a_\alpha$, we get

$$\Psi \exp x = c(-\lambda)^n \exp(\lambda z_+ + \lambda^{-1} z_-) \prod_{1 \le \alpha \le n} (-\lambda^{-1} + a_\alpha^{-1}),$$

i.e. the function $\Psi \exp x$ has the same structure with respect to z_- as that of Ψ with respect to z_+ with the obvious change $a_\alpha \to a_\alpha^{-1}$, $\lambda \to \lambda^{-1}$, $x \to -x$. Therefore, from (2.3b) we get a system similar to (2.4):

$$(a_\alpha^{-1})' = -\frac{\tilde{P}_n(a_\alpha^{-2})}{\prod\limits_{\beta \ne \alpha} (a_\alpha^{-2} - a_\beta^{-2})}, \quad \tilde{P}_n(a_\alpha^{-2}) \equiv \prod_{\beta=1}^n (a_\alpha^{-2} - \bar{\mu}_\beta^{-2}),$$

or

$$a_\alpha' = -\prod_{\beta \ne \alpha} a_\beta^2 \frac{a_\alpha^{2n} \tilde{P}_n(a_\alpha^{-2}) \cdot (-1)^n}{\prod\limits_{\gamma \ne \alpha} (a_\alpha^2 - a_\gamma^2)}. \tag{2.7}$$

The equation (2.3c) establishes on the above substitutions the additional constraints

$$\dot{a}_\alpha \prod_{\beta \ne \alpha} a_\beta^2 = a_\alpha', \quad \prod_\alpha a_\alpha^2 = 1 - \sum_\beta a_\beta', \quad \prod_\alpha a_\alpha^{-2} = 1 - \sum_\beta (a_\beta^{-1})',$$

from which we derive with the help of (2.5) and (2.7)

$$P_n(a_\alpha^2) = (-1)^n a_\alpha^{2n} \tilde{P}_n(a_\alpha^{-2}).$$

Therefore, $\dot{a}_\alpha$, a_α' are determined by the same polynomials of degree n whose roots μ_α^2 depend neither on z_+ nor on z_- and satisfy $\prod\limits_\alpha \mu_\alpha^2 = (-1)^n$. Finally, we get the following solution of the scalar LA-pair:

$$\Psi = c \exp(\lambda z_+ + \lambda^{-1} z_-) \prod_{1 \le \alpha \le n} (-\lambda + a_\alpha),$$

$$\prod_\alpha \frac{a_\alpha - \mu_\beta}{a_\alpha + \mu_\beta} = \exp 2(-\mu_\beta z_+ - \mu_\beta^{-1} z_- + \nu_\beta) \equiv \exp 2z_\beta, \tag{2.8}$$

$$\exp(-x) = \prod_\alpha a_\alpha, \quad \prod_\alpha \mu_\alpha^2 = (-1)^n.$$

The parameters μ_β, ν_β entering (2.8) do not depend on z_+ and z_- and the n relations (2.8) enable us to determine the elementary symmetric functions $s_\alpha \equiv \sum\limits_{\beta \ne \gamma \ne \ldots \ne \sigma} a_\beta a_\gamma \ldots a_\sigma$ from the solution of the system

$$\text{sh } z_\beta \cdot s_n - \mu_\beta \text{ ch } z_\beta \cdot s_{n-1} + \mu_\beta^2 \text{ sh } z_\beta \cdot s_{n-2} - \ldots = 0,$$

$$1 \le \beta \le n, \quad s_0 \equiv 1. \tag{2.9}$$

In particular, finding from (2.9) that

$$s_n = \prod_\alpha a_\alpha = \exp(-x)$$

we get the known n-soliton solution of the sine-Gordon equation as the ratio of two determinants of order n.

We have intentionally reproduced here this rather cumbersome deduction of soliton solutions of the sine-Gordon equation to make the reader familiar with the details. Now, in order to generalize it to A_r let us give the shortest solution of this problem. Since (2.3) is invariant with respect to the change $\lambda \to -\lambda$, it contains together with $\Psi_1 \equiv \Psi(\lambda)$ the second independent solution $\Psi_2 \equiv \Psi(-\lambda)$. Thanks to (2.3a), $\det \begin{pmatrix} \Psi_1 & \Psi_2 \\ \dot\Psi_1 & \dot\Psi_2 \end{pmatrix}$ does not depend on z_+ whereas from (2.3c) we get

$$\det \begin{pmatrix} \Psi_1 & \Psi_2 \\ \dot\Psi_1 & \dot\Psi_2 \end{pmatrix} = \lambda^2 \exp 2x \cdot \det \begin{pmatrix} \Psi_1 & \Psi_2 \\ \Psi_1' & \Psi_2' \end{pmatrix}$$

$$= \lambda^2 \det \begin{pmatrix} \exp x \cdot \Psi_1 & \exp x \cdot \Psi_2 \\ (\exp x \cdot \Psi_1)' & (\exp x \cdot \Psi_2)' \end{pmatrix} \qquad (2.10)$$

$$\equiv \lambda^2 \det \begin{pmatrix} \tilde\Psi_1 & \tilde\Psi_2 \\ \tilde\Psi_1' & \tilde\Psi_2' \end{pmatrix},$$

or in the obviously simplified notations

$$(\Psi, \dot\Psi) = \lambda^2 \exp 2x (\Psi, \Psi') = \lambda^2 (\exp x \cdot \Psi, (\exp x \cdot \Psi)')$$
$$\equiv \lambda^2 (\tilde\Psi, \tilde\Psi'). \qquad (2.11)$$

It follows from (2.3b) that the last determinant does not depend on z_- whereas the first one does not depend on z_+. Therefore, each of the terms in the chain of equations (2.11) is constant.

Let us calculate the determinants entering these equations explicitly assuming, as earlier, that

$$\Psi = \exp(\lambda z_+ + \lambda^{-1} z_-) \prod_\alpha (-\lambda + a_\alpha) \quad (c = \text{const} = 1).$$

Since $(\Psi, \dot\Psi)$ depends neither on z_+ nor on z_- we get

$$(\Psi, \dot\Psi) \equiv \begin{vmatrix} \Psi_1 & \Psi_2 \\ \dot\Psi_1 & \dot\Psi_2 \end{vmatrix} = \Psi_1 \Psi_2 \begin{vmatrix} 1 & 1 \\ (\ln \Psi_1)^\cdot & (\ln \Psi_2)^\cdot \end{vmatrix}$$

$$= \prod_\alpha (a_\alpha^2 - \lambda^2) \begin{vmatrix} \lambda + \sum_\alpha \dfrac{\dot a_\alpha}{-\lambda + a_\alpha} & -\lambda + \sum_\alpha \dfrac{\dot a_\alpha}{a_\alpha + \lambda} \end{vmatrix} \qquad (2.12)$$

$$= -\prod_\alpha (a_\alpha^2 - \lambda^2) \cdot 2\lambda \left(1 + \sum_\beta \dfrac{\dot a_\beta}{a_\beta^2 - \lambda^2} \right) = -2\lambda \prod_\alpha (\lambda_\alpha^2 - \lambda^2)$$

and therefore, in the n points of the λ^2-plane $\lambda^2 = \lambda_\alpha^2$, $1 \le \alpha \le n$, we get

$$1 = -\sum_\alpha \frac{\dot{a}_\alpha}{a_\alpha^2 - \lambda_\beta^2}, \quad 1 = -\sum_\alpha \frac{(a_\alpha^{-1})'}{a_\alpha^{-2} - \lambda_\beta^{-2}}.$$

For the same values of λ we get $(\Psi, \dot{\Psi}) = (\Psi, \Psi') = 0$, i.e., $\Psi_1 = c\Psi_2$, $c = $ const. Therefore, $\Psi_1 - c(\lambda)\Psi_2|_{\lambda^2 = \lambda_\beta^2} = 0$ and a_α are determined from (2.8), as the first integrals of this system are obtained as the coefficients of the powers of λ^2 from (2.12).

Notice that we did not need any algebraic operations except for the solution of a system of linear equations and the calculation of the corresponding determinants to get a soliton solution of (2.9).

Notice one more fact. Recall that $\Psi = \exp(-x)$ appeared after considering the Cartan element $\exp h\tau$ in the Gauss decomposition, where actually $\exp h\tau \in A_1$. In a degenerate representation of A_1 the element appears only with even powers of λ. Therefore we can deduce that Ψ should be even as function in λ, i.e., $\Psi = c(\lambda)\Psi(\lambda) + c(-\lambda)\Psi(-\lambda)$ and the condition for Ψ to vanish at n points of the λ^2-plane means that $\exp(-\tau)$ has a zero and $\exp\tau$ a pole.

Now let us investigate what will change if we will start from the scalar LA-pair of the "vector" (three-dimensional) representation of A_1 with the basis

$$|2\rangle, X_+|2\rangle, X_+X_+|2\rangle, \ (X_-|2\rangle = 0, \ h|2\rangle = -|2\rangle),$$

i.e., construct a system of equations for $\Psi = \exp(-2\tau)$. The technique of the preceding section leads us to the following linear systems:

a) $$\exp(-2x)\,(\Psi \exp 2x)^\cdot = \lambda^2 \Psi' \exp 2x,$$

b) $$(\exp(-2x)\,(\exp(-2x)\,(\Psi \exp 2x)^\cdot)^\cdot)^\cdot = 4\lambda^2 \dot{\Psi} \exp(-2x),$$

c) $$\lambda^{-2}(\exp(-2x)\,(\Psi \exp 2x)^\cdot)^\cdot - \lambda^2 \exp 2x(\Psi' \exp 2x)' \qquad (2.13)$$
$$= 2\Psi(1 - \exp 4x),$$

d) $$(\exp 2x(\Psi' \exp 2x)')' = 4\lambda^{-2} \exp 2x(\Psi \exp 2x)'.$$

The sine-Gordon equation is a corollary of only two of these equations. Indeed, substituting, e.g., $(\dot{\Psi} \exp 2x)$ and $(\Psi' \exp 2x)$ from (2.13a) into (2.13c) we get

$$(\Psi' \exp 2x)^\cdot - \exp 2x(\exp(-2x)\,(\Psi \exp 2x)^\cdot)' = 2\Psi(1 - \exp 4x),$$

implying $\dot{x}' = \exp 2x - \exp(-2x)$. The chain of determinants similar to (2.11) for this pair of equations (2.13) is the following:

$$(\Psi, \dot{\Psi}, \ddot{\Psi}) = \lambda^4 \exp 4x \cdot (\Psi, \dot{\Psi}, \Psi'') \quad = \lambda^6 \exp 6x \cdot (\Psi, \Psi', \Psi'')$$
$$= \lambda^2 \exp 2x(\Psi, \Psi', \dot{\Psi}) \quad = \lambda^2(\tilde{\Psi}, \tilde{\Psi}', \tilde{\Psi}'').$$

The product of any two solutions Ψ_1, Ψ_2 is a wave function of the scalar LA-pair with $l = 2$ (cf. 6.1) whose spectral equations are of maximal order 5. Therefore, they have exactly 5 linearly independent solutions and among the 6 quadratic combinations Ψ_1^2, Ψ_2^2, Ψ_3^2, $\Psi_1\Psi_2$, $\Psi_1\Psi_3$, $\Psi_2\Psi_3$ of the three fundamental solutions of (2.13) there should exist one linear dependence $0 = \sum\limits_{1 \le i,j \le 3} c_{ij} \Psi_i \Psi_j$ which after the reduction of the quadratic form to the diagonal form takes the form $\Psi_3^2 = \Psi_+\Psi_-$, i.e., the square of one of the solutions of (2.13) equals to the product of two others.

Therefore, to construct a complete solution of (2.13) it suffices to know two of the solutions of this equation, Ψ_+, Ψ_-. The first two equations of the scalar LA-pair (2.13) are automatically satisfied by $\Psi = (\Psi_+\Psi_-)^{1/2}$, where Ψ_+, Ψ_- are its solutions. The spectral equations (2.13b, d) lead to the relations relating $\Psi_\pm$, and it is more convenient to express them for the functions $\theta_\pm$ given by the formulas $\Psi_+ \equiv \theta_+^2$, $\Psi_- \equiv \theta_-^2$. Namely,

$$\dot\theta_+\theta_- - \dot\theta_-\theta_+ = f_-(z_-), \ \exp 2x(\theta_+'\theta_- - \theta_-'\theta_+) = f_+(z_+),$$

which, taking (2.13a) into account, yields

$$(\theta,\dot\theta) = \lambda^2 \exp 2x(\theta,\theta') = c; \ \dot c = c' = 0,$$

or

$$\frac{\ddot\theta_+}{\theta_+} = \frac{\ddot\theta_-}{\theta_-}, \ \frac{\tilde\theta_+''}{\tilde\theta_+} = \frac{\tilde\theta_-''}{\tilde\theta_-}; \ \ \tilde\theta \equiv \theta\exp x.$$

The fact that $\Psi_\pm \equiv \theta_\pm^2$ are solutions of the spectral equation gives additionally

$$\frac{\ddot\theta_+}{\theta_+} + \ddot x - \dot x^2 - \lambda^2 = c\theta_+^{-4},$$

which is compatible with the above relations only for $c = 0$. Again we get for $\theta_\pm$ the system of equations of the scalar LA-pair of the spinor representation (2.3).

Therefore, whatever representation of A_1 is taken to express it, the LA-pair reduces to the LA-pair of the fundamental (2-dimensional spinor) representation and its $2l + 1$ solutions are representable in the form $\Psi_1^{2l-n}\Psi_2^n$, $n = 0, 1, \ldots, 2l$, where Ψ_1, Ψ_2 are any solutions of the LA-pair of the spinor representation.

The structure of the solutions of the LA-pair of the l-th representation mirrors the construction of the basic vectors of this representation constructed from the spinor components.

Notice that from the group theoretical point of view the condition $\Psi_3^2 = \Psi_+\Psi_-$ on wave functions of the scalar LA-pair of the vector representation means that a vector in the 3-dimensional representation is isotropic. It guarantees the irreducibility of the representation with $l = 2$ constructed from it with 5 independent vectors (instead of 6 when isotropy fails).

§ 6.3 Generalized Bargmann potentials*)

Bargmann's work gives conditions on the potentials of the one-dimensional Schrödinger equation $\ddot{\Psi} = (\lambda^2 + u(z))\Psi$ for the existence of a solution with the following form of analytic dependence on λ:

$$\Psi = \exp(\lambda z) \cdot \prod_{1 \leq \alpha \leq n} (a_\alpha - \lambda).$$

These results are reproduced in another form in the preceding section where we have completely integrated the system of equations in unknown functions a_α and found out that $u = 2 \sum_\alpha \dot{a}_\alpha$. Now let us generalize these results for linear equations of arbitrary order

$$\overset{[\cdot(r+1)]}{\Psi} + \sum_{s=0}^{r-1} u_s \overset{[\cdot s]}{\Psi} = \lambda^{r+1}\Psi \quad (\overset{[\cdot s]}{\Psi} \equiv \partial^s \Psi/\partial z^s) \tag{3.1}$$

and find the conditions on coefficient functions u_s, $0 \leq s \leq r-1$, under which (3.1) has solutions of the form

$$\Psi = \exp(\lambda z) \prod_\alpha (a_\alpha - \lambda).$$

As was mentioned in 6.1, the spectral equations of the scalar LA-pair of the fundamental representations of the lowest dimensions lead exactly to equations of the form (3.1). Therefore a solution of the above problem is necessary for the construction of the simplest soliton solutions for these series. The following theorem holds:

Equation (3.1) possesses a solution of the form $\Psi = \exp(\lambda z) \prod_\alpha (a_\alpha - \lambda)$ *if the unknowns* a_α *are determined from the condition that*

$$\tilde{\Phi} \equiv \sum_{J=1}^{r+1} d(\lambda_J) \exp(\lambda_J z) \cdot \prod_{1 \leq \alpha \leq n} (a_\alpha - \lambda_J) \tag{3.2}$$

vanishes at n different points $\lambda^{r+1} = \lambda_J^{r+1}$; *the generalized Bargmann potentials are expressed in elementary symmetric functions in* a_i *and their derivatives via* (3.11).

We will not give these formulas here since it is more convenient for us to introduce the terms entering them in the process of our calculations.

Introduce $r + 1$ linearly independent functions determining Ψ via (3.2), setting

$$\Psi_J = \exp(\lambda_J z) \prod_\alpha (a_\alpha - \lambda_J), \quad \prod_{1 \leq J \leq 1+r} \Psi_J = \prod_{1 \leq \alpha \leq r} (a_\alpha^{r+1} - \lambda^{r+1}), \tag{3.3}$$

---

*) [125, 126, 133, 139]

where λ is an arbitrary number and a_α are determined by the conditions of the Theorem. For the first derivatives of the functions Ψ_J we get, thanks to (3.3),

$$\dot{\Psi}_J = \left(\lambda_J + \sum_\alpha \frac{\dot{a}_\alpha}{(a_\alpha - \lambda_J)} \right) \Psi \equiv \varphi_J^1 \Psi_J.$$

The general recursive relations for the derivatives of an arbitrary order, say, s, are of the form

$$\overset{[\cdot s]}{\Psi}_J = \varphi_J^s \Psi_J, \quad \varphi_J^{s+1} = \dot{\varphi}_J^s + \varphi_J^s \varphi_J^1, \quad \varphi_J^0 \equiv 1, \quad \varphi_J^1 \equiv \lambda_J + \sum_\alpha \frac{\dot{a}_\alpha}{(a_\alpha - \lambda_J)}. \quad (3.4)$$

As follows from the definition, the Wronskian of $r + 1$ functions Ψ_J

$$G(\lambda_1, \ldots, \lambda_{r+1}; a) \equiv (\Psi, \dot{\Psi}, \ddot{\Psi}, \ldots, \overset{[\cdot r]}{\Psi})$$
$$= \prod_{1 \leq \alpha \leq n} (a_\alpha^{r+1} - \lambda^{r+1})(1, \varphi^1, \varphi^2, \ldots, \varphi^r) \quad (3.5)$$

is an antisymmetric function in the roots λ_J. The only function of this type is the Vandermonde determinant $W = \prod_{I>J} (\lambda_I - \lambda_J)$, and therefore $G = Wf(\lambda^{r+1})$, where f is symmetric with respect to permutation of the roots and therefore coincides with $\prod_{1 \leq J \leq r+1} \lambda_J = (-1)^r \lambda^{r+1}$. From the relations (3.4) determining φ_J^s we easily find their analytic structure with respect to λ_J:

$$\varphi_J^s = \lambda_J^s + \sum_{0 \leq j \leq s-2} B_s^j \lambda_s^j + \sum_{1 \leq \alpha \leq n} \frac{A_\alpha^s}{a_\alpha - \lambda_J}$$

or

$$\varphi^s = \lambda^s + \sum_j B_s^j \lambda^j + \sum_\alpha{}' \frac{A_\alpha^s}{a_\alpha - \lambda}, \quad (3.6)$$

where B_s^j, A_α^s depend polynomially on the a_α and their derivatives up to order s inclusive, and $A_\alpha^1 \equiv \dot{a}_\alpha$.

The technique of calculation of determinants (3.5) with functions φ^s given by (3.6) is rather simple. Namely, we subtract the first column from all the columns of the determinant thus getting the common factor $\prod_{J \neq 1} (\lambda_J - \lambda_1)$, then subtract the second column from all the columns except the first one of the obtained determinant thus getting the factor $\prod_{J \neq 1,2} (\lambda_J - \lambda_2)$, etc.

For example, for the determinant of order 3 we get

$$(\Psi, \dot{\Psi}, \ddot{\Psi}) = \prod_{\alpha=1}^{n}(a_\alpha^3 - \lambda^3) \cdot (1, \varphi^1, \varphi^2) = \prod_{\alpha}(a_\alpha^3 - \lambda^3)$$

$$\times \det \begin{pmatrix} 1, & 1, & 1 \\ \lambda_1 + \sum_\beta \dfrac{A_\beta^1}{a_\beta - \lambda_1}, & \lambda_2 + \sum_\beta \dfrac{A_\beta^1}{a_\beta - \lambda_2}, & \lambda_3 + \sum_\beta \dfrac{A_\beta^1}{a_\beta - \lambda_3} \\ \lambda_1^2 + B_2^0 + \sum_\beta \dfrac{A_\beta^2}{a_\beta - \lambda_2}, & \lambda_2^2 + B_2^0 + \sum_\beta \dfrac{A_\beta^2}{a_\beta - \lambda_2}, & \lambda_3^2 + B_2^0 + \sum_\beta \dfrac{A_\beta^2}{a_\beta - \lambda_3} \end{pmatrix}$$

$$= \prod_\alpha(a_\alpha^3 - \lambda^3)(\lambda_2 - \lambda_1)(\lambda_3 - \lambda_1)$$

$$\times \det \begin{pmatrix} 1 + \sum_\beta \dfrac{A_\beta^1}{(a_\beta - \lambda_1)(a_\beta - \lambda_2)}, & 1 + \sum_\beta \dfrac{A_\beta^1}{(a_\beta - \lambda_1)(a_\beta - \lambda_3)} \\ \lambda_2 + \lambda_1 + \sum_\beta \dfrac{A_\beta^2}{(a_\beta - \lambda_1)(a_\beta - \lambda_2)}, & \lambda_3 + \lambda_1 + \sum_\beta \dfrac{A_\beta^2}{(a_\beta - \lambda_1)(a_\beta - \lambda_3)} \end{pmatrix}$$

$$= \prod_\alpha(a_\alpha^3 - \lambda^3)W(\lambda_1, \lambda_2, \lambda_3) \det \begin{pmatrix} 1 + \sum_\beta \dfrac{A_\beta^1 a_\beta}{a_\beta^3 - \lambda^3} & \sum_\beta \dfrac{A_\beta^1}{a_\beta^3 - \lambda^3} \\ \sum_\beta \dfrac{A_\beta^2 a_\beta}{a_\beta^3 - \lambda^3} & 1 + \sum_\beta \dfrac{A_\beta^2}{a_\beta^3 - \lambda^3} \end{pmatrix}.$$

For an arbitrary r we get

$$G = W(\lambda_1, \ldots, \lambda_{r+1}) \prod_\alpha(a_\alpha^{r+1} - \lambda^{r+1}) \det\left(\delta_{ij} + B_i^j + \sum_\alpha \frac{A_\alpha^i a_\alpha^{r-j}}{a_\alpha^{r+1} - \lambda^{r+1}} \right)$$

$$\equiv W P_n(\lambda^{r+1}),$$

(3.7)

where B_i^j is nonzero if $i \geq j + 2$, i.e., $B_i^i = B_{i+1}^i = 0$ and the nonzero entries B_j^i are determined from (3.6).

By definition G is a polynomial in roots and thanks to (3.7) we have $G = W P_n(\lambda^{r+1})$, where P_n is a polynomial of degree n in λ^{r+1} with leading coefficient $(-1)^n$.

Furthermore, the conditions of Theorem imply that $\tilde{\Phi}$ vanishes at n points of the λ^{r+1}-plane and therefore there is a nontrivial linear relation between the $r+1$ functions Ψ_J. Hence $(\Psi, \dot{\Psi}, \ldots, \overset{[\cdot r]}{\Psi})|_{\lambda^{r+1} = \lambda_J^{r+1}} = 0$ which means that $G = W(\lambda_1, \ldots, \lambda_{r+1}) \prod_J(\lambda_J^{r+1} - \lambda^{r+1})$. Therefore, the coefficients of $P_n(\lambda^{r+1})$ are integrals constructed from the a_α and their derivatives up to order r and determined by the conditions of Theorem.

The above arguments imply that

$$\dot{G} = 0, \quad (\Psi, \dot{\Psi}, \ldots, \overset{[\cdot(r-1)]}{\Psi}, \overset{[\cdot(r+1)]}{\Psi}) = 0.$$

If we consider this relation as an equation with respect to any of the unknown functions Ψ_J, say $\Psi_{r+1} \equiv \mathcal{Y}$, with $\Psi_1, \ldots, \Psi_r$ known we get the linear equation

$$(\Psi, \dot{\Psi}, \ldots, \overset{[\cdot(r-1)]}{\Psi}, \overset{[\cdot(r+1)]}{\Psi})$$

$$= \det \begin{pmatrix} \Psi_1 & \Psi_2 & \ldots & \Psi_r & \mathcal{Y} \\ \dot{\Psi}_1 & \dot{\Psi}_2 & \ldots & \dot{\Psi}_r & \dot{\mathcal{Y}} \\ \vdots & \vdots & & \vdots & \vdots \\ \overset{[\cdot(r-1)]}{\Psi_1} & \overset{[\cdot(r-1)]}{\Psi_2} & \ldots & \overset{[\cdot(r-1)]}{\Psi_3} & \overset{[\cdot(r-1)]}{\mathcal{Y}} \\ \overset{[\cdot(r+1)]}{\Psi_1} & \overset{[\cdot(r+1)]}{\Psi_2} & \ldots & \overset{[\cdot(r+1)]}{\Psi_r} & \overset{[\cdot(r+1)]}{\mathcal{Y}} \end{pmatrix} = 0 \qquad (3.8)$$

and the set Ψ_J is its fundamental solution.

Therefore, the coefficients of the equation after dividing by a common factor (the coefficient at the highest derivative $\mathcal{Y}^{[\cdot(r+1)]}$) should be symmetric with respect to the permutation of the roots λ_J, i.e. the functions in λ^{r+1}.

Expanding (3.8) with respect to the elements of the last column of the determinant we get

$$\overset{[\cdot(r+1)]}{\mathcal{Y}} - \frac{(\Psi, \ldots, \overset{[\cdot(r-2)]}{\Psi}, \overset{[\cdot(r+1)]}{\Psi})}{(\Psi, \ldots, \overset{[\cdot(r-2)]}{\Psi}, \overset{[\cdot(r-1)]}{\Psi})} \overset{[\cdot(r-1)]}{\mathcal{Y}}$$

$$+ \frac{(\Psi, \ldots, \overset{[\cdot(r-3)]}{\Psi}, \overset{[\cdot(r-1)]}{\Psi}, \overset{[\cdot(r+1)]}{\Psi})}{(\Psi, \ldots, \overset{[\cdot(r-2)]}{\Psi}, \overset{[\cdot(r-1)]}{\Psi})} \overset{[\cdot(r-2)]}{\mathcal{Y}} - \cdots = 0. \qquad (3.9)$$

Before we calculate the coefficients of (3.9) explicitly in general we consider the particular case $r = 2$. Making use of (3.6) we get

$$\frac{(\Psi, \ddot{\Psi})}{(\Psi, \dot{\Psi})} = \frac{(1, \phi^3)}{(1, \varphi^1)}$$

$$= \frac{\det \begin{pmatrix} 1, & 1 \\ \lambda_1^3 + B_3^1 \lambda_1 + B_3^0 + \sum_\alpha \frac{A_\alpha^3}{a_\alpha - \lambda_1}, & \lambda_2^3 + B_3^1 \lambda_2 + B_3^0 + \sum_\alpha \frac{A_\alpha^3}{a_\alpha - \lambda_2} \end{pmatrix}}{\det \begin{pmatrix} 1 & 1 \\ \lambda_1 + \sum_\alpha \frac{A_\alpha^1}{a_\alpha - \lambda_1} & \lambda_2 + \sum_\alpha \frac{A_\alpha^1}{a_\alpha - \lambda_2} \end{pmatrix}}$$

$$= \frac{B_3^1 + \sum_\alpha \frac{A_\alpha^3 (a_\alpha - \lambda_3)}{(a_\alpha^3 - \lambda^3)}}{1 + \sum_\alpha \frac{A_\alpha^1 (a_\alpha - \lambda_3)}{(a_\alpha^3 - \lambda^3)}} = B_3^1; \quad A_\alpha^3 = B_3^1 A_\alpha^1,$$

where the latter relations hold since $\det \begin{pmatrix} \Psi & \ddot{\Psi} \\ \dot{\Psi} & \dot{\Psi} \end{pmatrix}$ is a symmetric function in three roots and cannot depend on λ_3.

Notice that the relation $A_\alpha^3 = B_3^1 A_\alpha^1$ is exactly the system of third order differential equations satisfied by the functions a_α from the conditions of the theorem. Further,

$$\frac{(\dot{\Psi}, \ddot{\Psi})}{(\Psi, \dot{\Psi})} = \frac{(\varphi^1, \varphi^3)}{(1, \varphi^1)}$$

$$= \frac{\det \begin{pmatrix} \lambda_1 + \sum_\alpha \dfrac{A_\alpha^1}{a_\alpha - \lambda_1} & 1 + \sum_\alpha \dfrac{A_\alpha^1(a_\alpha - \lambda_3)}{a_\alpha^3 - \lambda^3} \\ \lambda^3 + B_3^1\lambda_1 + B_3^0 + \sum_\alpha \dfrac{A_\alpha^3}{a_\alpha - \lambda_1} & B_3^1 + \sum_\alpha \dfrac{A_\alpha^3(a_\alpha - \lambda_3)}{a_\alpha^3 - \lambda^3} \end{pmatrix}}{1 + \sum_\alpha \frac{A_\alpha^1(a_\alpha - \lambda_3)}{a_\alpha^3 - \lambda^3}}$$

$$= -(\lambda^3 + B_3^0),$$

where we have made use of the above-mentioned relation. Therefore, if $r = 2$, then equation (3.9) takes the form

$$\dddot{y} - B_3^1\dot{y} - (\lambda^3 + B_3^0)y = 0.$$

This example hints how we should calculate in the general case. First starting from (3.8) we get equations defining A_α^r. We have

$$0 = (\Psi, \dot{\Psi}, \ldots, \overset{[\cdot(r-1)]}{\Psi}, \overset{[\cdot(r+1)]}{\Psi}) = (1, \varphi^1, \ldots, \varphi^{r-1}, \varphi^{r+1})$$

$$= \left(1, \lambda + \sum_\alpha \frac{A_\alpha^1}{a_\alpha - \lambda}, \lambda^2 + B_2^0 + \sum_\alpha \frac{A_\alpha^2}{a_\alpha - \lambda}, \ldots, \lambda^{r-1}\right.$$

$$+ \sum_{0 \le s \le r-3} B_{r-1}^s \lambda^s + \sum_\alpha \frac{A_\alpha^{r-1}}{a_\alpha - \lambda}, \lambda^{r+1} + \sum_{0 \le s \le r-1} B_{r+1}^s \lambda^s + \sum_\alpha \frac{A_\alpha^{r+1}}{a_\alpha - \lambda}\right)$$

$$= \left(1, \lambda + \sum_\alpha \frac{A_\alpha^1}{a_\alpha - \lambda}, \lambda^2 + \sum_\alpha \frac{A_\alpha^2}{a_\alpha - \lambda}, \lambda^3 + \sum_\alpha \frac{\tilde{A}_\alpha^3}{a_\alpha - \lambda}, \ldots, \lambda^{r-1}\right.$$

$$\left.+ \sum_\alpha \frac{\tilde{A}_\alpha^{r-1}}{a_\alpha - \lambda}, \sum_\alpha \frac{\tilde{A}_\alpha^{r+1}}{a_\alpha - \lambda}\right)$$

$$= W(\lambda) \det{}_r \begin{pmatrix} 1 + \sum_\alpha \dfrac{\tilde{A}_\alpha^1 a_\alpha^{r-1}}{a_\alpha^{r+1} - \lambda^{r+1}} & \cdots & \sum_\alpha \dfrac{\tilde{A}_\alpha^1}{a_\alpha^{r+1} - \lambda^{r+1}} \\ \vdots & \cdots & \vdots \\ \sum_\alpha \dfrac{\tilde{A}_\alpha^{r+1} a_\alpha^{r-1}}{a_\alpha^{r+1} - \lambda^{r+1}} & \cdots & \sum_\alpha \dfrac{\tilde{A}_\alpha^{r+1}}{a_\alpha^{r+1} - \lambda^{r+1}} \end{pmatrix};$$

$$\tilde{A}_\beta^s = A_\beta^s - \sum_{1 \le j \le s-2} B_s^j \tilde{A}_\beta^j. \tag{3.10}$$

The chain (3.10) has been obtained after consecutive subtractions from the columns of the determinant its columns with smaller numbers multiplied by certain factors and with formula (3.7) used to put the determinant into symmetric (with respect to the permutations of the roots) form.

Equations (3.10) should be identities in λ and hence the equations in a_α are of the form $\tilde{A}^{r+1} = 0$. Making use of the equations of motion $\tilde{A}^{r+1} = 0$ for the coefficients (3.9) we get

$$-u_s = \tilde{B}^s_{r+1} \equiv B^s_{r+1} - \sum_j B^j_{r+1}\tilde{B}^s_j, \quad 1 \le s \le r-1;$$

$$-u_0 = \tilde{B}^0_{r+1} = B^0_{r+1} - \sum_j B^j_{r+1}\tilde{B}^0_j \qquad (3.11)$$

(as earlier, $B^p_q \ne 0$ for $p \le q+2$).

The Bargmann potentials (3.11) are easiest to be obtain if we notice that the equations of motion for A^s are of the form $A^{r+1} = - \sum_{1 \le s \le r-1} u_s A^s$ (as follows from (3.1), (3.8), (3.9)) and compare them with the above equations $\tilde{A}^{r+1} = 0$.

We get the same result by calculating Bargmann potentials directly from their definition as the ratio of two determinants from (3.9). In doing so we find the coefficient of the 0-th power of the derivative in (3.9) from the following chain of equations

$$-\frac{(\varphi^{r+1}, \varphi^1, \varphi^2, \dots, \varphi^{r-1})}{(1, \varphi^1, \varphi^2, \dots, \varphi^{r-1})} = -\left(\lambda^{r+1} + \sum_{s=0}^{r-1} B^s_{r+1}\lambda^s + \sum_\alpha \frac{\tilde{A}^{r+1}_\alpha}{a_\alpha - \lambda},\right.$$

$$\lambda + \sum_\alpha \frac{\tilde{A}^1_\alpha}{a_\alpha - \lambda}, \lambda^2 + \tilde{B}^0_2 + \sum_\alpha \frac{\tilde{A}^2_\alpha}{a_\alpha - \lambda}, \lambda^3 + \tilde{B}^0_3 + \sum_\alpha \frac{\tilde{A}^3_\alpha}{a_\alpha - \lambda}, \dots$$

$$\left. \dots, \lambda^{r-1} + \tilde{B}^0_{r-1} + \sum_\alpha \frac{\tilde{A}^{r-1}_\alpha}{a_\alpha - \lambda}\right) \cdot \left(1, \lambda + \sum_\alpha \frac{\tilde{A}^1_\alpha}{a_\alpha - \lambda}, \lambda^2 + \tilde{B}^0_2\right.$$

$$\left. + \sum_\alpha \frac{\tilde{A}^2_\alpha}{a_\alpha - \lambda}, \dots, \lambda^{r-1} + \tilde{B}^0_{r-1} + \sum_\alpha \frac{\tilde{A}^{r-1}_\alpha}{a_\alpha - \lambda}\right)^{-1}$$

$$= -\left(\lambda^{r+1} + \tilde{B}^0_{r+1} - \sum_{s=0}^{r-1} B^s_{r+1}\tilde{B}^0_s, \lambda + \sum_\alpha \frac{\tilde{A}^1_\alpha}{a_\alpha - \lambda}, \dots, \lambda^{r-1}\right.$$

$$\left. + \tilde{B}^0_{r-1} + \sum_\alpha \frac{\tilde{A}^{r-1}_\alpha}{a_\alpha - \lambda}\right)\left(1, \lambda + \sum_\alpha \frac{\tilde{A}^1_\alpha}{a_\alpha - \lambda}, \dots, \lambda^{r-1} + \tilde{B}^0_{r-1}\right.$$

$$\left. + \sum_\alpha \frac{\tilde{A}^{r-1}_\alpha}{a_\alpha - \lambda}\right)^{-1} = -(\lambda^{r+1} + \tilde{B}^0_{r+1}). \qquad (3.12)$$

This implies the above expression for u_0 with recursive relations which determine $\tilde{B}_{r+1}^0 = B_{r+1}^0 - \sum_{0 \leq s \leq r-1} B_{r+1}^s \tilde{B}_s^0$. The expressions for generalized Bargmann potentials u_s, $s \neq 0$, can be obtained from a chain of equations similar to (3.12).

Let us give one more expression for the Bargmann potentials of equation (3.1) that will be convenient in what follows. Notice that the $r + 1$ linearly independent solutions of the $(r + 1)$-th order equation

$$\overset{[\cdot(r+1)]}{\Psi} + \sum_{0 \leq j \leq r-1} q_j \overset{[\cdot j]}{\Psi} = 0 \tag{3.13}$$

are related by the condition that their Wronskian is constant:

$$V \equiv (\Psi, \dot{\Psi}, \ddot{\Psi}, \ldots, \overset{[\cdot(r)]}{\Psi}) = 1. \tag{3.14}$$

Equation (3.14) considered as an ordinary differential equation in an unknown function Ψ_{r+1} with Ψ_J known, $1 \leq J \leq r$, has been solved in 4.1; namely,

$$\Psi_1 = \varphi_1, \quad \Psi_2 = \varphi_1 \int \varphi_2, \quad \Psi_2 = \varphi_1 \int \varphi_2 \int \varphi_3, \ldots,$$
$$\Psi_{r+1} = \varphi_1 \int \varphi_2 \int \cdots \int \varphi_{r+1}, \tag{3.15}$$

where the $(r + 1)$ functions φ_J are subject to the unique relation

$$\prod_{1 \leq J \leq r+1} \varphi_J^{r+2-J} = 1. \tag{3.16}$$

Obviously, a set of $(r + 1)$ solutions of the equation

$$(\varphi_{r+1}^{-1}(\varphi_r^{-1}(\ldots(\varphi_2^{-1}(\varphi_1^{-1}\Psi)^{\cdot})^{\cdot}\ldots)^{\cdot})^{\cdot}\underbrace{\ldots)^{\cdot}}_{r+1} = 0 \tag{3.17}$$

coincides with (3.15) and therefore the coefficient functions q_j of (3.13) are expressed via φ_J after the differentiations in (3.17) are performed. The condition (3.16) guarantees the vanishing of the coefficients at $\overset{[\cdot r]}{\Psi}$.

It remains to relate the functions φ_J with the fundamental solutions of (3.13). We have $\varphi_1 = \Psi_1 \equiv D_1(V)$ and

$$D_2(V) = \det \begin{pmatrix} \Psi_1 & \Psi_2 \\ \dot{\Psi}_1 & \dot{\Psi}_2 \end{pmatrix} = \det \begin{pmatrix} \varphi_1 & \varphi_1 \int \varphi_2 \\ \dot{\varphi}_1 & \dot{\varphi}_1 \int \varphi_2 + \varphi_1 \varphi_2 \end{pmatrix} = \varphi_1^2 \varphi_2.$$

Continuing the reduction process we get

$$\prod_{1 \leq j \leq s} \varphi_j^{s+1-j} = D_s(V), \tag{3.18}$$

where $D_s(V)$ denotes the principal minors of order s of the Wronskian determinant (3.14).

The condition (3.16) is a corollary of (3.18) for $s = r + 1$ and is equivalent to (3.14). Substituting φ_J into (3.17) let us rewrite it in the form

$$(D_r^2 D_{r-1}^{-1}(\dots(D_3^{-1}D_2^2 D_1^{-1}(D_2^{-1}D_1^2(D_1^{-1}\underbrace{\Psi})^{\cdot})^{\cdot})^{\cdot}\dots)^{\cdot} = 0, \qquad (3.19)$$
$$\underbrace{}_{r+1}$$

which is the Frobenius generalization of Viète's theorem establishing a relation between the coefficients and a set of $r + 1$ linearly independent solutions of a linear differential equation of an arbitrary order.

Now apply (3.19) to the considered case, equation (3.1), which possesses $r + 1$ linearly independent solutions

$$\Psi_J = \exp(\lambda_J z) \prod_{1 \le \alpha \le n} (a_\alpha - \lambda_J). \qquad (3.20)$$

As follows from (3.11), the generalized Bargmann potentials u_s do not depend on λ and it is convenient to calculate them with the help of a complete set of solutions for $\lambda = 0$.

According to the general scheme we have

$$\varphi_1 = D_1(V)|_{\lambda=0} = \prod_\alpha a_\alpha, \quad \Psi_2|_{\lambda=0} = \Psi_1|_{\lambda=0} = D_1,$$

and therefore the second fundamental solution for $\lambda = 0$ should be chosen in the form $\tilde{\Psi}_2 = (\Psi_2 - \Psi_1)/(\lambda_2 - \lambda_1)$. Then

$$\varphi_1^2 \varphi_2 = D_2(V)|_{\lambda_1,\lambda_2=0} = \det \begin{pmatrix} \Psi_1 & \dfrac{\Psi_2 - \Psi_1}{\lambda_2 - \lambda_1} \\ \dot{\Psi}_1 & \dfrac{\dot{\Psi}_2 - \dot{\Psi}_1}{\lambda_2 - \lambda_1} \end{pmatrix}\Bigg|_{\lambda_1,\lambda_2=0}$$

$$= \frac{1}{\lambda_2 - \lambda_1} \det \begin{pmatrix} \Psi_1 & \Psi_2 \\ \dot{\Psi}_1 & \dot{\Psi}_2 \end{pmatrix}\Bigg|_{\lambda_1,\lambda_2=0}$$

$$= \exp[(\lambda_1 + \lambda_2)z] \prod_{1 \le \alpha \le n} (a_\alpha - \lambda_1)(a_\alpha - \lambda_2) \cdot (\lambda_2 - \lambda_1)^{-1}$$

$$\times \det \begin{pmatrix} 1, & 1 \\ \lambda_1 + \displaystyle\sum_\alpha \dfrac{A_\alpha^1}{a_\alpha - \lambda_1}, & \lambda_2 + \displaystyle\sum_\alpha \dfrac{A_\alpha^2}{a_\alpha - \lambda_2} \end{pmatrix}\Bigg|_{\lambda_1,\lambda_2=0}$$

$$= \exp[(\lambda_1 + \lambda_2)z] \prod_\alpha (a_\alpha - \lambda_1)(a_\alpha - \lambda_2)$$

$$\times \left[1 + \sum_\alpha \frac{A_\alpha^1}{(a_\alpha - \lambda_1)(a_\alpha - \lambda_2)}\right]\Bigg|_{\lambda_1,\lambda_2=0} = \prod_\alpha a_\alpha^2 \left(1 + \sum_\alpha \frac{A_\alpha^1}{a_\alpha^2}\right).$$

Continuing the reduction and making use of a technique similar to (3.7) we get the required expressions

$$
\prod_{1 \leq j \leq s} \varphi_j^{s+1-j} = \lim_{\lambda_1,\ldots,\lambda_s \to 0} W^{-1}(\lambda_1,\ldots,\lambda_s) V(\Psi, \dot{\Psi}, \ddot{\Psi}, \ldots, \overset{[\cdot(s-1)]}{\Psi})
$$

$$
= \prod_{1 \leq \alpha \leq n} a_\alpha^s D_{s-1}(\delta_{ij} + B_j^i + \sum_\alpha A_\alpha^i a_\alpha^{-(j+1)}), \quad 1 \leq i, j \leq r,
$$

(3.21)

where D_s are the principal minors of order s of $r \times r$ matrix, $D_0 \equiv 1$, and B_j^i are nonzero for $i \leq j+2$, cf. (3.11). In what follows the $r \times r$ matrix

$$
J_{ij} \equiv \delta_{ij} + B_i^j + \sum_\alpha \frac{A_\alpha^i a_\alpha^{r-j}}{(a_\alpha^{r+1} - \lambda^{r+1})}
$$

(3.22)

will be called the *integrals of motion matrix*, since the conserved quantities are determined by the coefficients of the n-th degree polynomial in λ^{r+1} related to the determinant of (J_{ij}) via

$$
\prod_{1 \leq \alpha \leq n} (a_\alpha^{r+1} - \lambda^{r+1}) \det J_{ij} = \prod_{1 \leq \alpha \leq n} (\lambda_\alpha^{r+1} - \lambda^{r+1}).
$$

(3.23)

The above results imply that a linear differential equation of order $(r+1)$ whose coefficient functions coincide with generalized Bargmann potentials can be expressed in the form

$$
(\exp \rho_r (\exp \rho_{r-1}(\ldots (\exp \rho_1 (\exp(-x_1)\Psi)^\cdot)^\cdot \ldots)^\cdot)^\cdot = \lambda^{r+1} \exp x_r \cdot \Psi \quad (3.24)
$$
$$\underbrace{\qquad\qquad\qquad\qquad\qquad\qquad\qquad}_{r+1}$$

where

$$
\exp x_1 = \prod_{\alpha=1}^{n} a_\alpha, \quad \exp x_2 = J_1 \exp 2x_1, \ldots, \exp x_s = J_{s-1} \exp(sx_1).
$$

Here the J_s are the principal minors of the matrix (3.22) for $\lambda = 0$, $\rho_s \equiv (kx)_s$, and k is the Cartan matrix of A_r.

The form (3.24) of the equation whose coefficient functions coincide with the generalized Bargmann potentials suggests their relation with A_r. Indeed, as we will see in the following section the spectral equations of the 1st fundamental representation of A_r coincide with (3.24). This fact is crucial in the finding the simplest soliton solutions of the periodic Toda lattice connected with fundamental representations of A_r.

§ 6.4 Soliton solutions for the vector representation of A_r*)

This section generalizes the results of 6.2 on the sine-Gordon equation (related with A_1) to the case of the *simplest* (standard or identity or vector) representation of A_r. In this case we are lucky to get explicit equations for the scalar LA-pair and derive from them the information needed to find soliton solutions. To make the general expression of the LA-pair concrete let us make use of the explicit form of the Cartan matrix of A_r and the decomposition of its maximal root with the respect to the simple ones:

$$m = \sum_{1 \le j \le r} \pi_j, \quad (m\rho) = \rho_1 + \rho_r.$$

Let us express the root vector corresponding to m in the form

$$X_{+m} = [X_{+1}[X_{+2}[\dots [X_{+(r-1)}, X_{+r}]\dots]]],$$

$$X_{-m} = X^{\dagger}_{+m} = [X_{-r}[X_{-(r-1)}[\dots [X_{-2}, X_{-1}]\dots]]].$$

The highest vector $|1\rangle$ of the 1st fundamental representation of A_r satisfies

$$X_{+j}|1\rangle = 0, \quad h_j|1\rangle = \delta_{j1}|1\rangle.$$

Clearly, $r + 1$ vectors of this representation are

$$|1\rangle, X_{-1}|\rangle, \ X_{-2}X_{-1}|1\rangle, \dots, X_{-r}\dots X_{-2}X_{-1}|1\rangle;$$

$$(\langle 1|, \ \langle 1|X_{+1}, \dots, \langle 1|X_{+1}\dots X_{+r}, \text{ resp.})$$

$$\langle 1|X_{-j} = 0, \quad \langle 1|h_j = \delta_{j1}\langle 1|.$$

According to the general scheme of 6.1 let us first express the matrix elements $\langle j|g|1\rangle$ via the derivatives of the wave function $\Psi = \langle 1|g|1\rangle$ of this representation with respect to z_+, namely

$$\exp x_1 (\Psi \exp(-x_1))^{\cdot} = \langle 1|X_{+1}g|1\rangle,$$

$$\exp(x_2 - x_1)(\exp \rho_1 (\Psi \exp(-x_1))^{\cdot})^{\cdot} = \langle 1|X_{+1}X_{+2}g|1\rangle,$$

$$\dots\dots\dots\dots\dots\dots\dots\dots\dots\dots\dots\dots\dots\dots\dots\dots\dots\dots\dots \quad (4.1)$$

$$\exp(x_s - x_{s-1})(\exp \rho_{s-1}(\exp \rho_{s-2}(\dots (\underbrace{\Psi \exp(-x_1))^{\cdot}\dots)^{\cdot})^{\cdot}}_{s}$$

$$= \langle 1|X_{+1}X_{+2}\dots X_{+s}g|1\rangle; \quad x_0 \equiv 0, \ 1 \le s \le r.$$

Differentiating the last of equations (4.1) yields the spectral equation

$$(\exp \rho_r(\exp \rho_{r-1}(\dots (\exp \rho_1 (\Psi \exp(-x_1))^{\cdot})^{\cdot}\dots)^{\cdot})^{\cdot})^{\cdot}}_{(r+1)} = \lambda^{r+1}\Psi \exp x_r. \quad (4.2)$$

---

*) [126]

Similar calculation of the matrix elements $\langle j|g|1\rangle$ with the help of derivatives of the wave function with respect to z_- results in

$$\Psi' = \exp(x_1 + x_r)\lambda^{-(r+1)}\langle|X_{+1}X_{+2}\ldots X_{+r}g|1\rangle,$$

$$(\Psi'\exp(-x_1 - x_r))' = \lambda^{-(r+1)}\exp(-\rho_r)\langle 1|X_{+1}X_{+2}\ldots X_{+(r-1)}g|1\rangle,$$

$$\cdots\cdots\cdots\cdots\cdots\cdots\cdots\cdots\cdots\cdots\cdots\cdots\cdots\cdots\cdots\cdots$$

$$(\exp\rho_{s+1}(\exp\rho_{s+2}(\ldots(\exp\rho_r(\underbrace{\Psi'\exp(-x_1 - x_2))')'\ldots)')'}_{r-s}$$

$$= \lambda^{-(r+1)}\langle 1|X_{+1}X_{+2}\ldots X_{+s}g|1\rangle, \quad 1 \le s \le r.$$

(4.3)

Comparing (4.3) and (4.2) we get r equations of the scalar LA-pair

$$\exp\rho_{s+1}(\exp\rho_{s+2}(\ldots(\exp\rho_r(\underbrace{\Psi'\exp(-x_1 - x_r))')'\ldots)'}_{r-s}$$

$$= \lambda^{-(r+1)}\exp(x_s - x_{s-1})(\exp\rho_{s-1}(\exp\rho_{s-2}(\ldots$$

$$\ldots(\exp\rho_1(\underbrace{\Psi\exp(-x_1))^{\cdot})^{\cdot}\ldots)^{\cdot})^{\cdot}}_{s}, \quad 1 \le s \le r.$$

(4.4₁)

The two missing equations are of the form

a) $(\exp\rho_r(\exp\rho_{r-1}\ldots$

$$(\exp\rho_1(\underbrace{\Psi\exp(-x_1))^{\cdot})^{\cdot}\ldots)^{\cdot})^{\cdot}}_{r+1} = \lambda^{r+1}\Psi\exp x_r,$$

b) $(\exp\rho_2(\exp\rho_3(\ldots$

$$(\exp\rho_r(\underbrace{\Psi'\exp(-x_1 - x_r))')'\ldots)')'}_{r} = \lambda^{-(r+1)}\Psi\exp(-\rho_1).$$

(4.4₂)

It follows from the explicit form (4.4) of the equations of the scalar LA-pair for the 1st fundamental (vector-) representation of A_r that they are invariant with respect to the changes

$$\ll\cdot\gg\;\rightleftharpoons\;\ll'\gg\;(z_+ \rightleftharpoons z_-),\quad \Psi \to \Psi\exp(-x_1),\quad \lambda \to \lambda^{-1},$$

$$x_1 \to -x_1,\; x_2 \to x_r - x_1,\; x_3 \to x_{r-1} - x_1,\ldots,\; x_s \to x_{r-s+2} - x_1,$$

$$1 \le s \le r;\quad x_{r+1} \equiv 0,$$

i.e., the Weyl reflections with respect to the simple root π_1.

In the first spectral equation (4.4₂) there is no term with $\overset{[\cdot r]}{\Psi}$. Indeed, the coefficient of $\overset{[\cdot r]}{\Psi}$, as is easy to prove by induction, equals

$$-(r+1)\dot{x}_1 + r\dot{\rho}_1 + (r-1)\dot{\rho}_2 + \ldots + \dot{\rho}_r = 0.$$

A similar statement holds also for the second of the spectral equations (4.4_2) if for the wave function we take $\tilde{\Psi} \equiv \Psi \exp(-x_1)$. (The coefficient at $\overset{['(r-1)]}{\Psi}$ in the equation on $\tilde{\Psi}$ vanishes.)

As a corollary of (4.4_1) we get a chain of equations for the $(r+1)$-th order determinants:

$$(\Psi, \dot{\Psi}, \ldots, \overset{[\cdot r]}{\Psi}) = \lambda^{r+1} \exp(-x_1 - x_r)(\Psi, \dot{\Psi}, \ldots, \overset{[\cdot(r-1)]}{\Psi}, \Psi')$$

$$= \lambda^{2(r+1)} \exp(\rho_r - 2x_1 - 2x_r)(\Psi, \dot{\Psi}, \ldots, \overset{[\cdot(r-2)]}{\Psi}, \Psi'', \Psi')$$

$$= \lambda^{3(r+1)} \exp(\rho_{r-1} + 2\rho_r - 3x_1 - 3x_r)(\Psi, \dot{\Psi}, \ldots, \overset{[\cdot(r-3)]}{\Psi}, \Psi''', \Psi'', \Psi')$$

$$= \lambda^{s(r+1)} \exp(\rho_{r-s+2} + 2\rho_{r-s+1} + \ldots + (s-1)\rho_r - sx_1 - sx_r)(\Psi, \dot{\Psi}, \ldots,$$

$$\overset{[\cdot(r-s)]}{\Psi}, \overset{['s]}{\Psi}, \ldots, \Psi') = \ldots$$

$$\ldots = \lambda^{r(r+1)} \exp(-(r+1)x_1)(\Psi, \overset{['r]}{\Psi}, \ldots, \Psi'', \Psi')$$

$$= \lambda^{r(r+1)}(\tilde{\Psi}, \tilde{\Psi}', \ldots, \overset{['r]}{\Psi})(-1)^{\frac{r(r-1)}{2}}, \quad \rho_{r+1} \equiv 0.$$

$$(4.5)$$

By the above-mentioned properties of the spectral equations the first term in (4.5) does not depend on z_+, while the last one does not depend on z_-. Therefore, each term in the chain (4.5) is a constant which in general depends on λ.

Before we pass to the study of corollaries of (4.4) and (4.5) let us consider in detail the particular case $r = 2$. Notice that associated with A_2 is a first cousin of the sine-Gordon equation, the Bulough-Dodd equation, is associated whose soliton solutions are obtained from the soliton solutions for the periodic Toda lattice for A_2 under certain restrictions on the parameters.

The LA-pair for A_2 consists of the four equations (1.6) and the system (4.5) takes the form

$$(\Psi, \dot{\Psi}, \ddot{\Psi}) = \lambda^3 \exp(-x_1 - x_2)(\Psi, \dot{\Psi}, \Psi')$$

$$= \lambda^6 \exp(-3x_1), (\Psi, \Psi'', \Psi') \qquad (4.6)$$

$$\equiv \lambda^6(\tilde{\Psi}, \tilde{\Psi}'', \tilde{\Psi}') \quad (\tilde{\Psi} \equiv \Psi \exp(-x_1)).$$

After the appropriate transformations the spectral equations $(1.6_{a,d})$ turn into

$$\ddot{\Psi} - W_2\dot{\Psi} - W_3\Psi = \lambda^3\Psi, \qquad (4.7)$$

$$(\exp(-x_1)\Psi)''' - \tilde{W}_2(\exp(-x_1)\Psi)' - W_3(\exp(-x_1)\Psi)$$
$$= \lambda^{-3}(\exp(-x_1)\Psi), \qquad (4.8)$$

where

$$W_2 = \ddot{x}_1 + \ddot{x}_2 + \dot{x}_1^2 - \dot{x}_1\dot{x}_2 + \dot{x}_2^2 \text{ and } W_3 = \dddot{x}_1 + \dot{x}_1\ddot{p}_1 + \dot{x}_1\dot{x}_2(\dot{x}_1 - \dot{x}_2)$$

are the 2-nd and 3-rd order integrals (solutions of the characteristic equations associated with an exactly integrable system of equations of the generalized Toda lattice with fixed end-points, cf. 5.2).

The equation (4.8) is obtained from (4.7) under the Weyl group transformation

$$\Psi \to (\exp(-x_1)\Psi), \ \lambda \to \lambda^{-1}, \ \ll \cdot \gg \to \ll' \gg, x_1 \to -x_1, x_2 \to x_2 - x_1,$$

i.e., $\tilde{W}_j(x_1, x_2) = W_j(-x_1, x_2 - x_1)$. Let us seek a wave function of (1.6) in soliton form $\Psi = c \cdot \exp(\lambda z_+ + \lambda^{-1} z_-) \prod_{1 \leq \alpha \leq n} (a_\alpha - \lambda)$, where a_α is the set of unknown functions. The spectral equations $(1.6_{a,d})$ are typical of the equations considered in 6.3 and for their integration we will make use of the methods of 6.3. Define the functions a_α entering Ψ from the condition that $\tilde{\Phi}$ should vanish at n points of the λ^3-plane $\lambda^3 = (\lambda_1^3, \lambda_2^3, \dots, \lambda_n^3)$ such that $\prod_\alpha \lambda_\alpha^3 = 1$, where

$$\tilde{\Phi} \equiv \sum_{1 \leq J \leq 3} d(\lambda_J) \exp(\lambda_J z_+ + \lambda_J^{-1} z_-) \prod_{1 \leq \alpha \leq n} (a_\alpha - \lambda_J). \qquad (4.9)$$

Then the equations $(1.6_{a,d})$ will be satisfied and their coefficient functions will be expressed via the generalized Bargmann potentials. As a corollary of (4.9) and the results of 6.3 we see that $\Psi = c \exp(\lambda z_+ + \lambda^{-1} z_-) \prod_\alpha (a_\alpha - \lambda)$ satisfies (1.6_a) relative the differentiation with respect to z_+ under the identifications

$$\exp x_1 = \prod_\alpha a_\alpha, \ \exp x_2 = J_1 \exp 2x_1 = \prod_\alpha a_\alpha^2 \left(1 + \sum_\beta \dot{a}_\beta a_\beta^{-2}\right).$$

The spectral equation (1.6_d) in the wave function $\tilde{\Psi} \equiv \Psi \exp(-x_1)$ is invariant under the change

$$\lambda \to \lambda^{-1}, \ a_\alpha \to a_\alpha^{-1}, \ \ll \cdot \gg \to \ll' \gg, \ x_1 \to -x_1, \ x_2 \to x_2 - x_1$$

in (1.6_a) and therefore leads to

$$\exp(-x_1) = \prod_\alpha a_\alpha^{-1}, \ \exp(x_2 - x_1) = \prod_\alpha a_\alpha^{-2} \left(1 - \sum_\beta a_\beta'\right),$$

i.e. $\quad \exp x_1 = \prod_\alpha a_\alpha,$

$$\exp x_2 = \prod_\alpha a_\alpha^{-1} \left(1 - \sum_\beta a_\beta'\right) = \prod_\alpha a_\alpha^2 \left(1 + \sum_\beta \dot{a}_\beta a_\beta^{-2}\right).$$

Now let us calculate the determinants from (4.6):

$$(\Psi, \dot{\Psi}, \ddot{\Psi}) = W(\lambda_1, \lambda_2, \lambda_3) \prod_\alpha (\lambda_\alpha^3 - \lambda^3) \quad \text{(see (3.7))},$$

$$\lambda^3(\Psi, \dot{\Psi}, \Psi') = \lambda^3 \prod_\alpha (a_\alpha^3 - \lambda^3) \cdot \left(1, \lambda + \sum_\alpha \frac{\dot{a}_\alpha}{(a_\alpha - \lambda)}, \ \lambda^{-1} + \sum_\alpha \frac{a_\alpha'}{(a_\alpha - \lambda)}\right)$$

$$= W \prod_\alpha (a_\alpha^3 - \lambda^3) \cdot \left(1, \lambda + \sum_\alpha \frac{\dot{a}_\alpha}{a_\alpha - \lambda}, \ \lambda^2 \left(1 - \sum_\alpha a_\alpha'\right)\right.$$

$$\left. - \lambda \sum_\alpha a_\alpha a_\alpha' + \sum_\alpha a_\alpha^2 a_\alpha' + \sum_\alpha \frac{a_\alpha^3 a_\alpha'}{(a_\alpha - \lambda)}\right)$$

$$= W(\lambda_1, \lambda_2, \lambda_3) \prod_\alpha (a_\alpha^3 - \lambda^3)$$

$$\times \det \begin{pmatrix} 1 + \sum_\alpha \dfrac{a_\alpha \dot{a}_\alpha}{(a_\alpha^3 - \lambda^3)} & , & \sum_\alpha \dfrac{\dot{a}_\alpha}{(a_\alpha^3 - \lambda^3)} \\[3mm] -\sum_\alpha a_\alpha a_\alpha' + \sum_\alpha \dfrac{a_\alpha' a_\alpha^4}{(a_\alpha^3 - \lambda^3)}, & & 1 - \sum_\alpha a_\alpha' + \sum_\alpha \dfrac{a_\alpha^3 a_\alpha'}{(a_\alpha^3 - \lambda^3)} \end{pmatrix}.$$

The latter equation implies that $\lambda^3(\Psi, \dot{\Psi}, \Psi') \exp(-x_1 - x_2)$ is an n-th degree polynomial in λ^3 (up to W) with zeros at n points of the λ^3-plane thanks to (4.9), which therefore coincides up to a constant factor with $\prod_\alpha (\lambda_\alpha^3 - \lambda^3)$. Comparing the values of these polynomials at $\lambda = 0$ and $\lambda = \infty$ we get

$$1 = \exp(-x_1 - x_2)\left(1 - \sum_\alpha a_\alpha'\right),$$

$$\exp(-x_1 - x_2)\left(1 + \sum_\alpha \frac{\dot{a}_\alpha}{a_\alpha^2}\right) \prod_\beta a_\beta^3 = \prod_\alpha \lambda_\alpha^3,$$

i.e., the system of equations holds for $\prod_\alpha \lambda_\alpha^3 = 1$. We get the same condition requiring the equality of the first and the third terms in the chain (4.7).

Thus, a solution of the periodic Toda lattice for A_2 is

$$\exp x_1 = \prod_\alpha a_\alpha, \ \exp x_2 = \prod_\alpha a_\alpha^{-1}\left(1 - \sum_\beta a_\beta'\right) = \prod_\alpha a_\alpha^2 \left(1 + \sum_\beta (a_\beta^{-1}) \cdot\right),$$

$$\tag{4.10}$$

where the elementary symmetric functions a_i which enter (4.10) are determined from the linear system (4.9). The expressions (4.10) depend on $3n - 1$ parameters: $2n$ values $d(\lambda_2)/d(\lambda_1)$, $d(\lambda_3)/d(\lambda_1)$ at the points $\lambda_\alpha^3 = \lambda^3$ $(1 \le \alpha \le n)$ and $n - 1$ more parameters since $\prod_\alpha \lambda_\alpha^3 = 1$.

In the general case the following theorem holds for the solutions of the system of equations for the scalar LA-pair of the first fundamental representation of A_r:

There is a solution of (4.4) *whose wave function analytically depends on λ as follows*

$$\Psi = c \exp(\lambda z_+ + \lambda^{-1} z_-) \prod_{1 \leq \alpha \leq n} (a_\alpha - \lambda),$$

where $a_\alpha(z_+, z_-)$ are determined from the condition that

$$\tilde{\Phi} \equiv \sum_{1 \leq J \leq r+1} d(\lambda_J) \exp(\lambda_J z_+ + \lambda_J^{-1} z_-) \prod_{1 \leq \alpha \leq n} (a_\alpha - \lambda_J)$$

vanishes at n points $\lambda_1^{r+1}, \ldots, \lambda_n^{r+1}$ of λ^{r+1}-plane such that $\prod_{1 \leq \alpha \leq n} \lambda_\alpha^{r+1} = 1$.

The functions x_j satisfying the equations of the periodic Toda lattice for A_r are then expressed via the elementary symmetric functions a_α and their derivatives by the formula

$$\exp x_1 = \prod_{1 \leq \alpha \leq n} a_\alpha, \ldots, \exp x_s = J_{s-1} \prod_{1 \leq \alpha \leq n} a_\alpha^s = \tilde{J}_{r+1-s} \prod_{1 \leq \alpha \leq n} a_\alpha^{-(r+1-s)}$$

where $J_j (\tilde{J}_j)$ are the principal minors of the matrix of the integrals of motion (with respect to differentiations over $z_+(z_-)$) at $\lambda = 0$ ($\lambda = \infty$) and $J_0 \equiv 1$ ($\tilde{J}_0 \equiv 1$). The matrix $\tilde{J}$ is obtained from J under a Weyl group transformation.

In order for Ψ from the above theorem to satisfy the spectral equation (4.4_{2a}) it suffices thanks to (3.24) to identify

$$\exp x_s = J_{s-1} \prod_{1 \leq \alpha \leq n} a_\alpha^s. \tag{4.11}$$

For fulfillment of the spectral equation (4.4_{2b}) the relation between x_s and a_α should be expressed as

$$\exp x_s = \tilde{J}_{r+1-s} \prod_{1 \leq \alpha \leq n} a_\alpha^{-(r+1-s)}, \tag{4.12}$$

which is obtained from (4.11) by a transformation from the Weyl group.

It remains to establish that (4.11) and (4.12) are noncontradicted and (4.4_1) or, equivalently, (4.5) holds. The determinant $(\Psi, \dot{\Psi}, \ldots, \overset{[\cdot r]}{\Psi})$ has been calculated and $(\Psi, \dot{\Psi}, \ldots, \overset{[\cdot r]}{\Psi}) = W_{r+1} \prod_\alpha (\lambda_\alpha^{r+1} - \lambda^{r+1})$.

To calculate the last determinant in (4.5) we notice that

$$\tilde{\Psi} \equiv \Psi \exp(-x_1) = c \exp(\lambda z_+ + \lambda^{-1} z_-) \prod_\alpha (1 - \lambda a_\alpha^{-1})$$

$$\equiv (-1)^n \lambda^n c \exp(\lambda z_+ + \lambda^{-1} z_-) \prod_\alpha (a_\alpha^{-1} - \lambda^{-1}),$$

i.e., $\tilde{\Psi}$ is obtained from Ψ (up to a factor $(-1)^n \lambda^n$) under the formal change

$$\lambda \to \lambda^{-1}, \quad a_\alpha \to a_\alpha^{-1}, \quad \ll \cdot \gg \to \ll' \gg .$$

This immediately enables us to write the formula for this determinant:

$$(\tilde{\Psi}, \tilde{\Psi}', \ldots, \overset{['r]}{\tilde{\Psi}} = (-1)^{n(r+1)} \lambda^{n(r+1)} W_{r+1}(\lambda^{-1}) \prod_\alpha (\lambda_\alpha^{-(r+1)} - \lambda^{-(r+1)})$$

$$= (-1)^{2n(r+1)} W_{r+1}(\lambda) (-1)^{\frac{r(r-1)}{2}} \lambda^{-r(r+1)} \prod_\alpha (\lambda_\alpha^{r+1} - \lambda^{r+1}) \cdot \lambda_\alpha^{-(r+1)}.$$

Substituting these expressions into (4.5) we get constraints for the admissible values of λ_α^{r+1} entering the above Theorem: $\prod_\alpha \lambda_\alpha^{r+1} = 1$.

Calculating the second determinant from (4.5) we get

$$\lambda^{r+1}(\Psi, \dot{\Psi}, \ldots, \overset{[\cdot(r-1)]}{\Psi}, \Psi')$$

$$= \lambda^{r+1} \prod_\alpha (a_\alpha^{r+1} - \lambda^{r+1}) \cdot (1, \varphi^1, \varphi^2, \ldots, \varphi^{r-1}, \tilde{\varphi}^1)$$

$$= \prod_\alpha (a_\alpha^{r+1} - \lambda^{r+1}) \cdot \left(1, \lambda + \sum_\alpha \frac{A_\alpha^1}{a_\alpha - \lambda}, \ldots, \lambda^s + \sum_{0 \le p \le s-2} B_s^p \lambda^p \right.$$

$$+ \sum_\alpha \frac{A_\alpha^s}{a_\alpha - \lambda}, \ldots, \lambda^{r-1} + \sum_{0 \le p \le r-3} B_{r-1}^p \lambda^p + \sum_\alpha \frac{A_\alpha^{r-1}}{a_\alpha - \lambda}, \lambda^r + \left. \sum_\alpha \frac{a_\alpha'}{a_\alpha - \lambda} \cdot \lambda^{r+1} \right)$$

$$= W_{r+1} P_n(\lambda^{r+1}) = W_{r+1} \prod_\alpha (\lambda_\alpha^{r+1} - \lambda^{r+1}) \cdot \exp(x_1 + x_r).$$

$$(4.13)$$

By the conditions of the theorem, $\lambda^{r+1}(\Psi, \dot{\Psi}, \ldots, \Psi^{[\cdot(r-1)]}, \Psi') = 0$ at n-points of the λ^{r+1}-plane and therefore can differ from $\prod_\alpha (\lambda_\alpha^{r+1} - \lambda^{r+1})$ only by a constant factor. To find this factor it suffices to calculate the ratio of these polynomials at some point of the λ^{r+1}-plane.

The general scheme for calculating determinants of this type (cf. 6.3) shows that $P_n(\lambda^{r+1})$ is determined by the determinant of an $r \times r$ matrix whose first $r - 1$ rows are

$$\delta_{ij} + B_i^j + \sum_\alpha \frac{A_\alpha^i a_\alpha^{r-j}}{(a_\alpha^{r+1} - \lambda^{r+1})},$$

and the r-th row is given by

$$\delta_{jr} + \sum_\alpha \frac{a_\alpha' a_\alpha^{r-j} \lambda^{r+1}}{(a_\alpha^{r+1} - \lambda^{r+1})}.$$

We have

$$P_n(0) = \det_{r-1}\left(\delta_{ij} + B_i^j + \sum_\alpha A_\alpha^i a_\alpha^{-(j+1)}\right) = J_{r-1},$$

and

$$P_n(\lambda) \xrightarrow[\lambda\to\infty]{} \left(1 - \sum_\alpha a_\alpha'\right)\lambda^{n(r+1)}(-1)^n = \lambda^{n(r+1)}(-1)^n \tilde{J}_1,$$

implying

$$\exp(x_1 + x_r) = \left(1 - \sum_\alpha a_\alpha'\right) = \tilde{J}_1 = J_{r-1}\cdot\prod_\alpha a_\alpha^{r+1}\lambda_\alpha^{-(r+1)},$$

i.e.,

$$\exp x_r = \tilde{J}_1\prod_\alpha a_\alpha^{-1} = J_{r-1}\prod_\alpha a_\alpha^r.$$

Now let us calculate the third term of the chain (4.5):

$$\lambda^{2(r+1)}\exp(-x_{r-1} - 2x_1)(\Psi, \dot\Psi, \ldots, \overset{[\cdot(r-2)]}{\Psi}, \Psi'', \Psi')$$

$$= \lambda^{2(r+1)}\exp(-x_{r-1})(\Psi, \dot\Psi, \ldots, \overset{[\cdot(r-2)]}{\Psi}, \tilde\Psi'', \tilde\Psi')$$

$$= \lambda^{2(r+1)}\exp(-x_{r-1})\prod_\alpha(a_\alpha^{r+1} - \lambda^{r+1})\cdot(1, \varphi^1, \ldots, \varphi^{r-2},$$

$$\left(\lambda^{-2} + \bar B_\alpha^0 + \sum_\alpha\frac{\bar A_\alpha^2}{a_\alpha^{-1} - \lambda^{-1}}\right)\exp(-x_1), \left(\lambda^{-1} + \sum_\alpha\frac{\bar A_\alpha^1}{a_\alpha^{-1} - \lambda^{-1}}\right)\exp(-x_1))$$

$$= \exp(-x_{r-1})\prod_\alpha(a_\alpha^{r+1} - \lambda^{r+1})\cdot(1, \varphi^1, \ldots, \varphi^{r-2}, (\lambda^{r-1} + \bar B_2^0\lambda^{r+1}$$

$$+ \sum_\alpha\frac{\bar A_\alpha^2\lambda^{r+1}}{a_\alpha^{-1} - \lambda^{-1}}\right)\exp(-x_1), \left(\lambda^r + \sum_\alpha\frac{\bar A_\alpha^1\lambda^{r+1}}{a_\alpha^{-1} - \lambda^{-1}}\right)\exp(-x_1)).$$

$$(4.14)$$

(Here $\bar A_\alpha^s$, $\bar B_j^i$ are obtained from A_α^s, B_j^i under the change $a_\alpha \to a_\alpha^{-1}$ and $\ll\cdot\gg\,\to\,\ll'\gg$.)

The determinant in the last expression reduces to the determinant of a matrix whose first $r - 2$ rows are given by the matrix of integrals of motion (3.22) and whose lowest two rows are given by the matrix

$$\tilde J_{ij} = \exp(-x_1)\left(\delta_{ij} + \bar B_j^i + \sum_\alpha\frac{A_\alpha^i a_\alpha^{-j+1}}{(a_\alpha^{-(r+1)} - \lambda^{-(r+1)})}\right). \qquad (4.15)$$

(Recall that B^i_j, $\bar{B}^i_j$ are nonzero only for $j \geq i + 2$.) We have

$$J_{ij}|_{\lambda=0} = \delta_{ij} + B^j_i + \sum_\alpha A^i_\alpha a_\alpha^{-j-1},$$

whereas

$$\tilde{J}_{ij}|_{\lambda=0} = \exp(-x_1)\left(\delta_{ij} + \bar{B}^i_j\right)$$

and (J_{ij}) turns into a lower triangular matrix at $\lambda = \infty$, whereas

$$\tilde{J}_{ij}|_{\lambda=\infty} = \exp(-x_1)\left(\delta_{ij} + \bar{B}^i_j + \sum_\alpha A^i_\alpha a_\alpha^{r-j+2}\right).$$

Returning to calculating the determinant (4.14) we see that it is an n-th degree polynomial in λ^{r+1} with zeros at n points of the λ^{r+1}-plane and therefore differs from $(\Psi, \dot{\Psi}, \ldots, \overset{[\cdot r]}{\Psi})$ by a constant factor. Considering the ratio of these polynomials at $\lambda = 0$ and $\lambda = \infty$ we finally get

$$\exp x_{r-1} = J_{r-2} \prod_\alpha a_\alpha^{r-1} = \tilde{J}_2 \prod_\alpha a_\alpha^{-2},$$

where J_{r-2} ($\tilde{J}_2$) is the principal upper left (lower right) corner minor of the matrix J (3.22) or $\tilde{J}$ (4.15). It is not difficult to see that the matrices J and $\tilde{J}$ are obtained from each other under an appropriate permutation of indices and the simultaneous change $a_\alpha \rightleftarrows a_\alpha^{-1}, \ll \cdot \gg \rightleftarrows \ll' \gg$.

Now it is obvious how to calculate all the determinants from (4.5). From this calculation we get

$$\exp x_s = J_{s-1} \prod_\alpha a_\alpha^s = \tilde{J}_{r+1-s} \cdot \prod_\alpha a_\alpha^{-(r+1-s)},$$

proving the above theorem.

Let us give one more form of expression of the function $\tilde{\Phi}$ from the above theorem and with its help construct elementary symmetric functions s_α. Set

$$\mathcal{F}(\lambda) \equiv \sum_{J=1}^{r+1} d(\lambda_J)\exp(\lambda_J z_+ + \lambda_J^{-1} z_-);$$

$$\overset{[\cdot(r+1)]}{\mathcal{F}}(\lambda) = \lambda^{r+1}\mathcal{F}(\lambda).$$

Then

$$\tilde{\Phi} = s_n \mathcal{F} - s_{n-1}\dot{\mathcal{F}} + s_{n-2}\ddot{\mathcal{F}} - \ldots = \sum_{0 \leq \alpha \leq n} s_{n-\alpha}(-1)^\alpha \overset{[\cdot \alpha]}{\mathcal{F}}; \quad s_0 \equiv 1.$$

The system of equations determining s_α is of the form

$$\sum_{0 \leq \alpha \leq n} (-1)^\alpha s_{n-\alpha} \overset{[\cdot \alpha]}{\mathcal{F}}_\beta = 0, \quad 1 \leq \beta \leq n,$$

where $\mathcal{F}_\beta \equiv \mathcal{F}(\lambda_\beta)$ are n independent functions each of which satisfies the equations

$$\overset{[\cdot(r+1)]}{\mathcal{F}}_\beta = \lambda_\beta^{r+1}\mathcal{F}, \quad \overset{['(r+1)]}{\mathcal{F}}_\beta = \lambda_\beta^{-(r+1)}\mathcal{F}.$$

Between the elements of the rows of the matrix of integrals of motion there exists a recursive relation at $\lambda = 0$ which allows one to recover the whole matrix starting from its first row. To derive these relations notice that J_{ij} appears when we expand the functions φ^s (see 6.3) into the Taylor series at $\lambda = 0$:

$$\varphi^s = \varphi^s|_{\lambda=0} + \sum_{1\leq b<\infty} J_{sb}\lambda^b.$$

In other words, the φ^s are generating functions for the matrix of integrals of motion at $\lambda = 0$. Indeed, we have

$$\varphi^s = \lambda^s + \sum_{0\leq j\leq s-2} \lambda^j B_s^j + \sum_{1\leq\alpha\leq n} \frac{A_\alpha^s}{a_\alpha - \lambda}$$

$$= B_s^0 + \sum_\alpha \frac{A_\alpha^s}{a_\alpha} + \lambda_s + \sum_{1\leq b<\infty}\lambda^b\sum_\alpha A_\alpha^s a_\alpha^{-(b+1)} + \sum_{1\leq j\leq s-2}\lambda^j B_s^j = \varphi_0^s + \sum_{1\leq b<\infty} J_{sb}\lambda^b,$$

$$\varphi_0^s \equiv \varphi^s|_{\lambda=0}.$$

Make use of the recursive relation (see 6.3) relating the functions φ^s with neighbouring indices:

$$\varphi^{s+1} = \dot\varphi^s + \varphi^s\varphi^1$$

$$= \dot\varphi_0^s + \sum_{1\leq b<\infty} \dot J_{sb}\lambda^b + \left(\varphi_0^s + \sum_a J_{sa}\lambda^a\right)\cdot\left(\varphi_0^1 + \sum_b J_{1b}\lambda^b\right)$$

$$= \dot\varphi_0^s + \varphi_0^s\varphi_0^1 + \varphi_0^s\sum_b J_{1b}\lambda^b + \varphi_0^1\sum_b J_{sb}\lambda^b + \sum_{1\leq a<\infty}\lambda^a\sum_{1\leq b\leq a-1} J_{s,a-b}J_{1b},$$

i.e.,

$$J_{s+1,b} = \dot J_{sb} + \sum_{1\leq l\leq b-1} J_{s,l-b}J_{1l} + \varphi_0^s J_{1b} + \varphi_0^1 J_{sb}.$$

The last two terms do not contribute to the principal minors since they are proportional to the elements of the 1st and the s-th rows.

Finally, to calculate the principal minors of J we can make use of the recursive relations of the form

$$J_{s+1,b} = \dot J_{sb} + \sum_{1\leq l\leq b-1} J_{s,l-b}J_{1l}, \quad J_{1l} = \delta_{1l} + \sum_{1\leq\alpha\leq n}\dot a_\alpha a_\alpha^{-(l+1)}$$

$$\equiv \delta_{1l} - l^{-1}\left(\sum_\alpha a_\alpha^{-l}\right).$$

$$(4.16)$$

In particular, $J_{s+1,1} = \dot{J}_{s1} = \overset{[\cdot s]}{J_{1r}} = (1 + \sum_\alpha \dot{a}_\alpha a_\alpha^{-2})^{[\cdot s]}$. Similar relations appear in calculations of the principal minors of the lower right corner of $\tilde{J}$ in (4.16) under the standard change $a_\alpha \to a_\alpha^{-1}$, $\ll \cdot \gg \to \ll' \gg$.

In conclusion notice that to get solutions of the periodic Toda lattice for the series A_r we only had to solve linear algebraic equations and calculate determinants of some known matrices. We believe that this suggests that the problem has a purely algebraic solution much simpler than the above.

Chapter 7
Exactly integrable quantum dynamical systems

In this chapter we study the quantum systems described by equations (3.2.8). On the quantum level the dynamical systems can be considered in different representations, in particular in the Schrödinger or Heisenberg ones. For the systems under study in the one-dimensional case we can solve the problem in both these representations. In the Schrödinger representation the wave functions are the matrix elements of the principal continuous series of unitary representations of noncompact real forms of complex semisimple Lie groups taken between the states with definite quantum numbers (generalized Whittaker vectors). At the same time the existence of a Hamiltonian formalism for the considered systems (5.8.1) enables us to apply the usual methods of perturbation theory as for the classical case, cf. 5.8. In such an approach the first term in the Hamiltonian (3.2.14) plays the role of the free Hamiltonian, whereas the second one with a factor λ describes the interaction in the system with the coupling constant λ. In complete analogy with the classical consideration the series of the perturbation theory are polynomials in λ and reproduce an exact solution of the corresponding system. In the one-dimensional case our constructions are essentially based on the representation theory of Lie groups and algebras and the final results are formulated completely in terms of this theory.

In the quantum field (two-dimensional) case the adequacy of the group-theoretical description of these systems is lost, at least in the conventional sense. Though for the majority of exactly solvable systems of type (3.2.8), in particular, for the two-dimensional generalized Toda lattice with fixed endpoints, we can construct explicit expressions for Heisenberg operators as polynomials in λ, their description is no longer given in terms of representation theory. (In modern notions, the problem is solved in terms of the quantum group representations, see e.g., our recent reviews: *Sov. J. Part. Nucl.* **16** (1), 1985, 81–101, and *Acta Appl. Math.*, **16**, 1989, 1–74.) Notice also that unlike the classical case the solution of the one-dimensional problem cannot be obtained from solutions of the corresponding two-dimensional case. The reason is that forms of commutation relations of generalized coordinates and momenta in the spaces of different dimensions are different.

The solutions of the corresponding classical systems in both the one- and two-dimensional cases can be obtained from the quantum ones as Planck's constant $\hbar$ tends to zero. Then the expressions obtained in Chapters 3 and 4 are reproduced.

§ 7.1 The Hamiltonian (canonical) formalism and the Yang-Feldmann method*)

The quantization procedure of systems of the form (3.2.8) developed in this chapter in both the Schrödinger and Heisenberg representations is based on the Hamiltonian formalism. We make use of perturbation theory for the construction of explicit expressions for Heisenberg operators on the corresponding dynamical quantities. As we have mentioned above, we speak not about approximate results but about exact expressions resulting from summation of the series of perturbation theory in powers of the coupling constant λ explicitly introduced in (3.2.8) as a factor of its right-hand side. (In what follows, referring to (3.2.8) and (3.2.13), we will assume that λ is introduced in these equations.)

There are several equivalent formulations of the perturbation theory. In the framework of the canonical formalism the explicit expressions for the Heisenberg operators are determined by the known Schwinger formula (2.44). In a number of cases the Yang-Feldmann method turns out to be more convenient. In this method we explicitly integrate differential equations describing the corresponding quantum system, rewrite them in the integral form and automatically take into account the boundary conditions with subsequent expansion of the desired solutions into a power series in λ.

Usually the Hamiltonian of quantum mechanics is identified with the radial part of the quadratic Casimir operator of a semisimple Lie group G in an appropriate parametrization. The choice of the functions constituting a basis in the representation space of G plays the same role as the choice of the initial conditions in the phase space of the functional algebra $\mathfrak{g}^f$ for the classical problem (cf. 4.6). The Poisson brackets defining the classical system are replaced by the commutators of dynamical variables of the corresponding quantum system. There is a deep relation between the solution of the quantum problem and the representation theory first established by Kostant for systems of Toda lattice type for which he obtains an integral representation

---

*) [2, 23, 24, 53–55, 100, 108]

of one-component wave functions via Whittaker vectors. The latter are the basis functions mentioned.

In accordance with the results of 2.6, the Hamiltonian (3.2.14) of (5.8.1) coincides with the quadratic Casimir operator of the principal continuous series of the corresponding real form in the basis constituted by the generalized Whittaker vectors (2.6.7). Therefore its one-component wave functions $\Psi^{\{\rho,l\}}(\tau,\chi)$ are expressed as follows:

$$
\Psi^{\{\rho,l\}}(\tau,\chi) = \exp\left(\frac{1}{2}\sum_j \rho_j^0 \tau_j\right) \cdot \mathcal{N}(\sigma,l) \int d\mu(k) \overset{*\{l\}}{\mathcal{D}}(k_0)
$$
$$
\times \prod_j [R_j(\tau,\chi;k)]^{\rho_j} W^{\{\rho,l\}}(\tilde{k}).
$$
(1.1)

Here $\mathcal{D}^{\{l\}}(k_0)$ is the matrix element of C_0 satisfying $(2.6.6_1)$. The presence of $\mathcal{N}(\sigma,l)$ and the norming factors $\exp(\rho^0\tau)/2$, where $\rho_j = i\sigma_j - (1/2)\rho_j^0$, $-\infty < \sigma_j < \infty$ and $\sigma_j = \overset{*}{\sigma}_j$ for the principal continuous series is caused by the reduction of the Hamiltonian to a form self-adjoint with respect to the standard scalar product in the representation space considered, the form corresponding to the physical Hamiltonian of the model. Thanks to these factors the latter Hamiltonian has the correct spectrum.

The scattering states are realized in the asymptotic domain $(\underline{\alpha}(\tau) \to \infty)$ of the positive Weyl chamber $(\underline{\alpha}(\tau) > 0$ for all $\alpha)$ since the "particles" are free there.

Now consider how to apply the Yang-Feldmann method for constructing exact solutions of the quantum system (5.8.1) in the Heisenberg representation. For this let us reformulate the construction from 5.8 for the expression of the quantum operator in the dynamical variable $g_0(t)$. (In what follows we will skip the index 0 for brevity, retaining it for the solution of this system at $\lambda = 0$.)

In accordance with the free Lagrangian of the quantum system (5.8.1)

$$
\mathcal{L}_0(g) = (1/2)\mathrm{tr}(g_0^{-1}\dot{g}_0)^2 \quad \text{for} \quad g_0(t) = \exp(-X^0 q),
$$

the momenta $p_\alpha = \sum_\beta c_{\alpha\beta} \dot{q}_\beta$ are canonically conjugate to the coordinates $q_\alpha(t)$, i.e., $[p_\alpha, q_\beta] = -i\hbar\delta_{\alpha\beta}$. Recall that $(X^0 q)$ takes values in $\mathfrak{g}_0^f$, i.e. $1 \leq \alpha$, $\beta \leq d_1 = \dim \mathfrak{g}_0^f$ and therefore $c_{\alpha\beta} \equiv (X_\alpha^0, X_\beta^0)$ is a symmetric invertible $d_1 \times d_1$-matrix. It is convenient to introduce the operators

$$
f_\alpha \equiv \sum_\beta \mathfrak{m}_{\alpha\beta} q_\beta, \quad \mathfrak{m}_{\alpha\beta} \equiv \sum_\gamma C_{\beta\alpha}^{\gamma+},
$$

whose general form is $f_\alpha = A_\alpha t + B_\alpha$, where A_α and B_α are arbitrary operators independent on t and satisfying the relations

$$[A_\alpha, B_\beta] = -i\hbar\Theta_{\alpha\beta}, \quad [A_\alpha, A_\beta] = 0,$$
$$[B_\alpha, B_\beta] = 0; \qquad\qquad \Theta \equiv \mathbf{m}c^{-1}\mathbf{m}^T. \tag{1.2}$$

The system (5.8.2) or its one-dimensional variant (5.7.4) holds also in the quantum case and the same is true for the solutions expressed in the form similar to (5.7.6).

However, the passage to a more compact form (5.8.3) fails on the quantum level since the quantities

$$J_\pm^{(t,-t_2,\ldots,\varepsilon_s t_s)} \equiv g_0^{(t_1,-t_2,\ldots,\varepsilon_s t_s)} J_\pm (g_0^{(t_1,-t_2,\ldots,\varepsilon_s t_s)})^{-1},$$

$$g_0^{(t_1,-t_2,\ldots,\varepsilon_s t_s)} \equiv g_0(t_1)g_0^{-1}(t_2)\ldots g_0^{\varepsilon_s}(t_s),$$

$$\varepsilon_s \equiv (-1)^{s-1}, \quad T_s \equiv t_1 - t_2 + \ldots + \varepsilon_s t_s$$

commute differently. The reduction of the final expression for the solution $g(t)$ to the form of type (5.8.4) requires the reduction of these operators to the form $\tilde{g}_0(T_s)J_\pm\tilde{g}_0^{-1}(T_s)$, where the elements $\tilde{g}_0(T_s)$ do not reduce any more to $g_0(T_s)$ and contain factors depending on Planck's constant. Such a reduction will be illustrated in 7.3 with the example of a generalized Toda lattice.

§ 7.2 Basics from perturbation theory*)

This section is auxiliary. It contains some commutation relations and other formulas needed for the integration of the dynamical systems studied in this chapter.

Consider a one-dimensional quantum-mechanical Lagrangian system of r particles with a potential interaction. Here the Lagrangian is of the form

$$\mathcal{L}(u, \dot{u}) = \frac{1}{2} \sum_{\alpha,\beta=1}^r \underline{k}_{\alpha\beta}\dot{u}_\alpha\dot{u}_\beta - \mu^2 U(u), \tag{2.1}$$

where u_α is the Heisenberg operator of the coordinate of the α-th particle, the dot means the derivation with respect to time, $(\underline{k}_{\alpha\beta})$ is an invertible symmetric real matrix, $\mu^2(\equiv \lambda)$ is the coupling constant and $U(u_1,\ldots,u_r)$ describes the interaction. The corresponding Hamiltonian is

$$\mathfrak{h}(u, p) = \frac{1}{2} \sum_{\alpha,\beta=1}^r (\underline{k}^{-1})_{\alpha\beta}p_\alpha p_\beta + \mu^2 U(u), \tag{2.2}$$

where the canonical momenta equal $p_\alpha = \sum_\beta \underline{k}_{\alpha\beta}\dot{u}_\beta$.

---

*) [9, 10, 106–108, 127]

The equations of motion in Lagrangian and Hamiltonian forms respectively are

$$\ddot{u}_\alpha + \mu^2 \sum_\beta (\underline{k}^{-1})_{\alpha\beta} U_\beta(u) = 0; \quad U_\beta \equiv \partial U/\partial u_\beta;$$

$$\dot{p}_\alpha + \mu^2 U_\alpha(u) = 0, \quad \dot{u}_\alpha = \sum_\beta (\underline{k}^{-1})_{\alpha\beta} p_\beta. \tag{2.3}$$

To apply the perturbation theory consider first the system of non-interacting particles with

$$\mathcal{L}_0(\varphi, \dot{\varphi}) = \frac{1}{2} \sum_{\alpha,\beta} \underline{k}_{\alpha\beta} \dot{\varphi}_\alpha \dot{\varphi}_\beta; \quad \mathfrak{h}_0(\varphi, p^{(0)}) = \frac{1}{2} \sum_{\alpha,\beta} (\underline{k}^{-1})_{\alpha\beta} p_\alpha^{(0)} p_\beta^{(0)}.$$

The general solution for the Heisenberg coordinates φ_α of free particles is a uniform motion along the straight lines:

$$\varphi_\alpha = \tilde{a}_\alpha t + \tilde{b}_\alpha, \quad p_\alpha^{(0)} = \sum_\beta \underline{k}_{\alpha\beta} \tilde{a}_\beta,$$

where $\tilde{a}_\alpha$ and $\tilde{b}_\alpha$ are operators that do not depend on time. The canonical commutation relations (at equal times) are

$$[p_\alpha^{(0)}, \varphi_\beta] = -i\hbar\delta_{\alpha\beta}; \quad [p_\alpha^{(0)}, p_\beta^{(0)}] = [\varphi_\alpha, \varphi_\beta] = 0$$

and are expressed in terms of $\tilde{a}$ and $\tilde{b}$ in the form

$$[\tilde{a}_\alpha, \tilde{b}_\beta] = -i\hbar(\underline{k}^{-1})_{\alpha\beta}, \quad [\tilde{a}_\alpha, \tilde{a}_\beta] = [\tilde{b}_\alpha, \tilde{b}_\beta] = 0.$$

In what follows we will be interested in interaction functions that depend on their arguments via the linear combinations $u'_\alpha = \sum_\beta k_{\alpha\beta} u_\beta$ with the matrix k being related to $\underline{k}$ by the formula

$$\underline{k}_{\alpha\beta} = v_\alpha k_{\alpha\beta}, \tag{2.4}$$

where v_α are some positive numbers. Therefore introduce the new operators

$$\phi_\alpha \equiv \sum_\beta k_{\alpha\beta} \varphi_\beta \equiv a_\alpha t + b_\alpha, \tag{2.5}$$

and express the old ones by the formulas

$$\varphi_\alpha = \sum_\beta (k^{-1})_{\alpha\beta} \phi_\beta; \quad p_\alpha^{(0)} = v_\alpha \cdot a_\alpha. \tag{2.6}$$

Then the commutation relations (at arbitrary times) take the form

$$[a_\alpha, b_\beta] = -i\hbar\hat{k}_{\alpha\beta}; \qquad\qquad [a_\alpha, a_\beta] = [b_\alpha, b_\beta] = 0;$$

$$[\phi_\alpha(t), \phi_\beta(t')] = -i\hbar\hat{k}_{\alpha\beta}(t - t'); \quad [\varphi_\alpha(t), \phi_\beta(t')] = -i\hbar v_\alpha^{-1}\delta_{\alpha\beta}(t - t'), \tag{2.7}$$

where

$$\hat{k}_{\alpha\beta} \equiv k_{\alpha\beta} v_\beta^{-1}. \tag{2.8}$$

Since in this chapter we consider mostly the interaction of exponential type, we will also need the rules of multiplication for the operators of type $\exp \phi_\alpha$ and for the operators constructed with the $\exp \phi_\alpha$.

Here are the general formulas for operations of this type. First

$$\exp(-A) \cdot B \cdot \exp A = B + \sum_{l=1}^{\infty} \frac{1}{l!} [B, \underbrace{A, A, \dots, A}_{l \text{ times}}], \tag{2.9}$$

where

$$[A, B, C, \dots, M] \equiv [\dots[[A, B], C], \dots], M]. \tag{2.10}$$

Furthermore,

$$\exp A \cdot \exp B = T \exp \left\{ \int_0^1 d\tau (A + \exp(\tau A) \cdot B \cdot \exp(-\tau A)) \right\}, \tag{2.11}$$

where the "$T\exp$" sign means the ordering with respect to the parameter τ in the exponential series:

$$T \exp \left\{ \int_a^b d\tau Y(\tau) \right\} = 1 + \sum_{n=1}^{\infty} \int_a^b d\tau_1 \int_a^{\tau_1} d\tau_2 \dots \int_a^{\tau_{n-1}} d\tau_n Y(\tau_1) Y(\tau_2) \dots Y(\tau_n). \tag{2.12}$$

Clearly, the sign of the T-ordering can be skipped if the integrands in (2.11) commute with each other for different τ. In particular, as is clear from (2.9), this is the case when all the multiple commutators $[B, A, \dots, A]$ commute each other and with $A + B$. If, moreover, $[A, B]$ commutes with both A and B then

$$\exp A \cdot \exp B = \exp(1/2)[A, B] \cdot \exp(A + B) \tag{2.13}$$
$$= \exp[A, B] \cdot \exp B \cdot \exp A.$$

As equalities of the corresponding formal series, the formulas (2.9) and (2.11) always hold. However, to apply them in concrete physical models requires investigation of convergence in order to get meaningful operators.

Taking (2.7) and (2.13) into account we can now get the formula for the product:

$$\exp(\phi_\alpha(t)) \exp(\phi_\beta(t')) = \exp(-i\hbar \hat{k}_{\alpha\beta}(t - t')) \exp(\phi_\beta(t')) \exp(\phi_\alpha(t));$$

$$\exp(\varphi_\alpha(t)) \exp \phi_\beta(t') = \exp(-i\hbar v_\alpha^{-1} \delta_{\alpha\beta}(t - t')) \exp(\phi_\beta(t')) \exp \varphi_\alpha(t), \tag{2.14}$$

and also the formula for the factorization into operators which depend only on a or b:

$$\exp(\phi_\alpha(t)) = \exp\left(a_\alpha + \frac{i\hbar}{2}\hat{k}_{\alpha\alpha}t\right)\exp b_\alpha = \exp b_\alpha \cdot \exp\left(a_\alpha - \frac{i\hbar}{2}\hat{k}_{\alpha\alpha}t\right),$$
$$(2.15)$$

or, more generally,

$$\exp(\phi_{\alpha_1}(t_1))\ldots\exp(\phi_{\alpha_n}(t_n))$$

$$= \prod_{1\leq m\leq n}\exp b_{\alpha_m}\prod_{1\leq s\leq n}\exp\left(a_{\alpha_s} - \frac{i}{2}\hbar\hat{k}_{\alpha_s\alpha_s} - i\hbar\sum_{s+1\leq l\leq n}\hat{k}_{\alpha_s\alpha_l}\right)t_s$$

$$= \prod_{1\leq m\leq n}\exp(\phi_{\alpha_m}(t))\prod_{1\leq s\leq n}\exp\left(a_{\alpha_s} - \frac{i}{2}\hbar\hat{k}_{\alpha_s\alpha_s} - i\hbar\sum_{s+1\leq l\leq n}\hat{k}_{\alpha_s\alpha_l}\right)(t_s - t).$$
$$(2.16)$$

Notice also the formula

$$\exp\left(\sum_\alpha l_\alpha\phi_\alpha\right)\cdot a_\mu\cdot\exp\left(-\sum_\alpha l_\alpha\phi_\alpha\right) = a_\mu + i\hbar\left(\sum_\alpha \hat{k}_{\mu\alpha}l_\alpha\right) \quad (2.17)$$

where l_α are arbitrary constants, which is obtained from (2.9) with the help of (2.7); the series in (2.9) is cut off after the first term.

This also implies a general formula for an arbitrary function $f(a)$ of operators $a_1, \ldots, a_r$. Expand it into Taylor series termwise with the help of (2.17) and then perform the formal summation:

$$\exp\left(\sum_\alpha l_\alpha\phi_\alpha\right)f(a_\mu)\exp\left(-\sum_\alpha l_\alpha\phi_\alpha\right) = f\left(a_\mu + i\hbar\sum_\alpha \hat{k}_{\mu\alpha}l_\alpha\right). \quad (2.18)$$

In exactly the same way we get

$$\exp\left(\sum_\alpha l_\alpha\varphi_\alpha\right)f(a_\mu)\exp\left(-\sum_\alpha l_\alpha\varphi_\alpha\right) = f(a_\mu + i\hbar v_\mu^{-1}l_\mu), \quad (2.19)$$

which implies that $\exp\varphi_\mu$ is a shift operator with respect to a_μ, i.e., $\exp\varphi_\mu = \exp(i\hbar v_\mu^{-1}\partial/\partial a_\mu)$, which, however, is already clear from (2.6).

For interacting particles we can construct a solution with the help of formal perturbation series in the Heisenberg representation. Suppose there exist asymptotic limits ("in-operators") $\varphi(t) \equiv u_{\text{in}}(t) \equiv \lim_{t\to-\infty} u(t)$ satisfying the relations for the operators of free particles. Introduce the "S-matrix for finite time" setting

$$S(t) \equiv T\exp\left\{\frac{\mu^2}{i\hbar}\int_{-\infty}^t d\tau U(\varphi_1, \ldots, \varphi_r)\right\}. \quad (2.20)$$

Then for any function $A(u)$ of operators $u_1, \ldots, u_r$ (taken at the same time t) we can write

$$A(u(t)) = S^\dagger(t) A(\varphi(t)) S(t).$$

Substituting $A(u)$ in the form of the series

$$A = \sum_{n \in I}^\infty \mu^{2n} A_n,$$

we get

$$A_0 \equiv A(\varphi(t)),$$

$$A_n(t) = \int\limits_{-\infty}^\infty \cdots \int\limits_{-\infty}^\infty \prod_{m=1}^n \frac{dt_m}{i\hbar} \theta(t_{m-1} - t_m)[A(\varphi(t)), U_1, \ldots, U_n], \tag{2.21}$$

where $\theta(t)$ is the Heaviside step-function, $t_0 \equiv t$, $U_m \equiv U(\varphi(t_m))$.

Considering the exponential interactions with U a linear function in $\exp \phi_\alpha$, we will calculate the operators of the form $\exp(lu_\alpha)$. As is clear from (2.21), we will need the multiple commutators of the form $[\exp(l\varphi_\alpha), \exp \phi_\beta, \ldots, \exp \phi_\gamma]$. To calculate them let us order the exponentials in each summand of the commutator so that the time arguments decrease from left to right with the help of formulas (2.14), thus acquiring numerical factors. Simplifying further we get

$$[\exp(l\varphi_\alpha(t)), \exp\phi_{\beta_1}(t_1), \ldots, \exp\phi_{\beta_n}(t_n)] = \exp(l\varphi_\alpha(t)) \exp(\phi_{\beta_1}(t_1)) \times \cdots$$

$$\ldots \times \exp(\phi_{\beta_n}(t_n)) \cdot \prod_{s=1}^n \left\{ 1 - \exp\left[\frac{i\hbar l}{v_{\beta_s}} \delta_{\alpha\beta_s}(t - t_s) + i\hbar \sum_{j=1}^{s-1} \hat{k}_{\beta_j \beta_s}(t_j - t_s) \right] \right\}. \tag{2.22}$$

In addition, with the help of (2.16) we can move all the dependence on parameters $t_1, \ldots, t_n$ to the right-hand side (into the product of commuting factors) after which the integration with respect to t_i is elementary.

Now, consider a similar model in the two-dimensional space-time $x^\mu = (t, x)$ with r scalar Hermitian fields u_α. It is convenient to introduce the variables $z^\pm \equiv \frac{1}{2}(t \pm x)$ so that

$$u_{\alpha,\mu} u_\beta^{\prime\mu} = \frac{1}{2}\left(\frac{\partial u_\alpha}{\partial z^+} \frac{\partial u_\beta}{\partial z^-} + \frac{\partial u_\alpha}{\partial z^-} \frac{\partial u_\beta}{\partial z^+} \right); \quad u_{\alpha,\mu}^{\prime\mu} \equiv \frac{\partial^2 u_\alpha}{\partial z^+ \partial z^-}.$$

Let us assume that the spatial coordinate either runs over the whole infinite interval or is cyclic, i.e., $u_\alpha(t, x + 2L) = u_\alpha(t, x)$. The first case can be considered formally as the limit of the second one as $L \to \infty$.

The Lagrangian is of the form

$$\mathcal{L} = \frac{1}{4} \sum_{\alpha,\beta=1}^{r} \underline{k}_{\alpha\beta} \left(\frac{\partial u_\alpha}{\partial z^+} \frac{\partial u_\beta}{\partial z^-} + \frac{\partial u_\alpha}{\partial z^-} \frac{\partial u_\beta}{\partial z^+} \right) - \mu^2 V(u_1, \ldots, u_r) \qquad (2.23)$$

(see (2.1)). The corresponding equations of motion are

$$\frac{\partial^2 u_\alpha}{\partial z^+ \partial z^-} + \mu^2 \sum_{\beta=1}^{r} (\underline{k}^{-1})_{\alpha\beta} V_\beta(u); \quad V_\alpha \equiv \frac{\partial V}{\partial u_\alpha}. \qquad (2.24)$$

To quantize with the help of perturbation theory first consider the free system whose Lagrangean is the kinetic part of (2.23):

$$\mathcal{L}_0 = \frac{1}{4} \sum_{\alpha,\beta=1}^{r} \underline{k}_{\alpha\beta} \left(\frac{\partial \varphi_\alpha}{\partial z^+} \frac{\partial \varphi_\beta}{\partial z^-} + \frac{\partial \varphi_\alpha}{\partial z^-} \frac{\partial \varphi_\beta}{\partial z^+} \right).$$

The general solution of the corresponding equation of motion is

$$\varphi_\alpha(z^+, z^-) = \varphi_\alpha^+(z^+) + \varphi_\alpha^-(z^-), \qquad (2.25)$$

where $\varphi_\alpha^\pm(z^\pm)$ are arbitrary operator-function. The periodicity condition in x implies that (2.25) takes the form

$$\varphi_\alpha = Q^\alpha + \frac{t}{2L} P^\alpha + \frac{i}{\sqrt{4\pi}} \sum_{n \neq 0} \frac{1}{n} \left[A_n^{+\alpha} \exp\left(-\frac{2\pi i n}{L} z^+ \right) \right.$$
$$\left. + A_n^{-\alpha} \exp\left(-\frac{2\pi i n}{L} z^- \right) \right], \qquad (2.26)$$

where Q^α, P^α and $A_n^{\pm\alpha}$ are operators which are still arbitrary. The hermiticity property of φ_α means that

$$(Q^\alpha)^\dagger = Q^\alpha, \quad (P^\alpha)^\dagger = P^\alpha, \quad (A_n^{\pm\alpha})^\dagger = A_{-n}^{\pm\alpha}. \qquad (2.27)$$

To the Lagrangian $\mathcal{L}_0$ corresponds the Hamiltonian

$$\mathfrak{h}_0 = \frac{1}{2} \sum_{\alpha,\beta} \underline{k}_{\alpha\beta} \int_{-L}^{L} dx \left[\frac{\partial \varphi_\alpha}{\partial t} \frac{\partial \varphi_\beta}{\partial t} + \frac{\partial \varphi_\alpha}{\partial x} \frac{\partial \varphi_\beta}{\partial x} \right]$$
$$= \sum_{\alpha,\beta} \underline{k}_{\alpha\beta} \left\{ \frac{P^\alpha P^\beta}{4L} + \frac{\pi}{2L} \sum_{n \neq 0} [A_{-n}^{+\alpha} A_{+n}^{+\beta} + A_{-n}^{-\alpha} A_{+n}^{-\beta}] \right\} \qquad (2.28)$$

and the canonical quantization leads to the commutation relations

$$[Q^\alpha, P^\beta] = i\hbar (\underline{k}^{-1})_{\alpha\beta};$$
$$[A_m^{\pm\alpha}, A_n^{\pm\beta}] = i\hbar m \delta_{m,-n} (\underline{k}^{-1})_{\alpha\beta}; \qquad (2.29)$$

with other brackets being zero. This implies that

$$[\varphi_\alpha(z), \varphi_\beta(z')] = -i\hbar(\underline{k}^{-1})_{\alpha\beta}D(z - z'),$$

where

$$D(z) = \frac{1}{4} \sum_{n=-\infty}^{\infty} [\text{sgn}(z^+ - nL) + \text{sgn}(z^- - nL)]. \qquad (2.30)$$

(Here the sum is understood in the sense of the principal value.) It is clear from (2.29) that the operators φ_α can be represented in the Hilbert space

$$T = T_0 \otimes T^+ \otimes T^-,$$

where T_0 is the space of one-dimensional r-particle quantum-mechanical problems (see the beginning of this section) and $T^\pm$ are the Fock spaces (with particles of r types) in which the creation (annihilation) operators are expressed via $A_{-n}^{\pm\alpha}(A_{+n}^{\pm\alpha})$, $n > 0$.

As in (2.5) introduce the new operators $\phi_\alpha = \sum_\beta k_{\alpha\beta}\varphi_\beta$. We have

$$[\phi_\alpha(z), \phi_\beta(z')] = -i\hbar\hat{k}_{\alpha\beta}D(z - z'),$$

$$[\phi_\alpha(z), \varphi_\beta(z')] = -i\frac{\hbar}{v_\alpha}\delta_{\alpha\beta}D(z - z'). \qquad (2.31)$$

The following formulas will be given for the one-component case only, i.e., $r = 1$ ($\underline{k} = v$). They are obtained in the same way as (2.14–2.20) and their generalization for multi-component systems does not present any difficulties:

$$\exp(-b\varphi(z))\exp(a\varphi(z'))\exp(b(\varphi(z))) = \exp(iab\hbar v^{-1}D(z - z'))\exp(a\varphi(z')); \qquad (2.32)$$

$$\exp(aP + b\varphi) = \exp\left(a\left(P + i\frac{\hbar b}{2v}\right)\right)\exp(b\varphi); \qquad (2.33)$$

$$\exp(-b\varphi)f(P)\exp(b\varphi) = f\left(P - i\frac{\hbar}{v}b\right). \qquad (2.34)$$

We will also need the operator

$$\Phi(z) \equiv \int_{z^+ - L}^{z^+} \int_{z^- - L}^{z^-} \exp(2\varphi(\tilde{z}))d\tilde{z}^+ d\tilde{z}^- \qquad (2.35)$$

for which from (2.32–2.34) we get

$$\exp(-b\varphi)\Phi\exp(b\varphi) = \exp\left(\frac{i\hbar}{v}b\right)\Phi; \qquad (2.36)$$

$$\Phi f(P) = f\left(P + \frac{2i\hbar}{v}\right)\Phi. \qquad (2.37)$$

Let us also give without proof the reduction formula

$$\int_{z^+-L}^{z^+} \int_{z^--L}^{z^-} \exp(2\varphi(\tilde{z}))\Phi^{n-1}(\tilde{z})d\tilde{z}^+d\tilde{z}^- \tag{2.38}$$

$$= \frac{(1-\exp(i\hbar v^{-1}))^2(1-\exp(-nP-i\hbar n^2 v^{-1}))^2}{(1-\exp(i\hbar n v^{-1}))^2(1-\exp(-P-i\hbar n v^{-1}))^2},$$

which is verified by substituting the explicit expression (2.35) for Φ and modifying the integration domains, taking into account the properties of $\varphi(z^+, z^-)$ when its arguments are shifted by L (see (2.26)).

For infinite space (formally as $L \to \infty$) the operator functions in (2.25) are arbitrary and the function $D(z)$ obtained by the canonical quantization takes the form

$$D(z) = \frac{1}{4}(\operatorname{sgn}(z^+) + \operatorname{sgn}(z^-)). \tag{2.39}$$

Though, as is known, in this case there are difficulties with the realization of the operators φ_α in a Hilbert space (see e.g. [9, 10]) we will, however, write some algebraic relations for them that we will need in 7.5:

$$[\phi_\alpha^\pm(z^\pm), \phi_\beta^\pm(\tilde{z}^\pm)] = -\frac{i\hbar}{4v}\hat{k}_{\alpha\beta}\operatorname{sgn}(z^\pm - \tilde{z}^\pm);$$

$$[\phi_\alpha^\pm(z^\pm), \phi_\beta^\mp(\tilde{z}^\mp)] = 0;$$

$$\exp(a\phi_\alpha^\pm(z^\pm))\exp(b\phi_\beta^\pm(\tilde{z}^\pm)) \tag{2.40}$$

$$= \exp\left(-\frac{i\hbar ab}{4v}\hat{k}_{\alpha\beta}\operatorname{sgn}(z^\pm - \tilde{z}^\pm)\right)\exp(b\phi_\beta^\pm(\tilde{z}^\pm))\exp(a\phi_\alpha^\pm(z^\pm)).$$

Introducing the multi-index operators

$$\Phi_{\alpha\beta\dots}^\pm(z^\pm) \equiv \int_{-\infty}^{z^\pm} dz_1^\pm \exp(\phi_\alpha^\pm(z_1^\pm)) \int_{-\infty}^{z_1^\pm} dz_2^\pm \exp(\phi_\beta^\pm(z_2^\pm))\dots, \tag{2.41}$$

which are similar to retarded functionals in φ_α we get a relation similar to (2.36):

$$\exp(a\phi_\beta^\pm)F(\Phi_{\alpha_1\dots\alpha_n}^\pm)\exp(-a\phi_\beta^\pm)$$

$$= F\left(\exp\left(-\frac{i\hbar a}{4v}\sum_{m=1}^n \hat{k}_{\beta\alpha_m}\right) \cdot \Phi_{\alpha_1\dots\alpha_n}^\pm\right). \tag{2.42}$$

With the help of formulas (2.40) we get the following relations for $\Phi_{\alpha\beta\dots}^\pm$

$$[\Phi_{\alpha_1\dots\alpha_n}^\pm]^\dagger = \exp\left(\frac{i\hbar}{4v}\sum_{l<m}^n \hat{k}_{\alpha_l\alpha_m}\right) \cdot \Phi_{\alpha_1\dots\alpha_n}^\pm. \tag{2.43}$$

Now passing to the construction of solutions of (2.24) as functionals in φ_α we can again make use of the general formula (2.21), where for A we take the local function of field operators and for the perturbation function $U(t)$ we take

$$U(t) = \int\limits_{-L}^{L} dx V(\varphi(t,x)).$$

In other words, the n-th order of the perturbation theory for a local operator $A(u)$ is given by Schwinger's formula:

$$A_n(t,x) = \int\limits_{-\infty}^{\infty} \int\limits_{-L}^{L} [A(\varphi(t,x)), V_1, \ldots, V_n] \prod_{m=1}^{n} \frac{dt_m dx_m}{i\hbar} \theta(t_{m-1} - t_m), \tag{2.44}$$

$$V_i \equiv V(\varphi(t_i, x_i)).$$

Another variant for constructing solutions via perturbation theory is to make use of the Yang-Feldmann equation, i.e., the integral form of (2.24):

$$u_\alpha(t,x) = \varphi_\alpha(t,x) + \int\limits_{-\infty}^{\infty} dt' \int\limits_{-L}^{L} dx' D^{\mathrm{ret}}(t - t', x - x') j_\alpha(t', x'), \tag{2.45}$$

where φ_α are free fields, $D^{\mathrm{ret}}(t,x) = \theta(t) D(t,x)$ and D is given by (2.30). In our case

$$j_\alpha = -\mu^2 \sum_{\beta} (\underline{k}^{-1})_{\alpha\beta} V_\beta(u).$$

The equation (2.45) for u_α can be solved by iteration, taking φ_α as the 0-th approximation. Therefore we get a perturbation theory series for u_α with the coupling constant μ^2.

§ 7.3 One-dimensional generalized Toda lattice with fixed end-points*)

In this section we will apply the general construction of 7.1 for an important particular case of the system (3.2.13) — the generalized (finite nonperiodic) Toda lattice described by the equations

$$\ddot{u}_i = \lambda \exp(ku)_i, \quad 1 \leq i \leq r \quad (\lambda \equiv \mu^2), \tag{3.1}$$

whose Hamiltonian is

$$\mathfrak{h} = \frac{1}{2} \sum_{i,j} (vk)_{ij}^{-1} p_i p_j + \lambda \sum_{i} v_i \exp(ku)_i. \tag{3.2}$$

---

*) [107–109]

Obviously, this system follows from (3.2.13) for a simple Lie algebra $\mathfrak{g}$ with Cartan matrix k and the principal grading with

$$g = \exp\left\{ -\sum_i h_i \left(u_i - \sum_j k_{ij}^{-1} \ln k_j \right) \right\}, \quad J_\pm = \sum_i k_i^{1/2} \mathcal{X}_{\pm i},$$

$$C_0 = \mathrm{tr}\left(\sum_i h^j p_i \right); \qquad\qquad k_i \equiv 2 \sum_j k_{ij}^{-1},$$

and the relations (2.1.12) and (2.1.13) being taken into account. Here p_i and u_i are generalized momenta and coordinates. We will consider both Schrödinger's and Heisenberg's pictures.

7.3.1 Schrödinger's picture. The 2nd order Casimir operator for the normal real form G in the Iwasawa decomposition is given by the formula $(2.2.5_2)$ where, setting $\partial/\partial\tau_i = p_i$ and performing the canonical transformation $C \to \exp((1/2)\rho^0\tau)K\exp((-1/2)\rho^0\tau)$, we get

$$C = \frac{1}{2}\sum_{i,j}(vk)_{ij}^{-1}p_ip_j + \sum_{\alpha>0}v_\alpha^{-1}[\exp(-\alpha(\tau))K_\alpha^R Z_\alpha^L \tag{3.3}$$

$$- \exp(-2\alpha(\tau))Z_\alpha^L Z_\alpha^L].$$

The action of the operator (3.3) on its eigenfunctions satisfying (2.6.6) leads, up to obvious changes, to (3.2). In other words, the role of the basic states for the considered problem is played by Whittaker vectors (2.6.4).

If for eigenfunctions of the operator (3.3) we take vectors of a finite-dimensional representation $\{l\}$ of dimension N_l of the subgroup $K \subset G$ satisfying (2.6.6) then this operator reduces to the form

$$C = \frac{1}{2}\sum_{i,j}(vk)_{ij}^{-1}p_ip_j + \sum_i k_i^{1/2}[\exp((-k\tau)_i)K_i^R \tag{3.4}$$

$$- v_i k_i^{1/2}\exp((-2k\tau)_i)],$$

where K_i^R are the generators of the representation $\{l\}$. Then (3.4) corresponds to an exactly integrable quantum system with an N_l-component wave function which in the one-component case (i.e. when the eigenfunctions of $\mathfrak{h}$ are one-dimensional with respect to Z and invariant with respect to the right K-action, i.e. $K_i^R\varphi = 0$) turns into the generalized Toda lattice after the change

$$\tau_i = -(1/2)u_i + (1/2)\sum_j k_{ij}^{-1}\ln(-k_j).$$

Notice that for the simplest case, i.e., for $SL(2,\mathbb{R})$ the formula (3.4) reproduces the Morse potential describing oscillations of a two-atom molecule and therefore provides us with its group-theoretical interpretations.

Therefore to study a multicomponent variant of the generalized quantum Toda lattice we should take the Whittaker vector (2.6.4) for the initial state $\varphi_+(k)$, while an arbitrary vector of the left-regular representation of K $\varphi_-(k) = \sum_{\{n\}} D^{\{l\}}_{\{m\},\{n\}}(k)\eta_{\{n\}}$, where $\eta \equiv \{\eta_{\{n\}}\}$ is an arbitrary numerical vector, serves as the final state. (In the one-component case, clearly $\varphi_-(k) = \eta_{\{0\}}$.) For the noncompact transformations $\exp\sum_j h_j\tau_j \in A$ from G, with the help of the formula $\xi^{\{\rho\}}(\tilde{k}) = \exp(\tau\rho)\prod_j R_j^{-\rho_j}\xi^{\{\rho\}}(k)$, which follows from (2.3.8), we get the following integral representation of the desired wave function

$$\Psi^{\{\rho,l\}}_{\{m\}}(\tau) = \exp(i(\tau\sigma))\mathcal{N}(\sigma)\sum_{\{n\}}\int d\mu(k)\xi^{\{\rho\}}(k)$$

$$\times \overset{*}{\mathcal{D}}{}^{\{l\}}_{\{m\}\{n\}}(k)\exp\left\{-i\sum_j K^L_j \widetilde{\ln\xi^{\{j\}}}(k)\right\}\cdot\eta_{\{n\}} \tag{3.5}$$

(cf. (1.1)). Here the tilde over the exponential means that all its constituents depend on the parameters $k \in K$ transformed under $g(\tau)$. The asymptotics of (3.5) are given as $\alpha(\tau) \to \infty$ by the formula

$$\overset{\infty}{\Psi}{}^{\{\rho,l\}}_{\{m\}}(\tau) = \sum_{\omega\in W}\exp(i(\tau\sigma_\omega))\cdot\mathcal{N}(\sigma_\omega)\sum_{\{n\}}R^{\{l\}}_{\{m\}\{n\}}(\rho_\omega)\eta_{\{n\}}, \tag{3.6}$$

where

$$R^{\{l\}}_{\{m\}\{n\}}(\rho) \equiv \int d\mu(k)\overset{*}{\mathcal{D}}{}^{\{l\}}_{\{m\}\{n\}}(k)\xi^{\{\rho\}}(k); \tag{3.7}$$

and W is the Weyl group of G.

We have to confess that we do not know a proof of the fact that this limit procedure is well-defined. The only reasons to believe it are the coincidence of the formula derived from it with the well-known formula for the Plancherel measure of the principal continuous series of representations of G and, most important, its analogy with Fock's method. The integrals in (3.5) are taken on the boundary of their domain of convergence and passage to the limit similar to the procedure from 2.5.3 using Fock's method requires a more accurate analysis of the integrand considered as the distribution of a certain class.

It is convenient to consider (3.7) as a matrix element of the operator

$$\hat{R} \equiv \int d\mu(k)\xi^{\{\rho\}}(k)k, \tag{3.8}$$

where $k \in \mathcal{K}$ is of the form (1.5.19) where $X_{\pm\alpha}$ are taken in the representation $\{l\}$. The unitary scattering matrix

$$\hat{T}(\rho) \equiv \hat{R}(\hat{R}^\dagger)^{-1} \tag{3.9}$$

coincides then with the intertwining operator of the total Weyl reflection of the corresponding group whereas $\hat{R}\hat{R}^\dagger$ is the unit operator inverse proportional to the weight functions $\omega(\rho)$ of the Plancherel's measure of the principal continuous series of G. The quantities $R^{\{l\}}_{\{m\}\{n\}}$ play the role of Jost functions for the one-component problem and

$$R^{\{0\}}_{\{0\}\{0\}} \equiv c(\rho) = \int d\mu(k)\xi^{\{\rho\}}(k).$$

Making use of the universal parameterization (1.5.19) of k and the formulas (1.6.20) and (1.5.20) for the highest matrix elements and the invariant measure on $\mathcal{K}$ we have, up to irrelevant numerical factors,

$$\hat{R}^{-1} = \prod_{\alpha>0}^{\Sigma^+} \left[\frac{2(\alpha, \rho + \rho_0)}{(\alpha\alpha)} - 1 \right] \exp\left[-\frac{\pi}{2}(X_\alpha - X_{-\alpha}) \right]$$

$$\times B\left[\frac{(\alpha, \rho + \rho_0)}{(\alpha\alpha)} - i(X_\alpha - X_{-\alpha})/2, \frac{(\alpha, \rho + \rho_0)}{(\alpha\alpha)} + i(X_\alpha - X_{-\alpha})/2 \right], \tag{3.10}$$

where B is Euler's B-function. This implies in particular the following expressions in the one-component case for the function $c(\rho)$, the scattering amplitude $t(\rho)$ and the weight function $\omega(\rho)$:

$$c(\rho) = \prod_\alpha B\left(\frac{1}{2}, \frac{(\alpha, \rho + \rho_0)}{(\alpha\alpha)} - \frac{1}{2} \right) \cdot B^{-1}\left(\frac{1}{2}, \frac{(\alpha\rho_0)}{(\alpha\alpha)} - \frac{1}{2} \right), \tag{3.11}$$

$$t(\rho) = \prod_\alpha B\left(\frac{1}{2}, \frac{(\alpha, \rho + \rho_0)}{(\alpha\alpha)} - \frac{1}{2} \right) \cdot B^{-1}\left(\frac{1}{2}, \frac{(\alpha, \overset{*}{\rho} + \rho_0)}{(\alpha\alpha)} - \frac{1}{2} \right),$$

$$\omega(\rho) = \prod_\alpha B^2\left(\frac{1}{2}, \frac{(\alpha\rho_0)}{(\alpha\alpha)} - \frac{1}{2} \right) B^{-1}\left(\frac{1}{2}, \frac{(\alpha, \rho + \rho_0)}{(\alpha\alpha)} - \frac{1}{2} \right) \tag{3.12}$$

$$\times B^{-1}\left(\frac{1}{2}, \frac{(\alpha, \overset{*}{\rho} + \rho_0)}{(\alpha\alpha)} - \frac{1}{2} \right).$$

There is an alternative approach to the construction of wave functions of the systems considered and the study of their analytic properties: we directly calculate the functions as eigenfunctions of a total number of functionally independent operators in involution (integrals of motion) given on the phase space of the functional algebra corresponding to G.

Let us illustrate this approach with the example of the generalized Toda lattice. In this case such operators are $\hat{I}_q(p, \rho) \equiv \text{tr} L^q(t)$ (see 4.1.5) and

$p_j = v_j k_{ji}^{-1} \dot{\rho}_i$, $[p_i \rho_j] = \delta_{ij}$ $(\hbar = 1)$. In particular, for A_r the eigenvalues I_q of operators $\hat{I}_q$ are equal to

$$\sum_{1 \le j_1 < j_2 < \ldots < j_q \le r+1} \prod_{1 \le j \le r} l_j, \quad \text{where } m_j \equiv l_j - l_{j+1}.$$

Therefore the problem is to solve the system of equations $\hat{I}_q \Phi = I_q \Phi$, $1 \le q \le r$. It is convenient to solve it in the momentum representation setting $\rho_j = -\partial/\partial p_j$ and getting therefore a system of differential-difference equations, e.g.,

A_1: $\qquad [p^2 - \exp(-\partial/\partial p)]\Phi = (1/4)m^2 \Phi;$

A_2: $\qquad [p_1^2 + p_2^2 - p_1 p_2 - \exp(-\partial/\partial p_1) - \exp(-\partial/\partial p_2)]\Phi$

$$= (1/3)(m_1^2 + m_2^2 + m_1 m_2)\Phi,$$

$$[p_1 p_2 (p_1 - p_2) - p_2 \exp(-\partial/\partial p_1) + p_1 \exp(-\partial/\partial p_2)]\Phi$$

$$= -(1/27)(2m_1 + m_2)(m_1 + 2m_2)(m_1 - m_2)\Phi;$$

C_2: $\qquad [(1/2)p_1^2 + p_2^2 - p_1 p_2 - 2\exp(-\partial/\partial p_1) - \exp(-\partial/\partial p_2)]\Phi$

$$= (m_1^2 + m_1 m_2 + 1/2 m_2^2)\Phi,$$

$$[-(1/16)p_1^2(p_1 - 2p_2)^2 + (1/2)p_1(p_1 - 2p_2)\exp(-\partial/\partial p_1)$$

$$+(1/4)p_1^2 \exp(-\partial/\partial p_2) - \exp(-2\partial/\partial p_1)]\Phi$$

$$= -(1/16)(2m_1 + m_2)^2 m_2^2 \Phi.$$

It is rather simple to solve these equations only for A_1 and A_2; and the solutions are obviously expressed in terms of the products and ratios of Γ-functions of the form $\Gamma(p_1 + \alpha_i + 1)$, $\Gamma(p_2 - \alpha_i + 1)$, $\alpha_i \equiv \alpha_i(m)$, while in other cases we get difference equations with shifts by greater than 1 for which there is no general method.

The knowledge of explicit forms of weight functions (3.5) enables us to calculate their quasiclassical limit and therefore get the exact expressions for the solutions of the corresponding classical problem in one more way. Let us illustrate this procedure with the example of a one-component variant of generalized Toda lattice with

$$\Psi^{\{\rho\}}(\tau) = \exp(i(\tau\sigma))\mathcal{N}(\sigma) \int d\mu(k)\xi^{\{\rho\}}(k) \exp\left\{-i\sum_j K_j^L \widetilde{\ln \xi}^{\{j\}}(k)\right\}.$$
$$(3.13)$$

We apply the saddle point method. The exponent of a quickly oscillating function in the integrand (3.13) is of the form

$$S = \sum_j \left[P_j \log \xi^{\{j\}} + K_j^L \widetilde{\ln \xi}^{\{j\}} + P_j \tau_j + \left(\frac{1}{4}\right)\sum_i (vk)_{ij}^{-1} P_j P_t \right]. \quad (3.14)$$

The stationary points of this expression are determined from the condition that all its derivatives with respect to all the parameters of $k(\theta_\alpha) \in \mathcal{K}$ should vanish:

$$K^R_{-\alpha}S = 0 \quad \text{and/or} \quad K^L_\alpha S = 0 \text{ for any } \alpha \in R_+. \tag{3.15}$$

To find the classical trajectories which realize the exact solutions of (3.1) ($u_i = -2\tau_i$) one should equate the derivatives of the action with respect to arbitrary parameters P_j determined via (3.15) to arbitrary constants $\ln c_j$, i.e.,

$$\ln c_j = dS/dP_j \equiv \partial S/\partial P_j + \sum_\alpha (\partial S/\partial\theta_\alpha)(\partial\theta_\alpha/\partial P_j)$$
$$= \partial S/\partial P_j = \ln \xi^{\{j\}} + (1/2)u_j + (1/2)\sum_i (vk)^{-1}_{ji} P_i t. \tag{3.16}$$

This implies that

$$\xi^{\{j\}}(k) = c_j \exp(\alpha_j t - u_j), \quad \text{where } \alpha_j \equiv -\sum_j (vk)^{-1}_{ji} P_i. \tag{3.17}$$

(Notice that $\partial S/\partial\theta_\alpha = 0$ thanks to the fact that the action is extremal.) Therefore we get the known expressions (see 4.1.5) for the solutions of the considered problem which satisfy the system (1.4) provided that (3.15) holds.

As an example consider in detail the Toda lattice associated with A_2. Denote by k^β_α, $1 \le \alpha, \beta \le 3$, the elements of a matrix from $SO(3)$; then $\xi^{\{1\}} = k^1_1$, $\xi^{\{2\}} = k^1_1 k^2_2 - k^1_2 k^2_1 \equiv k^3_3$. With (2.6.3) taken into account we rewrite (3.15) in the form

$$K^R_{-\pi_1}S = P_1 \frac{k^2_1}{k^1_1} - \exp\left(\frac{2u_1 - u_2}{2}\right)\frac{k^3_3}{(k^1_1)^2} = 0,$$

$$K^R_{-\pi_2}S = -P_2 \frac{k^2_3}{k^3_3} - \exp\left(\frac{2u_2 - u_1}{2}\right)\frac{k^1_1}{(k^3_3)^2} = 0,$$

$$K_{-(\pi_1+\pi_2)}S = P_1 \frac{k^3_1}{k^1_1} - P_2 \frac{k^1_3}{k^3_3} + \exp\left(\frac{2u_1 - u_2}{2}\right)\frac{k^2_3}{(k^1_1)^2}$$
$$+ \exp\left(\frac{2u_2 - u_1}{2}\right)\frac{k^2_1}{(k^3_3)^2} = 0, \tag{3.18}$$

(here π_1, $\pi_1 + \pi_2$, π_2 are the positive roots of A_2). Since

$$k^1_1 = c_1 \exp\left(\frac{\alpha_1 t - u_1}{2}\right), \quad k^3_3 = c_2 \exp\left(\frac{\alpha_2 t - u_2}{2}\right)$$

$$\left(\alpha_1 \equiv -\frac{1}{3}(2P_1 + P_2), \quad \alpha_2 \equiv -\frac{1}{3}(P_1 + 2P_2)\right)$$

and therefore, as follows from (3.18),

$$k_1^2 = P_1^{-1} c_1^{-1} c_2 \exp\left(\frac{(\alpha_2 - \alpha_1)t - u_1}{2}\right),$$

$$k_3^2 = -P_2^{-1} c_1 c_2^{-1} \exp\left(\frac{(\alpha_1 - \alpha_2)t - u_2}{2}\right),$$

making use of the orthogonality relation $\sum\limits_{1 \leq \alpha \leq 3} k_1^\alpha k_3^\alpha = 0$, we find

$$k_3^1 = (P_1 + P_2)^{-1} P_1^{-1} c_1^{-1} \exp\left(\frac{-\alpha_1 t - u_2}{2}\right),$$

$$k_1^3 = (P_1 + P_2)^{-1} P_2^{-1} c_2^{-1} \exp\left(\frac{-\alpha_2 t - u_1}{2}\right).$$

These formulas and the relations $\sum\limits_{1 \leq \alpha \leq 3}(k_1^\alpha)^2 = 1$, $\sum\limits_{1 \leq \alpha \leq 3}(k_3^\alpha)^2 = 1$ imply the explicit expressions for the desired classical trajectories for the Toda lattice associated with A_2 (cf. 4.1.5).

7.3.2 Heisenberg's picture (the canonical formalism). For an explicit construction of the Heisenberg operators $\exp(-u_i)$ in the Hamiltonian approach let us make use of Schwinger's formula (2.44) with $A(t) = \exp(-\varphi_j(t))$ and $U_j = \sum\limits_i v_i \exp\phi_i(t_j)$ of the form (2.5) and of the relations (2.16) and (2.22) which imply

$$[\exp(-u_i)]_0 = \exp(-\varphi_i),$$

$$[\exp(-u_i)]_{n\geq 1} = \exp(-\varphi_i) \sum_{j_1,\dots,j_n} \prod_{1\leq l\leq n} \frac{v_{j_l}}{i\hbar} \exp\phi_{j_l}(t) I_{j_1,\dots,j_n}, \qquad (3.19)$$

where

$$I_{j_1\dots j_n} \equiv \int_{-\infty}^{\infty} \dots \int \prod_{m=1}^{n} dt_m \theta(t_{m-1} - t_m) \exp\left[a_{j_m}^{(1)}(t_m - t)\right]$$

$$\times \left[1 - \exp i\hbar \sum_{l=0}^{m-1} \hat{k}_{j_l j_m}(t_l - t_m)\right]$$

$$= \sum_{m=0}^{n}(-1)^m \sum_{1\leq q_1 <\dots<q_m} \prod_{s=1}^{n}[c_{j_s}(q) + c_{j_{s+1}}(q) + \dots + c_{j_n}(q)]^{-1},$$

$$c_{j_s}(q) \equiv a_{j_s}^{(1)} + i\hbar \sum_{l=1}^{m}\left[\theta(q_l - s)\hat{k}_{j_s j_{q_l}} + \delta_{sq_l}\left(\hat{\delta}_{ij_s} - \sum_{p=1}^{s-1}\hat{k}_{j_p j_s}\right)\right],$$

$$\hat{\delta}_{ij} \equiv \delta_{ij} v_i^{-1}, \quad a_{j_m}^{(1)} \equiv a_{j_m} - i\frac{\hbar}{2}\hat{k}_{j_m j_m} - i\hbar \sum_{l=m+1}^{n} \hat{k}_{j_m j_l}. \qquad (3.20)$$

The further analysis and simplification of solutions given in the form (3.20) is rather difficult. Therefore let us rewrite them in a somewhat different form giving as an illustration the expressions for the first three orders:

$$I_j = (a_j^{(1)})^{-1} - (a_j^{(1)} + i\hbar\hat{\delta}_{ij})^{-1} \equiv i\hbar\hat{\delta}_{ij}[a_j^{(1)}(a_j^{(1)} + i\hbar\hat{\delta}_{ij})]^{-1};$$

$$I_{jl} = [(a_j^{(1)} + a_l^{(1)})a_l^{(1)}]^{-1} - [(a_j^{(1)} + a_l^{(1)} + i\hbar\hat{\delta}_{ij})a_l^{(1)}]^{-1}$$
$$+ [(a_j^{(1)} + a_l^{(1)} + i\hbar\hat{\delta}_{ij} + i\hbar\hat{\delta}_{il})(a_l^{(1)} + i\hbar\hat{\delta}_{il} - i\hbar\hat{k}_{il})]^{-1}$$
$$- [(a_j^{(1)} + a_l^{(1)} + i\hbar\hat{\delta}_{il})(a_l^{(1)} + i\hbar\hat{\delta}_{il} - i\hbar\hat{k}_{jl})]^{-1}$$
$$\equiv (i\hbar)^2 v_j^{-1} v_l^{-1} \delta_{ij}(2\delta_{il} - k_{jl})(a_j^{(1)} + a_l^{(1)})^{-1}(a_j^{(1)} + a_l^{(1)} + i\hbar\hat{\delta}_{ij}$$
$$+ i\hbar\hat{\delta}_{il})^{-1}(a_l^{(1)})^{-1}(a_l^{(1)} + i\hbar\hat{\delta}_{il} - i\hbar\hat{k}_{jl})^{-1};$$

$$I_{jlm}(i\hbar)^{-1}v_j = \delta_{ij}(a_{jlm}^{(1)})^{-1}(a_{jlm}^{(1)} + i\hbar\hat{\delta}_{ij})^{-1}(a_{lm}^{(1)})^{-1}(a_m^{(1)})^{-1}$$
$$- \delta_{ij}(a_{jlm}^{(1)} + i\hbar\hat{\delta}_{il})^{-1}(a_{jlm}^{(1)} + i\hbar\hat{\delta}_{il} + i\hbar\hat{\delta}_{ij})^{-1}(a_{lm}^{(1)} + i\hbar\hat{\delta}_{il} - i\hbar\hat{k}_{jl})^{-1}$$
$$\times (a_m^{(1)})^{-1} - \delta_{ij}(a_{jlm}^{(1)} + i\hbar\hat{\delta}_{im})^{-1}(a_{jl}^{(1)} + i\hbar\hat{\delta}_{ij} + i\hbar\hat{\delta}_{im})^{-1}$$
$$\times (a_{lm}^{(1)} + i\hbar\hat{\delta}_{im} - i\hbar\hat{k}_{jm})^{-1}(a_{lm}^{(1)} + i\hbar\hat{\delta}_{im} - i\hbar\hat{k}_{jm} - i\hbar\hat{k}_{lm})^{-1}$$
$$+ \delta_{ij}(a_{jlm}^{(1)} + i\hbar\hat{\delta}_{il} + i\hbar\hat{\delta}_{im})^{-1}(a_{jl}^{(1)} + i\hbar\hat{\delta}_{ij} + i\hbar\hat{\delta}_{il} + i\hbar\hat{\delta}_{im})^{-1}$$
$$\times (a_{lm}^{(1)} + i\hbar\hat{\delta}_{im} - i\hbar\hat{k}_{jm} + i\hbar\hat{\delta}_{il} - i\hbar\hat{k}_{jl})^{-1}$$
$$\times (a_m^{(1)} + i\hbar\hat{\delta}_{im} - i\hbar\hat{k}_{jm} - i\hbar\hat{k}_{lm})^{-1};$$

$$a_{jlm}^{(1)} \equiv a_j^{(1)} + a_l^{(1)} + a_m^{(1)}; \quad a_{jl}^{(1)} \equiv a_j^{(1)} + a_l^{(1)}.$$

Set

$$(-i\hbar)^{-2}v_j v_l v_m I'_{jlm} \equiv \delta_{ij}(2\delta_{il} - k_{jl})(\delta_{im} - k_{jm} - k_{lm})$$
$$\times (a_{jlm}^{(1)})^{-1}(a_{jlm}^{(1)} + i\hbar\hat{\delta}_{ij} + i\hbar\hat{\delta}_{il} + i\hbar\hat{\delta}_{im})^{-1}(a_{lm}^{(1)})^{-1}(a_m^{(1)})^{-1}$$
$$\times (a_{lm}^{(1)} + i\hbar\hat{\delta}_{il} - i\hbar\hat{k}_{jl} + i\hbar\hat{\delta}_{im} - i\hbar\hat{k}_{jm})^{-1}$$
$$\times (a_m^{(1)} + i\hbar\hat{\delta}_{im} - i\hbar\hat{k}_{jm} - i\hbar\hat{k}_{lm})^{-1}$$
$$+ \delta_{ij}(2\delta_{il} - k_{jl})(2\delta_{im} - k_{jm})(a_{jlm}^{(1)})^{-1}(a_{lm}^{(1)})^{-1}$$
$$\times (a_{jlm}^{(1)} + i\hbar\hat{\delta}_{ij} + i\hbar\hat{\delta}_{il} + i\hbar\hat{\delta}_{im})^{-1}$$
$$\times (a_m^{(1)})^{-1}(a_{lm}^{(1)} + i\hbar\hat{\delta}_{il} - i\hbar\hat{k}_{jl} + i\hbar\hat{\delta}_{im} - i\hbar\hat{k}_{jm})^{-1}$$
$$\times (a_l^{(1)} + i\hbar\hat{\delta}_{il} - i\hbar\hat{k}_{jl} - i\hbar\hat{k}_{lm})^{-1}.$$

The difference $I_{jlm} - I'_{jlm}$ is skew-symmetric with respect to the indices l and m. Therefore multiplying it by $v_l v_m \exp\phi_l(t)\exp\phi_m(t)$ and summing with respect to l and m we get 0. Hence

$$[\exp(-u_i)]_3 = \exp(-\varphi_i)\sum_{j,l,m}(i\hbar)^{-3}v_j v_l v_m \exp\phi_j \exp\phi_l \exp\phi_m I'_{jlm}.$$

The final form of the solution can be expressed in the following compact form:

$$\exp(-u_i(t)) = \langle i|\hat{\mathcal{M}}(t)\exp(-\varphi_i(t))\hat{\mathcal{M}}^\dagger(t)|i\rangle,$$

$$\hat{\mathcal{M}}(t) = \sum_{n=0}^{\infty}\lambda^n \sum_{j_1\ldots j_n} R_{j_1\ldots j_n} X_{j_1}\ldots X_{j_n};$$

$$R_{j_1\ldots j_n} \equiv \prod_{s=1}^{n}\exp\left(\frac{\phi_{j_s}}{2}\right)\left[\sum_{s\leq m\leq n} a_{j_m}\right]^{-1} \tag{3.21}$$

(cf. (5.8.4). Notice that there is no factor λ^n in the expansion of $\hat{\mathcal{M}}^\dagger$.)

The matrix element is expressed via formula (1.4.22) where X_{+j} is considered in the representation with $X^\dagger_{+j} = X_{-j}$. The first three orders in the expansion in the powers of λ coincide with the expressions given above.

For an arbitrary order n we get

$$[\exp(-u_i(t)]_n = \exp(-\varphi_i) \sum_{j_1,\ldots,j_n} \delta_{ij_1} \prod_{1\leq s\leq n} \exp\phi_{j_s} \mathcal{P}^{(1)}_{j_1\ldots j_n}$$

$$\times \sum_{\omega}\prod_{l=2}^{n}\left[\delta_{ij_l}(1+\xi_{j_l}) - k_{ij_l} - \sum_{m=2}^{l-1}k_{j_m j_l}\theta(j_{\omega(l)} - j_{\omega(m)})\right]\mathcal{P}^{(2)}_{j_1\ldots j_n}, \tag{3.22}$$

where ω runs over all the permutation of indices $(2,\ldots,n)$ and

$$\mathcal{P}^{(\varepsilon)}_{j_1\ldots j_n} \equiv \prod_{l=1}^{n}\left[\sum_{s=l}^{n}a^{(\varepsilon)}_{j_s}\right]^{-1}, \quad \varepsilon = 1, 2; \tag{3.23}$$

$$a^{(2)}_{j_s} \equiv a^{(1)}_{j_s} + i\hbar\delta_{ij_s} - i\hbar\sum_{l=1}^{s-1}\hat{k}_{j_l j_s} + i\hbar\sum_{l=s+1}^{n}\hat{k}_{j_s j_l}.$$

The passage to the classical limit as $\hbar \to 0$ reduces to the trivial change $a^{(\varepsilon)}_j \to a_j$ and when k coincides with Cartan's matrix of a finite-dimensional simple Lie algebra the final result reproduces the equivalent form of expression of the known classical solution (see 4.1.5) of the one-dimensional (finite nonperiodic) Toda lattice. The algebraic dependence of the solutions on the Cartan matrix is the same in the classical and quantum cases. Therefore the factors which depend only on k_{ij} in the classical domain also justify the cut off of the perturbation theory series in the powers of λ for the corresponding quantum problem. Thus the solutions (3.21)–(3.23) are representable by finite polynomials in λ.

7.3.3 Heisenberg's picture (Yang-Feldmann's formalism). The general considerations of 7.1 concerning the applicability of the Yang-Feldmann formalism for finding Heisenberg's operators can be reduced to explicit expressions (3.21)–(3.23) for the generalized Toda lattice. Here $\phi_i \equiv (k\varphi)_i$ and $g_0(t) = \exp[-(hk^{-1}a)t - (hk^{-1}b)]$, where $[\phi_i(t_1), \phi_j(t_2)] = -i\hbar \hat{k}_{ij}(t_1 - t_2)$ (see notations from 7.2). Then we can show that

$$J_+^{(t_1, -t_2, \ldots, \varepsilon_s t_s)} = g_0^{(\varepsilon_s)}(T_s) J_+ (g_0^{(\varepsilon_s)}(T_s))^{-1}, \tag{3.24}$$

where

$$g_0^{(+1)}(T) \equiv g_0(T),$$

$$g_0^{(-1)}(T) \equiv \exp[-(hk\tilde{a})T + i(\hbar/2)(hk^{-1}hv^{-1})T]; \tag{3.25}$$

$$\tilde{a}_j \equiv a_j + i\hbar v_j^{-1}.$$

A similar formula for $J_-^{(t_1, -t_2, \ldots, \varepsilon_s t_s)}$ follows from (3.24) under Hermitian conjugation $(\dagger)$, taking relations $X_{+j}^{\dagger} = X_{-j}$, $h_j^{\dagger} = h_j$ into account. (Recall that in the case considered $J_{\pm} = \sum_j k_j^{1/2} X_{\pm j}$.)

Proof of (3.24) can be obtained by the reduction procedure and the direct calculations for $s = 2$. Indeed,

$$J_+^{(t_1, -t_2)} = g_0(t_1) \left[\sum_j k_j^{1/2} \exp \phi_j(t_2) X_j \right] g_0^{-1}(t_1)$$

$$\equiv \sum_j k_j^{1/2} [g_0(t_1) \exp \phi_j(t_2) g_0^{-1}(t_1)][g_0(t_1) X_j g_0^{-1}(t_1)]$$

$$= \sum_j k_j^{1/2} \{\exp[i\hbar(hv^{-1})_j T_2] \exp \phi_j(t_2)\}\{\exp(-\phi_j(t_1)) X_j\}$$

$$= \sum_j k_j^{1/2} \exp[i\hbar(hv^{-1})_j T_2] \exp\{[\phi_j(t_2) - \phi_j(t_1)] - i\hbar v_j^{-1} T_2\} X_j$$

$$\equiv g_0^{(-1)}(T_2) J_+ (g_0^{(-1)}(T_2))^{-1}.$$

Substituting these relations for $J_{\pm}^{(t_1, -t_2, \ldots \varepsilon_s t_s)}$ into the corresponding formulas for $\hat{u}_m(t)$ yields (3.21)–(3.23) if we take into account the fact that $\exp(-u_i) \equiv \prod_j k_j^{-(k^{-1})_{ij}} \langle i|g(t)|i \rangle$.

§ 7.4 The Liouville equation*)

Consider a Heisenberg field $u(t, x)$ on the cylinder

$$u(t, x + 2L) = u(t, x), \quad -\infty < t < \infty. \tag{4.1}$$

Take the Lagrangian

$$\mathcal{L} = \frac{w}{2}(u_{,\alpha} u^{,\alpha} - \mu^2 \exp 2u) \equiv \frac{w}{2}\left(\frac{\partial u}{\partial z^+}\frac{\partial u}{\partial z^-} - \mu^2 \exp 2u\right)$$

$$\equiv \mathcal{L}_0 + \mu^2 \mathcal{L}_I, \, z^\pm \equiv \frac{t \pm x}{2}, \tag{4.2}$$

where the dimension of the constant w coincides with the dimension of action which makes u dimensionless. Here $\mu^2 \equiv \lambda$ is the constant of self-interaction. The corresponding equation of motion is the Liouville equation:

$$\frac{\partial^2 u}{\partial z^+ \partial z^-} + \lambda \exp 2u = 0. \tag{4.3}$$

Let us seek u as a function in the asymptotic field $\varphi(t, x) \equiv u_{in} = \lim_{t \to -\infty} u(t, x)$ whose explicit form is given by the expansion (2.26). For this let us make use of the Yang-Feldmann equation which in this case takes the form

$$u(z) = \varphi(z) - \lambda \int\limits_{-\infty}^{\infty} \left(\int\limits_{-L}^{L} D^{\mathrm{ret}}(z - z') \exp 2u(z') dx' \right) dt'$$

$$= \varphi(z) - \lambda \sum_{l=0}^{\infty}(l+1) \int\limits_{z^- - L}^{z^-} \left(\int\limits_{z^+ - (l+1)L}^{z^+ - lL} \exp 2u(\tilde{z}) d\tilde{z}^- d\tilde{z}^+ \right). \tag{4.4}$$

Here we have split the total domain of integration into subdomains on which the retarded Green function is constant and made use of the periodicity of u in x.

Let us solve this equation by iteration considering λ as a small parameter. In the first approximation substitute φ in the right-hand side instead of u. After that the domain of integration over $\tilde{z}$ can be divided into segments of length L and, making use of multiplication formulas from 7.2, one can reduce all the integrals to an integral over the standard interval $(z^\pm, z^\pm - L)$. An

*) [106]

elementary summation of the series obtained leads to the final formula

$$u_1 = -(1 - \exp(-P - i))^{-2}\Phi; \quad \Phi \equiv \int\limits_{z^- - L}^{z^-} \left(\int\limits_{z^+ - L}^{z^+} \exp 2\varphi(\tilde{z}) d\tilde{z}^- \, d\tilde{z}^+ \right).$$

(Hereafter we skip the factor $\hbar/w$ at i for simplicity and recover it only in the final answer.)

To get the second approximation insert $\varphi + u_1$ in the right-hand side of (4.4) instead of u. The formulas (2.13) and (2.38) show that u_2 is proportional to Φ^2 with the proportionality factor depending only on P. In general, we can write

$$u = \varphi + \tilde{u} = \varphi + \sum_{l=1}^{\infty} \lambda^l u_l = \varphi + \sum_{l=1}^{\infty} \lambda^l f_l(P)\Phi^l. \tag{4.5}$$

Substitute this expression into (4.4). We get

$$\exp(2u) = \exp(2\varphi)[\exp(-2\varphi)\exp(2u)]$$

$$= \exp(2\varphi) \cdot T \exp \int\limits_{0}^{1} dt \exp(-2\varphi t)\tilde{u} \exp(2\varphi t).$$

Making use of (2.34), (2.36) we calculate

$$\exp(-2\varphi t)u_l \exp(2\varphi t) = f_l(P - 2it)\exp(2ilt)\Phi^l.$$

We calculate the integrals over z with the help of the reduction formula (2.38). Finally, with the help of commutation relations for Φ with functions of P (2.37) we can move all Φ's in the right-hand side of (4.4) to the right. Then equating the coefficients of the same powers of λ we get the recurrence relation for $f_l(P)$

$$f_{l+1} = -\frac{(1 - \exp i)^2}{[1 - \exp i(l+1)]^2[1 - \exp(-P - i(l+1)]^2}$$

$$\times \sum_{m=1}^{l} 2^m \sum_{\substack{l_1,\ldots,l_m \\ (\Sigma l_i = l)}} \int\limits_{0}^{1} dt_1 \exp(2il_1t_1)f_{l_1}(P - 2it_1 + 2i) \int\limits_{0}^{t_1} dt_2 \exp(2il_2t_2)$$

$$\times f_{l_2}(P - 2it_2 + 2il_1 + 2i) \times \ldots$$

$$\times \int\limits_{0}^{t_{m-1}} dt_m \exp(2il_mt_m)f_{l_m}(P - 2it_m + 2il_1 + \ldots + 2il_m + 2i).$$

$$\tag{4.6}$$

First notice that $f_l(P)$ takes the form

$$f_l^{(0)} = \frac{(-1)^l}{i^{l-1}} \frac{(\exp(i) - 1)^l}{\exp(il) - 1} \quad \text{as } P \to \infty. \tag{4.7}$$

To see this without referring to the recurrence relations notice that the T-exponential in the expression for $\exp 2u$ turns into the usual exponential as $P \to \infty$ (since the integrands commute for any values of t) which enables us to calculate f_l in an elementary way. The solution of (4.6) is of the form

$$f_l = f_l^{(0)} \frac{1}{1 - \exp(-P - il)} \prod_{m=1}^{2l-1} \frac{1}{1 - \exp(-P - im)}, \tag{4.8}$$

which can be verified directly.

Now calculate

$$\frac{\partial^2 u}{\partial z^+ \partial z^-} = \frac{\partial^2 \varphi}{\partial z^+ \partial z^-} + \frac{\partial^2 \tilde{u}}{\partial z^+ \partial z^-} = \sum_{l=1}^{\infty} \lambda^l f_l(P) \frac{\partial^2 \Phi^l}{\partial z^+ \partial z^-}$$

$$= \sum_l \lambda^l f_l(P) \frac{(1 - \exp(il))^2}{(1 - \exp(i))^2} (1 - \exp(-P - il))^2 \exp(2\varphi) \cdot \Phi^{l-1}$$

$$= \exp(2\varphi) \sum_l \lambda^l f_l(P - 2i) \frac{(\exp(il) - 1)^2}{(\exp(i) - 1)^2} (1 - \exp(-P + 2i - li))^2 \Phi^{l-1}$$

$$= -\exp(2\varphi) \frac{\lambda}{(\exp(i) - 1)(1 - \exp(-P + i))}$$

$$\times \sum_l (\exp(il) - 1)(1 - \exp(-P + 2i - il))$$

$$\times [i\lambda(\exp(i) - 1)(1 - \exp(-P))^{-1}(1 - \exp(-P - i))^{-1}\Phi]^{l-1}$$

$$= \exp(2\varphi) \frac{\lambda}{(\exp(i) - 1)(1 - \exp(-P + i))} \left\{ (1 + \exp(-P + 2i)) \frac{1}{1 + X} \right.$$

$$\left. - \exp(i) \frac{1}{1 + X \exp(i)} - \exp(-P + i) \frac{1}{1 + X \exp(-i)} \right\}$$

$$= \exp(2\varphi) \frac{\lambda}{1 - \exp(-P + i)} \left\{ \frac{\exp(-P + i)}{1 + X \exp(-i)} - \frac{1}{1 + X \exp(i)} \right\} \frac{1}{1 + X}, \tag{4.9}$$

where

$$X \equiv \frac{\lambda}{i} (\exp(i) - 1)(1 - \exp(-P))^{-1}(1 - \exp(-P - i))^{-1} \Phi.$$

Here we have made use of the commutation relations for $F(P)$ and Φ and the explicit form (4.7), (4.8) of $f_l(P)$.

Now calculate $\exp(-u)$. We have

$$\exp(-u) = \exp(-\varphi)[\exp(\varphi)\exp(-u)] = \exp(-\varphi)R$$

where

$$R = R[\tilde{u}] = T\exp\left\{-\int_0^1 \exp(\varphi t)\cdot u\cdot\exp(-\varphi t)dt\right\}. \qquad (4.10)$$

The arguments used for the solution of Yang-Feldmann equation show that

$$R = \sum_{l=0}^{\infty}\lambda^l R_l(P)\Phi^l. \qquad (4.11)$$

Comparing with (4.9) we get

$$R_l = \sum_{m=1}^{l}(-1)^m \sum_{\substack{l_1,\dots,l_m \\ \Sigma l_i = l}}\int_0^1 dt_1 \exp(-il_1 t_1)f_{l_1}(P+it_1)\int_0^{t_1}dt_2$$

$$\times \exp(-il_2 t_2)f_{l_2}(P+it_2+2il_1)\times\dots \qquad (4.12)$$

$$\times \int_0^{t_{m-1}} dt_m\exp(-il_m t_m)f_{l_m}(P+it_m+2il_1+\dots+2il_m),\quad l\geq 1$$

and we find directly that

$$R_0 = 1;$$
$$R_1 = \frac{\exp(i)-1}{i\exp(i)}\cdot\frac{1}{(1-\exp(-P-i))(1-\exp(-P-2i))};\quad R_2 = 0. \qquad (4.13)$$

We have $R_l^{(0)} = 0$ $(l \geq 2)$ as $P \to \infty$, which is easy to see since the T-exponential turns into the usual exponential as $P \to \infty$.

Let us pass in (4.12) from integration over t_i to integration over $z_i \equiv P+it_i$. After extracting the factor $\exp(lP)$, the dependence on P still remains inside of the integration limits. After that it is easy to differentiate with respect to P and the remaining integrals reduce to R_l with lesser values of l:

$$\exp(lP)\frac{\partial}{\partial P}[R_l\exp(-lP)] = i\sum_{n=1}^{l}[\exp(-in)f_n(P+i)R_{l-n}(P+2in)$$

$$- f_n(P+2il-2in)R_{l-n}(P)].$$

Let us prove by induction that $R_l = 0$ for $l \geq 2$. Suppose $R_2 = \dots = R_{l-1} = 0$. Then

$$\exp(lP)\partial/\partial P[R_l\exp(-lP)] = \exp(-il)f_l(P+i) - f_l(P)$$

$$+ \exp(-i(l-1))f_{l-1}(P+i)R_1(P+2i(l-1)) - f_{l-1}(P+2i)R_1(P).$$

With the help of (4.8), (4.13) implies $\exp(lP)\partial/\partial P[R_l \exp(-lP)] = 0$, i.e., $R_l = c\exp(lP)$. Since we should get $R_l \to 0$ as $P \to \infty$, it follows that $c = 0$. Thus

$$\exp(-u) = \exp(-\varphi)\left[1 + \frac{2w \cdot \lambda}{\hbar}\sin\frac{\hbar}{2w}\exp\left(-\frac{i\hbar}{2w}\right)\right.$$
$$\left. \times \left(1 - \exp\left(-P - i\frac{\hbar}{w}\right)\right)^{-1}\left(1 - \exp\left(-P - 2i\frac{\hbar}{w}\right)\right)^{-1}\Phi\right].$$
$$(4.14)$$

Now we get

$$\exp 2u = R^{-1}\exp\varphi R^{-1}\exp\varphi = \exp(2\varphi)R^{-1}(P - 2i)R^{-1}(P - i).$$

Comparing this formula with (4.9) we see that (4.5), with $f_l(P)$ given by (4.7) and (4.8), satisfies Liouville's equation indeed.

An important characteristic feature of the solutions obtained is that the perturbation theory series for $\exp(-u)$ is finite and in this particular case $\exp(-x)$ is a linear function in λ.

We can perform the formal passage to a field in infinite-dimensional space introducing $P' = L^{-1/2}P$ and letting L tend to infinity; here $f_l = f_l^{(0)}$ and

$$\exp(-u) = \exp(-\varphi)\left[1 + \frac{2w\lambda}{\hbar}\sin\frac{\hbar}{2w}\exp\left(-\frac{i\hbar}{2w}\right)\Phi\right]. \qquad (4.15)$$

However, whereas the operators $\varphi(t, x)$ in a problem periodic with respect to x are well defined, the corresponding operators for the problem in infinite space have no physical interpretation: as is known, they cannot be realized in a Fock space [9, 10]. However, for an infinite space we can use the passage to a finite interval for a regularization procedure.

The expression (4.15) has a correct classical limit. Indeed, letting $\hbar$ tend to 0 we get one of the forms of the general classical solution of Liouville's equation expressed via two arbitrary functions given on characteristics.

We have solved the quantum Liouville equation with the help of Yang-Feldmann formalism. The same solution can also be obtained with the help of the canonical formalism and the "halved" S-matrix. We will illustrate this for an infinite space (which is technically simpler than the case of a finite interval).

We will construct the perturbation theory series for the operator $\exp(-u)$ making use of Schwinger's formula (2.44).

At zero-th order $\exp(-u) = \exp(-\varphi)$.

At first order

$$[\exp(-u)]_1 = i^{-1}\int dz'\theta(t - t')[\exp(-\varphi(z)), \exp(2\varphi(z'))].$$

Making use of (2.32) and the easily verified formula

$$\theta(t - t')[1 - a^{\text{sgn}(z-z')}] = (1 - a^2)\theta(z - z'),$$

we get

$$[\exp(-u)]_1 = \frac{2w}{\hbar} \sin \frac{\hbar}{2w} \exp\left(-\frac{i\hbar}{2w}\right) \exp(-\varphi) \cdot \Phi.$$

All subsequent orders vanish (see the general proof in 7.5). Thus we have indeed obtained formula (4.15) again.

§ 7.5 Multicomponent 2-dimensional models. 1*)

Now let us pass to generalized Liouville's equations, namely, to models with Lagrangian of the form (2.23), where interaction is of the exponential form

$$V(u) = \sum_{1 \leq \alpha \leq r} v_\alpha \exp \sum_{1 \leq \beta \leq r} k_{\alpha\beta} u_\beta. \tag{5.1}$$

Then the equations of motion are of the form

$$\frac{\partial^2 u_\alpha}{\partial z^+ \partial z^-} + \mu^2 \exp\left(\sum_{1 \leq \beta \leq r} k_{\alpha\beta} u_\beta\right) = 0 \tag{5.2}$$

(under the notations from 7.2). We will calculate Heisenberg's field operators with the help of formula (2.44). Though, as has been mentioned in 7.4, for x from an infinite coordinate line the interpretation of the fields φ_α is unclear, we will perform all the calculations for this example to avoid the additional algebraic complications which appear in the problem for a cyclic x.

The general structure on the interaction (5.1) enables one to transform formula (2.44) in the following remarkable way. Introduce the "two-point field operators" $\phi_{ij}^\alpha \equiv \phi_\alpha^+(z_i^+) + \phi_\alpha^-(z_j^-)$ which, thanks to (2.40), satisfy

$$[\phi_{ij}^\alpha, \phi_{nl}^\beta] = -\frac{i\hbar}{4} \hat{k}_{\alpha\beta}[\text{sgn}(z_i^+ - z_n^+) + \text{sgn}(z_j^- - z_l^-)]. \tag{5.3}$$

Define the operators

$$V_{ij} \equiv V(z_i^+, z_j^-) \equiv \sum_\alpha 2v_\alpha \exp \phi_{ij}^\alpha, \tag{5.4}$$

so that $V_i = V_{ii}/2$ in (2.44). Formulas (5.3) and (2.13) imply that

$$[V_{ij}, V_{nl}] = \sum_{\alpha,\beta} 4v_\alpha v_\beta \exp \phi_{ij}^\alpha \cdot \exp \phi_{nl}^\beta$$

$$\times \left\{1 - \exp\left[\frac{i\hbar}{4} \hat{k}_{\alpha\beta}(\text{sgn}(z_i^+ - z_n^+) + \text{sgn}(z_j^- - z_l^-))\right]\right\}. \tag{5.5}$$

---

*) [106]

Notice that the arguments to follow are actually based not on the exponential nature of the interactions but on the algebraic structure of relations (5.5) expressed by C-numerical factors in the right-hand side. In particular, it is essential that

$$[V_{ij}, V_{ql}] = 0 \text{ for } (z_i^+ - z_q^+)(z_j^- - z_l^-) < 0 \qquad (5.6)$$

since the arguments of the sgn functions are of different sign and all the exponentials turn into 1. For $A[u]$ take $\exp(gu_\alpha)$. Passing from integration over $dt_i dx_i$ to integration over $dz_i \equiv dz_i^+ dz_i^-$ we can rewrite (2.44) in the form

$$[\exp(gu_\alpha)]_n = \frac{1}{(i\hbar)^n} \int_{-\infty}^{\infty} \cdots \int \theta(t - t_1) \dots \theta(t_{n-1} - t_n)$$

$$\times [\exp(g\varphi_\alpha), V_{11}, \dots, V_{nn}] dz_1 \dots dz_n. \qquad (5.7)$$

At zero-th order we have $[\exp(gu_\alpha)]_0 = \exp(g\varphi_\alpha)$. At first order we get from (5.7):

$$[\exp(gu_\alpha)]_1 = \frac{1}{i\hbar} \int_{-\infty}^{\infty} dz_1 \theta(t - t_1)[\exp(g\varphi_\alpha), V_{11}]$$

$$= \frac{1}{i\hbar} \int_{-\infty}^{\infty} dz_1 \theta(z - z_1)[\exp(g\varphi_\alpha), V_{11}]; \qquad (5.8)$$

$$\theta(z - z_1) \equiv \theta(z^+ - z_1^+)\theta(z^- - z_1^-).$$

The latter equality follows from the fact that thanks to (5.6) the commutator in the integrand is nonzero only for $(z^+ - z_1^+)(z^- - z_1^-) < 0$ whereas $\theta(t - t_1)$ vanishes in the domain $z^+ < z_1^+$, $z^- < z_1^-$ since $t - t_1 = z^+ - z_1^+ + z^- - z_1^-$. This very condition determines the causal structure of the dependence of the terms of the first approximation (5.8).

Making use of (2.41) and (5.4) we get the final result

$$[\exp(gu_\alpha)]_1 = \exp(g\varphi_\alpha) \cdot \frac{2v_\alpha}{i\hbar} \left[1 - \exp\left(\frac{i\hbar g}{2v_\alpha}\right)\right] \Phi_\alpha^+ \Phi_\alpha^-. \qquad (5.9)$$

At second order we can write

$$[\exp(gu_\alpha)]_2 = \frac{1}{(i\hbar)^2} \int dz_1 dz_2 \theta(z - z_1)\theta(t_1 - t_2)[\exp(g\varphi_\alpha(z)), V_{11}, V_{22}]. \qquad (5.10)$$

Here again we have made use of the vanishing of $[\exp(g\varphi_\alpha), V_{11}]$ and the presence of $\theta(t - t_1)$ to introduce $\theta(z - z_1) \equiv \theta(z^+ - z_1^+)\theta(z^- - z_1^-)$ in the integrand of (5.10).

To reconstruct (5.10) in the Z-ordered form, introduce the following factor in the integrand:

$$1 \equiv [\theta(z_1^+ - z_2^+) + \theta(z_2^+ - z_1^+)][\theta(z_1^- - z_2^-) + \theta(z_2^- - z_1^-)]. \qquad (5.11)$$

One of the four summands which appear after simplification of (5.11), $\theta(z_2^+ - z_1^+)\theta(z_2^- - z_1^-)$, is not compatible with $\theta(t_1 - t_2)$ and is rejected; another one, $\theta(z_1^+ - z_2^+)\theta(z_1^- - z_2^-)$, immediately gives a Z-ordered structure whereas the second two contribute to the integrand in the form

$$\theta(z - z_1)\theta(z_1^+ + z_1^- - z_2^+ - z_2^-)[\exp(g\varphi_\alpha), V_{11}, V_{22}]$$

$$\times [\theta(z_2^+ - z_1^+)\theta(z_1^- - z_2^-) + \theta(z_1^+ - z_2^+)\theta(z_2^- - z_1^-)]. \tag{5.12}$$

Let us rename the variables of integration: $z_2^+ \rightleftharpoons z_1^+$ in the first and $z_2^- \rightleftharpoons z_1^-$ in the second of summands in (5.12). We get

$$\theta(z - z_1)\theta(z_1 - z_2)\{\theta(z_2^+ + z_1^- - z_1^+ - z_2^-)[\exp(g\varphi_\alpha), V_{21}, V_{12}]$$

$$+ \theta(z_1^+ + z_2^- - z_2^+ - z_1^-)[\exp(g\varphi_\alpha), V_{12}, V_{21}]. \tag{5.13}$$

Since now $(z_1^+ - z_2^+)(z_2^- - z_1^-) < 0$, (5.5) implies $[V_{12}, V_{21}] = 0$. It follows from the Jacobi identity that $[\exp(g\varphi_\alpha), V_{12}, V_{21}] = [\exp(g\varphi_\alpha), V_{21}, V_{12}]$ and after simplifying we get the following Z-ordered expression for the second order terms

$$[\exp(gu_\alpha)]_2 = \frac{1}{(i\hbar)^2} \int dz_1 \int dz_2 \theta(z - z_1)\theta(z_1 - z_2)$$

$$\times \{[\exp(g\varphi)_\alpha, V_{11}, V_{22}] + [\exp(g\varphi)_\alpha, V_{21}, V_{12}]\}. \tag{5.14}$$

The Z-ordered nature is justified by the fact that the retarded function of the two-dimensional Laplace operator is of the form

$$D^{\text{ret}}(z, \tilde{z}) = \frac{1}{2}\theta(z - \tilde{z}) \equiv \frac{1}{2}\theta(z^+ - \tilde{z}^+)\theta(z^- - \tilde{z}^-),$$

i.e., it does not vanish only in the domain $t > \tilde{t}$ (or $(z^+ - \tilde{z}^+) + (z^- - \tilde{z}^-) > 0$) inside of the light cone $(t - \tilde{t})^2 > (x - \tilde{x})^2$ (or $(z^+ - \tilde{z}^+)(z^- - \tilde{z}^-) > 0$).

Calculating at n-th order we start with the formula (5.15). To reduce it to the Z-ordered form multiply the integrand by two factors of the form

$$\sum_{\omega(i_1,\ldots,i_n)} \prod_{2 \le l \le n} \theta(z_{i_{l-1}}^\pm - z_{i_l}^\pm) = 1,$$

where the sum runs over all $(n!)$ the permutations of indices $i_1, \ldots, i_n$ (see a similar formula (5.11)). In addition, from the properties of commutators and thanks to the presence of the time-dependent θ-functions the integrand actually contains

$$\prod_{1 \le i \le n} \theta(z - z_i),$$

i.e., $z^+ > z_i^+$, $z^- > z_i^-$ for $i = 1, 2, \ldots, n$, as well as at first order. After simplification we get exactly $n! \times n!$ summands. Under an appropriate change of variables $z_i^+ \rightleftharpoons z_q^+$, $z_j^- \rightleftharpoons z_m^-$ (for each summand its own change) we reduce the integrand to the form with the "normal" order of integration $z^\pm \equiv z_0^\pm > z_1^\pm > \ldots > z_n^\pm$:

$$\prod_{1 \leq l \leq n} \theta(z_{l-1} - z_l) \sum_{\omega(i_1,\ldots,i_n)} \sum_{\omega(j_1,\ldots,j_n)} [\exp(g\varphi_\alpha), V_{i_1 j_1}, \ldots, V_{i_n j_n}]$$

$$\times \prod_{2 \leq s \leq n} \theta(z_{i_{s-1}}^+ + z_{j_{s-1}}^- - z_{i_1}^+ - z_{j_s}^-). \tag{5.15}$$

There are $n! \times n!$ summands here and the second of the products of θ-functions is the factor $\theta(t_1 - t_2) \ldots \theta(t_{n-1} - t_n)$ from (5.7) rewritten in new variables. The first product determines the Z-ordering. The expression (5.15) at second order corresponds to (5.13).

Consider some summand from (5.15). Let V_{qm} be any of the "Hamiltonians" standing in the multiple commutator to the left of the "Hamiltonian" V_{1j}. The time-dependent θ-functions give $z_1^+ + z_1^- - z_q^+ - z_m^- < 0$, but thanks to the Z-ordering $z_1^+ - z_q^+ > 0$ and hence $z_j^- - z_m^- < 0$.

Thus $(z_1^+ - z_q^+)(z_j^- - z_m^-) < 0$ and therefore $[V_{1j}, V_{qm}] = 0$ in (5.6).

It follows that thanks to the Jacobi identity we can put V_{1j} in the multiple commutator to the leftmost place, directly to the right of $\exp(g\varphi_\alpha)$. Similarly, V_{2q} can be placed to the right of V_{1j}, etc. Thus each of the multiple commutators in (5.15) can be reduced to the standard form

$$[\exp(g\varphi_\alpha), V_{1q_1}, V_{2q_2}, \ldots, V_{nq_n}].$$

Now (5.15) takes the form of a linear combination of independent n-multiple commutators with the coefficient of each of them being the sum of $n!$ products of θ-functions of the form

$$\sum_{\omega(i_1,\ldots,i_n)} \prod_{2 \leq q \leq n} \theta(z_{q-1}^+ + z_{i_{q-1}}^- - z_q^+ - z_{i_q}^-). \tag{5.16}$$

However, such a sum is identically equal to 1 since it reflects the division of the domain $z_0^+ > z_i^+, z_0^- > z_i^-$ into non-intersecting subdomains. As a result we get for the n-th order:

$$[\exp(gu_\alpha(z))]_n = \int \ldots \int \sum_{\omega(q_1,\ldots,q_n)}$$

$$[\exp(g\varphi_\alpha(z)), V_{1q_1}, \ldots, V_{nq_n}] \prod_{1 \leq m \leq n} \frac{\theta(z_{m-1} - z_m) dz_m}{i\hbar}. \tag{5.17}$$

This is the end of the first stage of the reduction to the Z-ordered structure.

Let us continue by substituting the explicit expressions (5.4) for V_{ij} into (5.17). We get the n-multiple summation over the indices β_i ($1 \leq i \leq n$). To calculate the multiple commutators we make use of (5.5). We get

$$[\exp(g\varphi_\alpha), V_{1q_1}, \ldots, V_{nq_n}] = \exp(g\varphi_\alpha) \sum_{\beta_1, \ldots, \beta_n} \exp(\varphi^+_{\beta_1}(1)) \ldots$$

$$\ldots \exp(\varphi^+_{\beta_n}(n)) \exp(\varphi^-_{\beta_1}(q_1)) \ldots \exp(\varphi^-_{\beta_n}(q_n))$$

$$\times \prod_{s=1}^n 2v_{\beta_s} \left\{ 1 - \exp\left[\frac{i\hbar g}{4v_\alpha} \delta_{\alpha\beta_s}(\varepsilon^+(0, s) + \varepsilon^-(0, q_s)) \right. \right. \tag{5.18}$$

$$\left. \left. + \frac{i\hbar}{4} \sum_{l=1}^{s-1} \hat{k}_{\beta_l \beta_s}(\varepsilon^+(l, s) + \varepsilon^-(q_l, q_s)) \right] \right\}.$$

In order to simplify (5.18) we have denoted $\varphi^\pm_{\beta_s}(q_i) \equiv \varphi^\pm_{\beta_s}(z^\pm_{q_i})$ and $\varepsilon^\pm(l, s) \equiv \operatorname{sgn}(z^\pm_l - z^\pm_s)$; $z^\pm_0 \equiv z^\pm$. Furthermore, we rearrange the factors $\exp\varphi^-_\mu(q_i)$ in (5.18) with respect to the normal order of the arguments $z^-_1, \ldots, z^-_n$. Making use of (2.40) we get

$$\exp(\varphi^-_{\beta_1}(q_1)) \ldots \exp(\varphi^-_{\beta_n}(q_n)) = \exp\left(\frac{i}{4} F_\omega \right) \exp(\varphi^-_{\beta_{j_1}}(1)) \ldots \exp(\varphi^-_{\beta_{j_n}}(n)),$$

where the permutation $\omega \colon (1, 2, \ldots, n) \to (j_1, j_2, \ldots, j_n)$ is obviously inverse to $(1, 2, \ldots, n) \to (q_1, \ldots, q_n)$ and

$$F_\omega(\beta, k) \equiv \sum_{l<m}^n \hat{k}_{\beta_{j_l} \beta_{j_m}} \theta(j_l - j_m); \tag{5.19}$$

the index ω means that F depends on a permutation; it also depends on β_i and the matrix elements of k.

Thanks to the presence of the factors $\prod_i \theta(z_{i-1} - z_i)$ which ensure the Z-ordering, we get the following values of some terms in the exponents in (5.18):

$$\varepsilon^+(0, s) + \varepsilon^-(0, q_s) = 2;$$

$$\varepsilon^+(l, s) + \varepsilon^-(q_l, q_s) = \begin{cases} 2, & \text{for } q_l < q_s \\ 0, & \text{for } q_l > q_s \end{cases} \text{ if } l < s.$$

The same can be rewritten in terms of the permutation of the indices β_i which is the inverse to the permutation of the z_i:

$$\varepsilon^+(l, s) + \varepsilon^-(q_l, q_s) = \begin{cases} 2 & \text{if } (\omega(l), \omega(s)) \text{ are ordered as } (l, s) \\ 0 & \text{otherwise,} \end{cases}$$

i.e.,

$$\varepsilon^+(l, s) + \varepsilon^-(q_l, q_s) = 2\theta(\omega(s) - \omega(l)).$$

Thanks to the above arguments we can rewrite (5.18) in the form

$$[\exp(g\varphi_\alpha), V_{1q_1}, \ldots, V_{nq_n}] = \exp(g\varphi_\alpha) \sum_{\beta_1,\ldots,\beta_n} \exp(\varphi^+_{\beta_1}(1))\ldots$$

$$\ldots\exp(\varphi^+_{\beta_n}(n))\exp(\varphi^-_{\beta_{\omega(1)}}(1))\ldots\exp(\varphi^-_{\beta_{\omega(n)}}(n))\exp\left(\frac{i\hbar}{4}F_\omega(\beta,k)\right)$$

$$\times \prod_{1\leq s\leq n} 2v_{\beta_s}\left\{1 - \exp\left[\frac{i\hbar g}{2v_\alpha}\delta_{\alpha\beta_s} + \frac{i\hbar}{2}\sum_{1\leq l\leq s-1}\hat{k}_{\beta_l\beta_s}\theta(\omega(s) - \omega(l))\right]\right\}.$$

$$(5.20)$$

Performing the Z-ordered integration of (5.20) and taking the definition (2.41) into account we get for the terms of the n-th order

$$[\exp(gu_\alpha)]_n = \exp(g\varphi_\alpha) \sum_{\beta_1\ldots\beta_n} \sum_{\omega(\beta_1,\ldots,\beta_n)} \exp\left(\frac{i\hbar}{4}F_\omega(\beta,k)\right)$$

$$\times \Phi^+_{\beta_1\ldots\beta_n}\Phi^-_{\omega(\beta_1)\ldots\omega(\beta_n)} \prod_{1\leq s\leq n}\frac{2v_{\beta_s}}{i\hbar}\left\{1 - \exp\left[\frac{i\hbar g}{2v_\alpha}\delta_{\alpha\beta_s}\right.\right. \qquad (5.21)$$

$$\left.\left. + \frac{i\hbar}{2}\sum_{1\leq l\leq s-1}\hat{k}_{\beta_l\beta_s}\theta(\omega(s) - \omega(l))\right]\right\}.$$

The expression (5.21) is constituted from $n!$ different structures corresponding to $n!$ permutations of the indices β_i, whereas in the final answer we will seek the sum of $(n-1)!$ structures. Let us first consider the calculation of the third order approximation to illustrate how we unite the summands for an arbitrary n. For simplicity set $g = -k_{\alpha\alpha}/2$.

Introduce the function $f(x) \equiv [1 - \exp(\frac{i\hbar x}{2})]/i\hbar$ and, skipping the summation sign over β_i, rewrite (5.21) for the third order (up to a common factor) in the form

$$\delta_{\alpha\beta_1}(\alpha\beta_2\beta_3)^+ \left\{(\beta_1\beta_2\beta_3)^- f\left(\hat{k}_{\beta_2\beta_3} + \hat{k}_{\beta_1\beta_3} - \frac{1}{2}\delta_{\alpha\beta_3}\hat{k}_{\alpha\alpha}\right) f\left(\hat{k}_{\beta_1\beta_2} - \frac{1}{2}\delta_{\alpha\beta_2}\hat{k}_{\alpha\alpha}\right)\right.$$

$$+\exp\left(\frac{i\hbar}{4}\hat{k}_{\beta_1\beta_2}\right)(\beta_2\beta_1\beta_3)^- f\left(\hat{k}_{\beta_1\beta_3} + \hat{k}_{\beta_2\beta_3} - \frac{1}{2}\delta_{\alpha\beta_3}\hat{k}_{\alpha\alpha}\right)$$

$$\times f\left(-\frac{1}{2}\delta_{\alpha\beta_2}\hat{k}_{\alpha\alpha}\right) +\exp\left(\frac{i\hbar}{4}\hat{k}_{\beta_2\beta_3}\right)(\beta_1\beta_3\beta_2)^- f\left(\hat{k}_{\beta_1\beta_3} - \frac{1}{2}\delta_{\alpha\beta_3}\hat{k}_{\alpha\alpha}\right)$$

$$\times f\left(\hat{k}_{\beta_1\beta_2} - \frac{1}{2}\delta_{\alpha\beta_2}\hat{k}_{\alpha\alpha}\right) +\exp\left(\frac{i\hbar}{4}(\hat{k}_{\beta_1\beta_2} + \hat{k}_{\beta_1\beta_3})\right)(\beta_2\beta_3\beta_1)^-$$

$$\times f\left(\hat{k}_{\beta_2\beta_3} - \frac{1}{2}\delta_{\alpha\beta_3}\hat{k}_{\alpha\alpha}\right) f\left(-\frac{1}{2}\delta_{\alpha\beta_2}\hat{k}_{\alpha\alpha}\right) +\exp\left(\frac{i\hbar}{4}(\hat{k}_{\beta_1\beta_3} + \hat{k}_{\beta_2\beta_3})\right)$$

$$\times (\beta_3\beta_1\beta_2)^- f\left(-\frac{1}{2}\delta_{\alpha\beta_3}\hat{k}_{\alpha\alpha}\right) f\left(\hat{k}_{\beta_1\beta_2} - \frac{1}{2}\delta_{\alpha\beta_2}\hat{k}_{\alpha\alpha}\right) + (\beta_3\beta_2\beta_1)^-$$

$$\times \exp\left(\frac{i\hbar}{4}(\hat{k}_{\beta_1\beta_2} + \hat{k}_{\beta_1\beta_3} + \hat{k}_{\beta_2\beta_3})\right) f\left(-\frac{1}{2}\delta_{\alpha\beta_3}\hat{k}_{\alpha\alpha}\right) f\left(-\frac{1}{2}\delta_{\alpha\beta_2}\hat{k}_{\alpha\alpha}\right)\Big\}. \quad (5.22)$$

Temporarily, set $\Phi^\pm_{\alpha\beta\gamma} \equiv (\alpha\beta\gamma)^\pm$.

First let us unite the two first terms of (5.22) taking into account that in the second summand we can replace β_2 by α and β_1 by β_2 (since $\alpha = \beta_1 = \beta_2$):

$$\delta_{\alpha\beta_1}(\alpha\beta_2\beta_3)^+(\alpha\beta_2\beta_3)^-\left(\hat{k}_{\beta_2\beta_3} + \hat{k}_{\beta_1\beta_3} - \frac{1}{2}\delta_{\alpha\beta_3}\hat{k}_{\alpha\alpha}\right)$$

$$\times \left[f\left(\hat{k}_{\beta_1\beta_2} - \frac{1}{2}\delta_{\alpha\beta_2}\hat{k}_{\alpha\alpha}\right) + \exp\left(\frac{i\hbar}{4}\hat{k}_{\beta_1\beta_2}\right) f\left(-\frac{1}{2}\delta_{\alpha\beta_2}\hat{k}_{\alpha\alpha}\right)\right]$$

$$= \delta_{\alpha\beta_1}(\alpha\beta_2\beta_3)^+(\alpha\beta_2\beta_3)^-$$

$$\times f\left(\hat{k}_{\beta_2\beta_3} + \hat{k}_{\beta_1\beta_3} - \frac{1}{2}\delta_{\alpha\beta_3}\hat{k}_{\alpha\alpha}\right) f(\hat{k}_{\beta_1\beta_2} - \delta_{\alpha\beta_2}\hat{k}_{\alpha\alpha}).$$

Similarly, the remaining four terms in (5.22) reduce after obvious changes of variables (indices) to the form

$$\delta_{\alpha\beta_1}(\alpha\beta_2\beta_3)^+(\alpha\beta_3\beta_2)^- \exp\left(\frac{i\hbar}{4}\hat{k}_{\beta_2\beta_3}\right) f(\hat{k}_{\beta_1\beta_3} - \delta_{\alpha\beta_3}\hat{k}_{\alpha\alpha})$$

$$\times f(\hat{k}_{\beta_1\beta_2} - \delta_{\alpha\beta_2}\hat{k}_{\alpha\alpha}).$$

Thus instead of $3! \equiv 6$ summands we have reduced (5.22), i.e., the third order, to 2 independent structures. The numerical coefficients of these structures coincide for $\beta_{\omega(1)} = \beta_1 = \alpha$ with the coefficients in (5.21) modified as follows: The factors whose exponent $\sum_{1\le l\le s-1} \hat{k}_{\beta_l\beta_s}\theta(\omega(s) - \omega(l))$ contains the unique term $\hat{k}_{\beta_1\beta_s} = \hat{k}_{\alpha\beta_s}$ (called *anomalous* in what follows) are replaced, thanks to (5.21)–(5.22), by the modified factors:

$$f\left(\hat{k}_{\alpha\beta_s} - \frac{1}{2}\delta_{\alpha\beta_s}\hat{k}_{\alpha\alpha}\right) \to f(\hat{k}_{\alpha\beta_s} - \delta_{\alpha\beta_s}\hat{k}_{\alpha\alpha})$$

while the other factors remain as they were.

Now, let us regroup the summands in (5.21) in the general case to get the final expression for the n-th order terms. First of all let us define the general number a_m of all permutations of the indices which do not affect β_1 and to which exactly m anomalous factors correspond. For $m = n - 1$ all the $n - 1$ factors should be anomalous, meaning that $\omega(s) - \omega(l) < 0$ for $l < s$ ($2 \le s, l \le n$); in other words, this condition is only satisfied for the mirror permutation $(2, 3, \ldots, n) \to (n, \ldots, 3, 2)$, hence $a_{n-1} = 1$.

The number a_{n-2} is the number of permutations which differ from the above (miror) in one of the indices which is placed to any place to the right of its initial place; for β_n there are $n-2$ such places, $n-3$ for β_{n-1}, etc. Thus

$$a_{n-2} = \sum_{1 \le l \le n-2} l = \frac{1}{2}(n-1)(n-2).$$

Similarly, a_{n-3} is the number of permutations that differ from the miror one by one index replaced to any place to the right of its initial position and by another index that initially stood to the left of the first one replaced to any place to any place to the right of its own initial position but to the left of the new position of the first index. We get

$$a_{n-3} = \sum_{1 \le l_1 \le l_2}^{n-2} l_1 l_2 = \frac{1}{24}(n-1)(n-2)(n-3)(3n-4).$$

Each next case is easily reduced to the preceding one and we get

$$a_{n-m} = \sum_{1 \le l_1 < l_2 < \ldots < l_{m-1}}^{n-2} \prod_{j=1}^{m-1} l_j.$$

In particular, $a_1 = (n-1)!$. Notice that a_m is the number of all permutations of $(n-1)$ elements for which the Young diagram contains exactly m rows. Therefore all the a_m exhaust all the permutations

$$\sum_{1 \le m \le n-1} a_m = (n-1)!,$$

which is also a direct corollary of their definition.

So far we have grouped the permutations which did not affect β_1 with respect to the number of anomalous factors. Now let is show how to group the other permutations. Consider the summand corresponding to the permutation $(\beta_1 \beta_n \ldots \beta_s \ldots \beta_2)$ with one anomalous factor corresponding to β_s, i.e., $f(\hat{k}_{\beta_1 \beta_s} - \frac{1}{2}\delta_{\alpha\beta_s}\hat{k}_{\alpha\alpha})$. Then the summand corresponding to $(\beta_s \beta_n \ldots \beta_1 \ldots \beta_2)$ differs from the one above by replacing the anomalous factor by $\exp(\frac{i\hbar}{4}\hat{k}_{\beta_1 \beta_s})$ $\times f(-\frac{1}{2}\delta_{\alpha\beta_s}\hat{k}_{\alpha\alpha})$. (Since this summand contains $\delta_{\alpha\beta_s}$ and the factor corresponding to β_1 contains $\delta_{\alpha\beta_1}$ we can redenote β_1 and β_s.) Let us unite these two summands in one, replacing in the term $(\beta_1 \beta_n \ldots \beta_s \ldots \beta_2)$ in the anomalous factor by the modified one (as in (5.22)):

$$f\left(\hat{k}_{\beta_1 \beta_s} - \frac{1}{2}\delta_{\alpha\beta_s}\hat{k}_{\alpha\alpha}\right) \rightarrow f(\hat{k}_{\beta_1 \beta_s} - \delta_{\alpha\beta_s}\hat{k}_{\alpha\alpha}).$$

Similarly, the summands

$$(\beta_l \beta_n \ldots \beta_1 \ldots \beta_s \ldots \beta_2), \ (\beta_s \beta_n \ldots \beta_l \ldots \beta_1 \ldots \beta_2), \ (\beta_l \beta_n \ldots \beta_s \ldots \beta_1 \ldots \beta_2)$$

differ from the summand $(\beta_1\beta_n\ldots\beta_l\ldots\beta_s\ldots\beta_2)$ with two anomalous factors only in the form of their factors. Unite all these four summands into one. In the general case, we unite the summands $(\beta_1\ldots\beta_{q_1}\ldots\beta_{q_m}\ldots)$ having m anomalous factors with m summands $(\beta_{q_1}\ldots\beta_1\ldots\beta_{q_m}\ldots)$, ..., $(\beta_{q_m}\ldots\beta_{q_1}\ldots\beta_1\ldots)$, where β_1 is interplaced with some of β_{q_i}. We then do that with $\binom{m}{2} = C_m^2$ summands obtained from the preceding ones by a permutation of β_1 with any of the anomalous indices to the right of it, etc., a total sum of all the summands is $1 + C_m^1 + C_m^2 + \ldots + C_m^m = 2^m$. This procedure exhausts all the permutations of n indices, i.e.,

$$\sum_{1 \leq m \leq n-1} a_m \cdot 2^m = n!.$$

(This formula corresponds to the reduction of the permutation group S_n to subgroup S_{n-1}.)

Thus we have the final expression for the n-th term of the perturbation theory in the form

$$\left[\exp\left(-\frac{1}{2}k_{\alpha\alpha}u_\alpha\right)\right]_n = \exp\left(-\frac{1}{2}k_{\alpha\alpha}\varphi_\alpha\right) \cdot 2v_\alpha f\left(-\frac{1}{2}\hat{k}_{\alpha\alpha}\right)$$

$$\times \sum_{\beta_2\ldots\beta_n}\sum_{\omega(\beta)} \exp\left(\frac{i\hbar}{4}F_\omega\right) \Phi^+_{\alpha\beta_2\ldots\beta_n} \Phi^-_{\alpha\beta_{\omega(2)}\ldots\beta_{\omega(n)}} \prod_{2\leq s\leq n} 2v_{\beta_s} \qquad (5.23)$$

$$\times f\left(-\delta^\omega\delta_{\alpha\beta_s}\hat{k}_{\alpha\alpha} + \hat{k}_{\alpha\beta_s} + \sum_{2\leq l\leq s-1}\hat{k}_{\beta_l\beta_s}\theta(\omega(s)-\omega(l))\right),$$

where $\delta^\omega = 1$ if the sum over l vanishes and $\delta^\omega = 1/2$ otherwise.

To analyse the properties of the series $\exp(-\frac{1}{2}k_{\alpha\alpha}u_\alpha)$, pass to the classical limit as $\hbar \to 0$:

$$\left[\exp\left(-\frac{1}{2}k_{\alpha\alpha}u_\alpha\right)\right]_n = \exp\left(-\frac{1}{2}k_{\alpha\alpha}\varphi_\alpha\right) \sum_{\beta_2\ldots\beta_n}\sum_{\omega(\beta)}$$

$$\times \Phi^+_{\alpha\beta_2\ldots\beta_n} \Phi^-_{\alpha\beta_{\omega(2)}\ldots\beta_{\omega(n)}} \prod_{s=2}^{n} [\delta^\omega\delta_{\alpha\beta_s}k_{\alpha\alpha} - k_{\alpha\beta_s} \qquad (5.24)$$

$$- \sum_{l=2}^{s-1} k_{\beta_l\beta_s}\theta(\omega(s)-\omega(l)).$$

This is another form of the classical solutions obtained in Chapter 4. However, there it is clear from the construction that $\exp(-\frac{1}{2}k_{\alpha\alpha}u_\alpha)$ is a polynomial in the coupling constant $\lambda(\equiv \mu^2)$ if and only if $(k_{\alpha\beta})$ is (equivalent to) the Cartan matrix of a simple finite-dimensional Lie algebra. In terms of (5.24) this means that all the coefficients of $\Phi^+_{\alpha\beta\ldots}\Phi^-_{\alpha\gamma\ldots}$ vanish starting from some

n. But taking the explicit form of f into account this means that starting from this very n the coefficients in (5.23) also vanish so that the series of the quantum problem is also finite.

Conversely, if the series (5.23) is finite, the same is true in the classical limit and $(k_{\alpha\beta})$ is the Cartan matrix.

Finally, the operators $\exp(-\frac{1}{2}k_{\alpha\alpha}u_\alpha)$ are polynomials in λ if and only if $(k_{\alpha\beta})$ is the Cartan matrix of a simple finite-dimensional Lie algebra.

To illustrate the above let us give the explicit expressions for the Heisenberg operators corresponding to the algebras A_2 and B_2:

$$A_2: \qquad k = \begin{pmatrix} 2 & -1 \\ -1 & 2 \end{pmatrix}, \quad v_\alpha = v/2,$$

$$\exp(-u_1) = \exp(-\varphi_1)[1 + \lambda\eta\Phi_1^+\Phi_1^- + \lambda^2\eta^2\Phi_{12}^+\Phi_{12}^-],$$

$$\exp(-u_2) = \exp(-\varphi_2)[1 + \lambda\eta\Phi_2^+\Phi_2^- + \lambda^2\eta^2\Phi_{21}^+\Phi_{21}^-].$$

$$B_2: \qquad l = \begin{pmatrix} 2 & -2 \\ -1 & 2 \end{pmatrix}, \quad \{v_\alpha\} = v \cdot (1/2, 1).$$

$$\exp(-u_1) = \exp(-\varphi_1)\Bigg[1 + \lambda\eta\Phi_1^+\Phi_1^- + 2\lambda^2\eta^2\Phi_{12}^+\Phi_{12}^-$$

$$+ 4\lambda^3\eta^3 \exp\left(\tfrac{i\hbar}{2v}\right)\Phi_{122}^+\Phi_{122}^- + 4\lambda^4\eta^4 \exp\left(\tfrac{i\hbar}{2v}\right)\Phi_{1221}^+\Phi_{1221}^-\Bigg],$$

$$\exp(-u_2) = \exp(-\varphi_2)[1 + \lambda\eta'\Phi_2^+\Phi_2^- + \lambda^2\eta\eta'\Phi_{21}^+\Phi_{21}^- + \lambda^3\eta(\eta')^2\Phi_{212}^+\Phi_{212}^-].$$

Here

$$\eta \equiv \frac{2v}{\hbar}\sin\left(\frac{\hbar}{2v}\right)\exp\left(-\frac{i\hbar}{2v}\right),$$

$$\eta' \equiv \frac{4v}{\hbar}\sin\left(\frac{\hbar}{4v}\right)\exp\left(-\frac{i\hbar}{4v}\right).$$

§ 7.6 Multicomponent 2-dimensional models. 2

Here we will briefly discuss how to quantize the system (5.2) with a cyclic spatial coordinate x. The particular case $(r = 1)$ was considered in detail in 7.4. The characteristic pecularities of this case are also preserved for an arbitrary r. Namely, we have seen that when we pass to the cyclic coordinate in Φ the integration is performed in finite limits and Φ acquires a factor which depends only on the "zero mode" operator P. The coefficient which explicitly contains $\hbar$ remains the same and its algebraic properties guarantee the fact that the series of perturbation theory for $\exp(-u)$ is actually a polynomial.

We will not prove here the corresponding formulas which are much more complicated than those in the preceding section; we give only the scheme of calculations of the two first orders of perturbation theory and indicate the correspondence with the other problems considered above via some limit procedures.

We will perform all the calculations with the help of the Yang-Feldmann equation (2.45) where in this case

$$j_\alpha = -\lambda \exp \left(\sum_{1 \le \beta \le r} k_{\alpha\beta} u_\beta \right).$$

The calculations in the first order do not differ from those performed in 7.4 for the Liouville equations; we have

$$[u_\alpha]_1 = \left[1 - \exp \left(-\frac{1}{2} \left(p_\alpha + \frac{i\hbar}{2} \hat{k}_{\alpha\alpha} \right) \right) \right]^{-2} \Phi^L_{\alpha;\alpha}, \tag{6.1}$$

where

$$\Phi^L_{\alpha;\alpha} \equiv \int\limits_{z^+-L}^{z^+} \left(\int\limits_{z^--L}^{z^-} \exp \phi_\alpha(\tilde{z}) d\tilde{z}^+ d\tilde{z}^- \right) \tag{6.2}$$

$$p_\alpha \equiv \sum_\beta k_{\alpha\beta} P_\beta; \quad q_\alpha \equiv \sum_\beta k_{\alpha\beta} Q_\beta; \quad a_n^{\pm\alpha} \equiv \sum_\beta k_{\alpha\beta} A_n^{\pm\alpha}.$$

The second order is calculated via the scheme of 7.5 where instead of (2.34)–(2.36) we make use of the formulas

$$\exp(-b\phi_\beta) f(p_\alpha) \exp(b\phi_\beta) = f(p_\alpha - i\hbar b \hat{k}_{\alpha\beta}),$$

$$\exp(ap_\alpha + b\phi_\beta) = \exp \left(ap_\alpha + \frac{i\hbar}{2} ab \hat{k}_{\alpha\beta} \right) \exp(b\phi_\beta) \tag{6.3}$$

which are proved in the same way. Then we divide the total domain of integration into subdomains where D^{ret} is constant. Then the integrals are reduced with the help of the identity

$$\exp(-b\phi_\beta) \Phi^L_{\alpha;\alpha} \exp(b\phi_\beta) = \exp \left(\frac{i\hbar}{2} b \hat{k}_{\alpha\beta} \right) \Phi^L_{\alpha;\alpha}, \tag{6.4}$$

which generalizes (2.36), to the integrals over the standard domain:

$$\int\limits_{z^+-L}^{z^+} \left(\int\limits_{z^--L}^{z^-} \exp \phi_\alpha(\tilde{z}) \cdot \Phi^L_{\beta;\beta}(\tilde{z}) d\tilde{z}^+ d\tilde{z}^- \right) \tag{6.5}$$

with coefficients depending only on the operators p. The series of coefficients obtained can be summed and by elementary transformations similar to those from 7.5 the integral (6.5) can be reduced to the form

$$\Phi^L_{\alpha\beta;\alpha\beta} = \int\limits_{z^+-L}^{z^+} \left(\int\limits_{z^--L}^{z^-} \exp\phi_\alpha(z_1) \right.$$

$$\left. \times \left(\int\limits_{z_1^+-L}^{z_1^+} \left(\int\limits_{z_1^--L}^{z_1^-} \exp\phi_\beta(z_2) dz_1^+ \right) dz_1^- \right) dz_2^+ \right) dz_2^-. \tag{6.6}$$

Finally we get

$$[u_\alpha]_2 = \frac{1}{i\hbar} \sum_\beta 2v_\beta \left[\exp\left(\frac{i\hbar}{2} \hat{k}_{\alpha\beta} \right) - 1 \right]$$

$$\times \left[1 - \exp\left(-\frac{1}{2} \left(p_\alpha + \frac{i\hbar}{2} \hat{k}_{\alpha\alpha} + p_\beta + \frac{i\hbar}{2} \hat{k}_{\beta\beta} + i\hbar\hat{k}_{\alpha\beta} \right) \right) \right]^{-2}$$

$$\times \left[1 - \exp\left(-\frac{1}{2} \left(p_\beta + \frac{i\hbar}{2} \hat{k}_{\beta\beta} + i\hbar\hat{k}_{\alpha\beta} \right) \right) \right]^{-1}$$

$$\times \left[1 - \exp\left(-\frac{1}{2} \left(p_\beta + \frac{i\hbar}{2} \hat{k}_{\beta\beta} \right) \right) \right]^{-1} \Phi^L_{\alpha\beta;\alpha\beta}.$$

As in the case of infinite space it is convenient to pass to the operator $\exp(-\frac{1}{2}k_{\alpha\alpha}u_\alpha)$. This is performed according to the general rules of 7.2 with the help of (2.27)–(2.28) and we get

$$\left[\exp\left(-\frac{1}{2}k_{\alpha\alpha}u_\alpha \right) \right]_1 = \exp\left(-\frac{1}{2}k_{\alpha\alpha}\varphi_\alpha \right) \frac{2v_\alpha}{i\hbar} \left[1 - \exp\left(-\frac{i\hbar}{2v_\alpha} \right) \right]$$

$$\times \left[1 - \exp\left(-\frac{1}{2} \left(p_\alpha + \frac{i\hbar}{2} \hat{k}_{\alpha\alpha} \right) \right) \right]^{-1} \left[1 - \exp\left(-\frac{1}{2} \left(p_\alpha + \frac{i\hbar}{v_\alpha} + \frac{i\hbar}{2} \hat{k}_{\alpha\alpha} \right) \right) \right]^{-1} \Phi^L_{\alpha;\alpha};$$

$$\left[\exp\left(-\frac{1}{2}k_{\alpha\alpha}u_\alpha \right) \right]_2 = \exp\left(-\frac{1}{2}k_{\alpha\alpha}\varphi_\alpha \right) \frac{2v_\alpha}{i\hbar} \left[1 - \exp\left(-\frac{i\hbar}{2v_\alpha} \right) \right]$$

$$\times \sum_{\beta\neq\alpha} \frac{2v_\beta}{i\hbar} \left[1 - \exp\left(\frac{i\hbar}{2} \hat{k}_{\alpha\beta} \right) \right] \left[1 - \exp\left(-\frac{1}{2} \left(p_\beta + \frac{i\hbar}{2} \hat{k}_{\beta\beta} \right) \right) \right]^{-1}$$

$$\times \left[1 - \exp\left(-\frac{1}{2} \left(p_\beta + \frac{i\hbar}{2} \hat{k}_{\beta\beta} + i\hbar\hat{k}_{\alpha\beta} \right) \right) \right]^{-1} \times \left[1 - \exp\left(-\frac{1}{2} \left(p_\alpha + \frac{i\hbar}{v_\alpha} \right. \right. \right.$$

$$\left. \left. \left. + \frac{i\hbar}{2} \hat{k}_{\alpha\alpha} + p_\beta + \frac{i\hbar}{2} \hat{k}_{\beta\beta} + i\hbar\hat{k}_{\alpha\beta} \right) \right) \right]^{-1} \left[1 - \exp\left(-\frac{1}{2} \left(p_\alpha + \frac{i\hbar}{2} \hat{k}_{\alpha\alpha} \right. \right. \right.$$

$$\left. \left. \left. + p_\beta + \frac{i\hbar}{2} \hat{k}_{\beta\beta} + i\hbar\hat{k}_{\alpha\beta} \right) \right) \right]^{-1} \Phi^L_{\alpha\beta;\alpha\beta}.$$

The calculation of the higher orders is similar and the final result is

$$\left[\exp\left(-\frac{1}{2}k_{\alpha\alpha}u_\alpha\right)\right]_n = \exp\left(-\frac{1}{2}k_{\alpha\alpha}\varphi_\alpha\right) \sum_{\beta_2,\ldots,\beta_n} \sum_{\omega(\beta)} \qquad (6.7)$$

$$\times K^\alpha_{\beta;\omega(\beta)} S^{+\alpha}_\beta S^{-\alpha}_{\omega(\beta)} \Phi^L_{\alpha\beta_2\ldots\beta_n;\alpha\omega(\beta_2)\ldots\omega(\beta_n)},$$

where the second sum runs over all the permutations of $(\beta_2,\ldots,\beta_n)$ and the sum over each β_s runs from 1 to r. Here the functionals

$$\Phi^L_{\alpha\beta\ldots;\mu\nu\ldots} \equiv \int\limits_{z^+-L}^{z^+} dz_1^+ \int\limits_{z^--L}^{z^-} dz_1^- \exp\phi_{\alpha\mu}(z_1) \int\limits_{z_1^+-L}^{z^+} dz_2^+ \int\limits_{z_1^--L}^{z^-} dz_2^-$$

$$\times \exp\phi_{\beta\nu}(z_2)\ldots;$$

$$\phi_{\alpha\mu} \equiv q_\alpha + \phi_\alpha^+(z^+) + \phi_\mu^-(z^-);$$

$$\phi_\alpha^\pm \equiv \frac{1}{2L}p_\alpha z^\pm + \frac{i}{\sqrt{4\pi}} \sum_{n\neq 0} \frac{1}{n}a_n^{\pm\alpha} \exp\left(-\frac{2\pi i n}{L}z^\pm\right),$$

are analogues of the functionals $\Phi^+_{\alpha\beta_2\ldots\beta_n} \Phi^-_{\alpha\omega(\beta_2)\ldots\omega(\beta_n)}$ from (5.23); the functionals

$$S^{\pm\alpha}_{\beta_2\ldots\beta_n} \equiv \prod_{s=1}^n \left[1 - \exp\left(-\frac{1}{2}\left(p^\pm_{\beta_s} + p^\pm_{\beta_{s+1}} + \ldots + p^\pm_{\beta_n}\right)\right)\right]^{-1}; \quad \beta_1 \equiv \alpha;$$

$$p^+_{\beta_s} \equiv p_{\beta_s} + \frac{i\hbar}{2}\hat{k}_{\beta_s\beta_s} + i\hbar\sum_{l=s+1}^n \hat{k}_{\beta_l\beta_s};$$

$$p^-_{\beta_s} \equiv p_{\beta_s} + \frac{i\hbar}{2}\hat{k}_{\beta_s\beta_s} + i\hbar\sum_{l=1}^{s-1} \hat{k}_{\beta_l\beta_s} + \frac{i\hbar}{v_\alpha}\delta_{\alpha\beta_s},$$

$$(6.8)$$

are functions in the "zero mode" operators p_μ only. Their coefficients

$$K^\alpha_{\beta;\omega(\beta)} \equiv 2v_\alpha f\left(-\frac{1}{2}\hat{k}_{\alpha\alpha}\right)\exp\left[\frac{i\hbar}{4}F_\omega\right] \cdot \prod_{s=2}^n 2v_{\beta_s} \times f(-\delta^\omega\delta_{\alpha\beta_s}\hat{k}_{\alpha\alpha} + \hat{k}_{\alpha\beta_s} + \Delta^\omega),$$

$$\Delta^\omega \equiv \sum_{l=2}^{s-1} \hat{k}_{\beta_l\beta_s}\theta(\omega(s)-\omega(l)), \delta^\omega \equiv \begin{cases} 1, & \Delta=0; \\ 0, & \Delta\neq 0; \end{cases} \qquad (6.9)$$

$$f(x) \equiv \left[1 - \exp\left(\frac{i\hbar}{2}x\right)\right]\Big/ i\hbar,$$

where F_ω is given by (5.19), coincide with the coefficients in (5.23).

Thus, as compared with infinite space, the formulas for Heisenberg's operators are modified as follows: the functionals Φ of free fields are replaced by similar functionals having the standard form of integrals over a finite space-time domain with factors which carry an explicit dependence on the "zero mode" operators. Since the coefficients which depend only on $\hbar$ and k are the same, we can immediately deduce that the series of the perturbation theory are finite in this problem in exactly the same cases as for infinite space, namely, if and only if k is (equivalent to) the Cartan matrix of a simple finite-dimensional Lie algebra.

The formal correspondence with the case of infinite space can be established as follows: introduce the new operators $p'_\alpha \equiv L^{-1/2}p_\alpha$ and let L tend to infinity. Then the Fourier series in (6.26) tends to the Fourier integral while p'_α now commutes with all the operators and therefore we may set $p'_\alpha = 0$. Now, the Hamiltonian (2.28) turns into the corresponding Hamiltonian in infinite space. Indeed, passing to the limit in (6.7) we return to (2.23).

Further, notice that the structure of the exponents of the exponentials in (6.8) is similar to that of the denominators that appear in the solution of the one-dimensional problem (7.3). This is not accidental. Indeed, as is clear from (2.26)–(2.29), setting $a_\alpha \equiv (2L)^{-1}p_\alpha$ and $v_\alpha \equiv 2Lv_\alpha$, we see that $a_n^{\pm\alpha}$ commute as $L \to \infty$ and therefore we may put $a_n^{\pm\alpha} = 0$. As a result, the Hamiltonian (2.28) turns into the Hamiltonian of the one-dimensional problem. To prove this let us move Φ to the left of S in (6.7) making use of (6.4) and, setting $p_\alpha = 2La_\alpha$, $v_\alpha = \frac{1}{2L}v'_\alpha$, pass to the limit as $L \to \infty$. Then Φ^L turns into $\Phi^+_{\alpha\beta_2...\beta_n} \Phi^-_{\alpha\omega(\beta_2)...\omega(\beta_n)}$ with factor L^{2n}, the exponentials in (6.8) and (6.9) turn into linear functions with factor L^{-2n} and for the Heisenberg operators we actually get (5.7).

The passage to the classical limit $\hbar \to 0$ in (6.7) does not differ from the corresponding passage in 7.5. The verification of the correspondence principle can be performed substituting the functions ϕ_α periodic with respect to x into the known classical solutions from Chapter 4 (see also (5.24)) and reducing the functionals Φ to the integrals over the standard domain with a factor depending on p_μ.

In conclusion let us give for an illustration the operators which modify formulas (5.25) for A_2:

$$S^{+1}S^{-1} = \left[1 - \exp\left(-\frac{1}{2}\left(p_1 + \frac{2i\hbar}{v}\right)\right)\right]^{-1}\left[1 - \exp\left(-\frac{1}{2}\left(p_1 + \frac{4i\hbar}{v}\right)\right)\right]^{-1};$$

$$S_2^{+1}S_2^{-1} = S^{+1}S^{-1}\bigg|_{p_1 \to p_2} \times \left[1 - \exp\left(-\frac{1}{2}\left(p_1 + p_2 + \frac{2i\hbar}{v}\right)\right)\right]^{-1}$$

$$\times \left[1 - \exp\left(-\frac{1}{2}\left(p_1 + p_2 + \frac{4i\hbar}{v}\right)\right)\right]^{-1};$$

$$S^{+2}S^{-2} = S^{+1}S^{-1}\big|_{p_1 \to p_2}; \, S_1^{+2}S_1^{-2} = S_2^{+1}S_2^{-1}\big|_{p_1 \rightleftharpoons p_2}.$$

Afterword

Let us emphasize that the Russian version of this book was completed in 1982. The delay in the publication of this monograph (which was not our fault) turned out to be very opportune, since only now a part of the original results of the authors, presented in it, is being understood and rediscovered in the literature.

The later results of the authors that are not in the book tightly connected with its contents, are covered in the following surveys:

1. Leznov, A.N. and M.V. Saveliev, Nonlinear equations and graded Lie algebras, *J. Soviet Math.* **36** (1987), 699–721.

2. Leznov, A.N. and M.V. Saveliev, Exactly and completely integrable nonlinear dynamical systems, *Acta. Appl. Math.* **16** (1989), 1–74.

3. Leznov, A.N., M.V. Saveliev and I.A. Fedoseev, Exactly solvable systems in quantum mechanics and two-dimensional quantum field theory, *Soviet J. Particles and Nuclei* **16** (1985), 81–101.

4. Leznov, A.N., V.I. Manko and M.V. Saveliev, Soliton solutions to nonlinear equations and group representation theory, in V.L. Ginzburg, ed., *Solitons and instantons and operator quantization*, Nova Sci. Publ., Commack, 1986, 83–267.

5. Leznov, A.N., V.I. Manko and S.M. Chumakov, Dynamical symmetries of nonlinear equations, in A.A. Komar, ed., *Group theory and gravitation and physics of elementary particles*, Nova Sci. Publ., Commack, 1986, 232–277.

6. Leites, D.A., M.V. Saveliev and V.V. Serganova, Embeddings of Lie superalgebras $osp(N/2)$ and the associated nonlinear equations, in M.A. Markov, V.I. Manko and V.V. Dodonov, eds., *Group-theoretical methods in physics*, VNU Sci. Press, Utrecht, 1986, vol 1, 255–297.

June 1991, Moscow

After 1982, a number of excellent reviews and books of other authors, concerning the integrability problem of nonlinear dynamical systems, were published. They have not been included in our brief Afterword, however, since they intersect very little with the methods and explicit results given here.

Bibliography

[1] Ablowitz, M.J., D.J. Kaup, A.C. Newell and H. Segur, *Studies in Appl. Math.* **53** (1974), 249–315.

[2] Adler, M., *Inv. Math.* **50** (1979), 219–248.

[3] De Alfaro V. and T. Regge, *Potential scattering*, J. Wiley, New York, 1965.

[4] Arnold, V.I., *Mathematical methods of classical mechanics*, Springer, Berlin, 1978.

[5] Atiyah, M.F., V.G. Drinfeld, N.J. Hitchin and Yu.I. Manin, *Phys. Lett.* **65**A (1978), 185–189.

[6] Barbashov, B.M., V.V. Nesterenko and A.M. Chervyakov, *Comm. Math. Phys.* **84** (1982), 471–479.

[7] Barut, A.O. and R. Rączka, *Theory of group representations and applications.* PWN — Polish Sci. Publ., Warsaw, 1977.

[8] Berezin, F.A., *Introduction to superanalysis*, Kluwer, Dordrecht, 1987. Acad. Press, New York, 1966.

[9] Bogolyubov, N.N., A.A. Logunov, A.I. Oksak and I.T. Todorov, *General principles of quantum field theory.* Kluwer, Dordrecht, 1989.

[10] Bogolyubov, N.N. and D.V. Shirkov, *Introduction to the theory of quantized fields*, Interscience, New York, 1969.

[11] Bogoyavlensky, O.I., *Comm. Math. Phys.* **51** (1976), 201–209.

[12] Bourbaki, N., *Groupes et algèbres de Lie*, Hermann, Paris, 1968, Ch. 4–6; 1975, Ch. 7–8.

[13] Vilenkin, N.Ya., *Special functions and the theory of group representations.* AMS, Providence, 1968.

[14] Volterra, V., *Leçons sur la théorie mathématique de la lutte pour la vie*, Gautier-Villars, Paris, 1931.

[15] Ganoulis, N., P. Goddard and D. Olive, *Nucl. Phys.* **B205** (1982), 601.

[16] Gantmakher, F.R., *Mat. Sb.* **5** (1939), 101–144.

[17] Gantmakher, F.R., *Mat. Sb.* **5** (1939), 217–249.

[18] Gardner, C.S., J.M. Greene, M.D. Kruskal and R.M. Miura, *Phys. Rev. Lett.* **19** (1967), 1095–1097.

[19] Gelfand, I.M., M.I. Graev and N.Ya. Vilenkin, *Generalized functions*, v. 5: *Integral geometry and representation theory.* Acad. Press, New York-London, 1966.

[20] Gelfand, I.M. and M.A. Naimark, *Unitäre Darstellungen der klassischen Gruppen.* Akademie, Berlin, 1957.

[21] Gindikin, S.G. and F.I. Karpelevich, *Soviet Math. Dokl.* **3** (1962), 862.

[22] Goddard, P. and D. Olive, *Rep. Progr. Phys.* **41** (1978), 1357.

[23] Goodman, R. and N.R. Wallach, *J. Functional Anal.* **39** (1980), 199–279.

[24] Goodman, R. and N.R. Wallach, *Comm. Math. Phys.* **83** (1982), 355–386.

[25] Jackiw, R., *Rev. Mod. Phys.* **49** (1977), 681–706.

[26] Jackiw, R., C. Nohl and C. Rebbi, *Classical and semiclassical solutions of the Yang-Mills theory*, Plenum Press, New York, 1979.

[27] Jacobson, N., *Lie algebras*, Interscience, New York, 1962.

[28] Dirac, P.A.M., *Principles of quantum mechanics*, Oxford Univ. Press, Oxford 1958.

[29] Drinfeld, V.G., I.M. Krichever, Yu.I. Manin and S.P. Novikov, *Soviet Sci. Rev., Phys. Rev.* 1978.

[30] Dubrovin, B.A., I.M. Krichever, and S.P. Novikov, Soviet Sci. Rev. Sect. C: Math. Phys. Rev. **3** (1982), pp. 1–150.

[31] Doubrovin, B.A., S.P. Novikov and A.T. Fomenko, *Géométrie contemporaine (en trois parties)*, Mir, Moscow, 1985–87.

[32] Dynkin, E.B., *Amer. Math. Soc. Transl.* **17** (1950), 159–227.

[33] Dynkin, E.B., *Amer. Math. Soc. Transl., Ser. 2*, **6** (1957), pp. 111–244.

[34] Dynkin, E.B., *Amer. Math. Soc. Transl., Ser. 2*, **6** (1957), 245–378.

[35] Zhelobenko, D.P., *Compact Lie groups and their representations*, AMS, Providence, 1973.

[36] Zhelobenko, D.P., *Harmonic analysis in semisimple Lie groups*, Nauka, Moscow, 1974 (in Russian).

[37] Zhiber, A.V. and A.B. Shabat, *Soviet Math. Dokl.* **247** (1979), 1103–1107.

[38] Zhiber, A.V., N.Kh. Ibragimov and A.B. Shabat, *Soviet Math. Dokl.* **249** (1979), 26–39.

[39] Zakharov, V.E., S.V. Manakov, S.P. Novikov and L.P. Pitaevsky, *Theory of solitons: The method of the inverse scattering problem*, Plenum Press, New York, 1984.

[40] Zakharov, V.E., S.L. Musher and A.M. Rubenchik, *Soviet Phys. JETP* **19** (1974), 249–253.

[41] Zakharov, V.E. and A.B. Shabat, *Functional Anal. Appl.* **8** (1974), 43–53.

[42] Zakharov, V.E. and A.B. Shabat, *Functional Anal. Appl.* **13** (1979), 13–22.

[43] Inönü, E. and E.P. Wigner, *Proc. Nat. Acad. Sci. USA* **39** (1953), 510–524.

[44] Kac, V.G., *Infinite dimensional Lie algebras*, 2nd ed., Cambridge Univ. Press, Cambridge 1985.

[45] Kac, V.G. and A.K. Raina, *Bombay lectures on highest weight representations of infinite-dimensional Lie algebras*. World Sci., Singapore 1987.

[46] Kac, V.G. and D.H. Peterson, Lectures on the infinite wedge representations and the MKP hierarchy. In: *Sém. Math. Sup.* **102**, Montreal Univ. 1986, 141–184.

[47] Kac, V.G., *Adv. in Math.* **30**, (1978), 85–136.

[48] Kac, V.G., *Adv. in Math.* **26** (1977), 8–96.

[49] Scheunert, M., *Lie superalgebras*, Lecture Notes in Math. **76**, Springer, Berlin, 1979.

[50] Kac, M. and P. van Moerbeke, *Adv. in Math.* **16** (1975), 160–169.

[51] Kirillov, A.A., *Elements of the theory of representations*, Springer, New York, 1976.

[52] Kostant, B., *Amer. J. Math.* **81** (1959), 973–1032.

[53] Kostant, B., *Invent. Math.* **48** (1978), 101–184.

[54] Kostant, B., London Math. Soc. Lecture Note Ser. **34** (1979), 287–298.

[55] Kostant, B., in B. Kostant, ed., *Lectures in Modern Analysis and Applications, III*, Lecture Notes in Math. **170** (1970), 80–94.

[56] Kostant, B., *Adv. in Math.* **34** (1979), 195–338.

[57] Krichever, I.M. and S.P. Novikov, *Russian Math. Surveys* **35** (1980), 47–69.

[58] Kruskal, M.D., in J. Moser, ed., *Dynamical Systems, Theory and Applications*, Lecture Notes in Phys. **38**, (1975), 310–354, Springer, Berlin.

[59] Lax, P.D., *Comm. Pure Appl. Math.* **21** (1968), 467–490.

[60] Leznov, A.N. and M.V. Saveliev, *Functional Anal. Appl.* **8** (1974), 87–89.

[61] Leznov, A.N. and M.V. Saveliev, *Soviet J. Particles and Nuclei* **7** (1976), 22–41.

[62] Leznov, A.N., I.A. Malkin and V.I. Manko, *Proc. of PhIAN SSSR* **96** (1977), 24–71, Moscow (in Russian).

[63] Leznov, A.N. and M.V. Saveliev, *Phys. Lett. B* **79** (1978), 294–296.

[64] Leznov, A.N. and M.V. Saveliev, *Phys. Lett. B* **83** (1979), 314–316.

[65] Leznov, A.N. and M.V. Saveliev, Preprint IHEP **78–177**, Serpukhov, 1978; *Lett. Math. Phys.* **3** (1979), 207–211.

[66] Leznov, A.N., *Soviet J. Theor. Math. Phys.* **42** (1980), 343–349.

[67] Leznov, A.N. and M.V. Saveliev, *Functional Anal. Appl.* **14** (1980), 87–89.

[68] Leznov, A.N. and M.V. Saveliev, *Physica D* **3** (1981), 62–72.

[69] Leznov, A.N. and M.V. Saveliev, *Comm. Math. Phys.* **74** (1980), 111–118.

[70] Leznov, A.N. and M.V. Saveliev, *Soviet J. Particles and Nuclei* **11** (1980), 14–34.

[71] Leznov, A.N., D.A. Leites and M.V. Saveliev, *Phys. Lett. B* **96** (1980), 97–100.

[72] Leznov, A.N. and M.V. Saveliev, *Soviet J. Particles and Nuclei* **12** (1981), 125–161.

[73] Leznov, A.N., M.V. Saveliev and V.G. Smirnov, *Soviet J. Theor. Math. Phys.* **48** (1981), 3-12.

[74] Leznov, A.N., M.V. Saveliev and V.G. Smirnov, *Soviet J. Theor. Math. Phys.* **47** (1981), 216–224.

[75] Leznov, A.N., V.G. Smirnov and A.B. Shabat, *Soviet J. Theor. Math. Phys.* **51** (1982), 10–21.

[76] Leznov, A.N., IHEP Preprint **83-7**, Serpukhov, 1983.

[77] Leznov, A.N. and V.V. Krushchev, in: *Group-theoretical Methods in Phys.*, M.A. Markov and V.I. Manko, eds., Nauka, Moscow, 1983.

[78] Leznov, A.N., IHEP Preprint **82-169**, Serpukhov, 1982.

[79] Leznov, A.N. and M.V. Saveliev, *Comm. Math. Phys.* **89** (1983), 59–75.

[80] Leites, D.A., *Russian Math. Surveys* **35** (1980), 3–57.

[81] Leites, D.A., *Supermanifold theory*, Karelia Branch of the USSR Acad. of Sci., Petrozavodsk, 1983, in Russian; an expanded English version is preprinted in Reports of Dept. of Math. of Stockholm Univ, 1987-90.

[82] Lund, F. and T. Regge, *Phys. Rev. D* **14** (1976), 1524.

[83] Macdonald, I.G., *Séminaire Bourbaki*, 33ème année, 1980/1981, No. 577, 1–19.

[84] Manakov, S.V., *Soviet Phys. JETP* **67** (1974), 543–555.

[85] Manin, Yu.I., *Soviet Probl. Mat. VINITI* **11** (1978), 5–152 (in Russian; Engl. transl. in J. Soviet Math.).

[86] Manin, Yu.I., ed., *Geometric ideas in physics*. Mir, Moscow, 1983 (in Russian).

[87] Manin, Yu.I., *Gauge fields and complex geometry*, Springer, New York, 1988.

[88] Mansfield, P., *Nuclear Phys. B* **208** (1982), 277–300.

[89] Miura, R.M., *J. Math. Phys.* **9** (1968), 1202.

[90] Moser, J., *Adv. in Math.* **16** (1975), 197–220.

[91] Moser, J., in J. Moser, ed., *Dynamical Systems, Theory and applications*, Lecture Notes in Phys. 38 (1975), 97–134.

[92] *Monopoles in quantum field theory*, Proc. Monopole Meeting, P. Goddard, W. Nahm and D. Olive, eds., Miramare, Trieste, 1981.

[93] Moody, R.V., *Bull. Amer. Math. Soc.* **73** (1967), 217–221.

[94] Moody, R.V., *J. Algebra* **10** (1968), 211–230.

[95] Naimark, M.A., *Theory of group representation*, Springer, New York, 1982.

[96] Ovsiannikov, L.V., *Group analysis of differential equations*, Acad. Press, New York, 1982.

[97] Polyakov, A.M., *Phys. Lett. B* **103** (1981), 207–210.

[98] Polyakov, A.M., *Phys. Lett. B* **103** (1981), 211–212.

[99] Pontryagin, L.S., *Topological groups*, Princeton Univ. Press, Princeton, 1946.

[100] Reyman, A.G., M.A. Semenov-Tian-Shansky and I.B. Frenkel, *Soviet Math. Dokl.* **247** (1979), 802–805.

[101] Serre, J-P., *Algèbres de Lie semisimples complexes*, Benjamin, New York, 1966.

[102] Scott, A.C., F.Y.E. Chu and D.W. McLaughlin, *Proc. IEEE* **61** (1973), 1449–1472.

[103] Longren K. and A. Scott, eds., *Solitons in Action*, Acad. Press, New York, 1978.

[104] Toda, M., *Phys. Rep. C* **18** (1975), 1.

[105] Faddeev, L.D., *Soviet Sci. Rev. Sect. C: Math. Phys. Rev.* **1** (1981), 13.

[106] Fedoseev, I.A. and A.N. Leznov, IHEP Preprints **81-15**, **81-52**, **82-45** and **82-169**, Serpukhov, 1981 and 1982.

[107] Fedoseev, I.A., A.N. Leznov and M.V. Saveliev, *Phys. Lett. B* **116** (1982), 49–52.

[108] Fedoseev, I.A., A.N. Leznov and M.V. Saveliev, *Nuovo Cimento A* **76** (1983), 596–612.

[109] Fock, V.A., *The Foundations of Quantum Mechanics*, Nauka, Moscow, 1976 (in Russian).

[110] Frenkel, I.B., in D. Winter, ed., *Lie Algebras and Related Topics*, Lecture Notes in Math. **933** (1982), 71–110.

[111] Frenkel, I.B. and V.G. Kac, *Invent. Math.* **62** (1980), 23.

[112] Harish-Chandra, *Automorphic forms on semisimple Lie groups*, Lecture Notes in Math. **62**, Springer, Berlin, 1968.

[113] Helgason, S., *Differential geometry, Lie groups and symmetric spaces*, Acad. Press, New York, 1978.

[114] Chevalley, C., *Theory of Lie groups*, vol. I, Princeton Univ. Press, Princeton. 1946;
id., *Théorie des groupes de Lie*, vols. II and III, Hermann, Paris, 1951 and 1955.

[115] Eisenhart, L.P., *Continuous groups of transformations*, Princeton Univ. Press, Princeton, 1933.

[116] Eisenhart, L.P., *Riemannian Geometry*, Princeton Univ. Press, Princeton, 1964.

[117] *Suppl. Progr. Theor. Phys.* **59** (1976).

[118] Fedoseev, I.A. and A.N. Leznov, *Phys. Lett. B* **141** (1984), 100–103.

[119] Braaten, E., T. Curtright and C. Thorn, *Phys. Lett. B* **118** (1982), 115–117.

[120] Gervais, J.L. and A. Neveu, *Nuclear Phys. B* **199** (1982), 59.

[121] Gervais, J.L. and A. Neveu, *Nuclear Phys. B* 209 (1982), 125.

[122] Gervais, J.L. and A. Neveu, *Phys. Lett. B* **123** (1983), 86.

[123] Curtright, T. and C. Thorn, *Phys. Rev. Lett.* **48** (1982), 1309–1312.

[124] Leznov, A.N., *Soviet J. Theor. Math. Phys.* **58** (1984), 156–160.

[125] Leznov, A.N., *Lett. Math. Phys.* **8** (1984), 5–8.

[126] Leznov, A.N. and M.V. Saveliev, *J. Soviet Math.* **36** (1987), 699–721.

[127] Leznov, A.N. and I.A. Fedoseev, *Soviet J. Theor. Math. Phys.* **53** (1982), 358–372.

[128] Leznov, A.N. and A.B. Shabat, in: A.B. Shabat, ed., *Integrable Systems*, Bashkiria Branch of the USSR Acad. Sci., Ufa, 1982 (in Russian).

[129] Saveliev, M.V., IHEP Preprints **83-32** and **83-58**, Serpukhov, 1983; *Soviet J. Theor. Math. Phys.* **60** (1984), 9–23; **69** (1986), 411–419; *Soviet Math. Dokl.* **292** (1987), 582–585.

[130] Ablowitz, M.J. and H. Segur, *Solitons and inverse scattering transform*, SIAM Appl. Math., Phil., 1981.

[131] Ibragimov, N.H., *Transformation groups applied to mathematical physics*, D. Reidel, Dordrecht, 1987.

[132] Leznov, A.N. and M.V. Saveliev, *Soviet J. Theor. Math. Phys.* **54** (1983), 323–337.

[133] Newton, R.G., *Scattering theory of waves and particles*. McGraw-Hill, New York, 1966.

[134] Bullough, R.K. and P.J. Caudrey, eds., *Solitons*, Springer, Berlin, 1980.

[135] Calogero, F. and A. Degasperis, *Spectral transform and solitons*, North-Holland, Amsterdam, 1982.

[136] Adler, M. and P. von Moerbeke, *Invent. Math.* **50** (1979) 219–248.

[137] Bais, F.A. and H.A. Weldon, *Phys. Lett. B* **79** (1978), 297–300.

[138] Bais, F.A. and H.A. Weldon, *Phys. Rev. Lett.* **41** (1978), 601–604.

[139] Bargmann, V., *Rev. Modern Phys.* **21** (1949), 488–491.

[140] Belavin, A.A., A.M. Polyakov, A.S. Schwarz and Yu.S. Tyupkin, *Phys. Lett. B* **59** (1975), 85–87.

[141] Calogero, F., *Nuovo Cimento B* **43** (1978), 177–242.

[142] Chaichian, M. and P.P. Kulish, *Phys. Lett. B* **78** (1978), 413–416.

[143] Crewther, R.J., *Acta Phys. Austriaca Suppl.* **19** (1978), 47–153.

[144] Farwell, R.S. and M. Minami, *Progr. Theoret. Phys.* **69** (1982), 1091–1099.

[145] Fordy, A.P. and J. Gibbons, *Comm. Math. Phys.* **77** (1980), 21–30.

[146] Hénon, M., *Phys. Rev. B* **9** (1974), 1421–1423.

[147] Leznov, A.N., in: R.Z. Sagdeev, ed., *Proc. of the Internat. Workshop on Nonlinear and Turbulent Processes*, Gordon & Breach, New York, 1983.

[148] Marchenko, V.A., in: R.Z. Sagdeev, ed., *Proc. of the Internat. Workshop on Nonlinear and Turbulent Processes*, Gordon & Breach, New York, 1983.

[149] Olive, D. and N. Turok, *Nuclear Phys. B* **215** (1983), 470–486.

[150] Saveliev, M.V., *Comm. Math. Phys.* **95** (1984) 199–216.

[151] Saveliev, M.V., in: R.Z. Sagdeev, ed., *Proc. of the Internat. Workshop on Nonlinear and Turbulent Processes*, Gordon & Breach, New York, 1983.

[152] Terng, C.L., *Ann. of Math.* **111** (1980), 491–510.

[153] t'Hooft, G., *Nuclear Phys. B* **79** (1974), 276–284.

[154] Toda, M. and M. Wadati, *J. Phys. Soc. Japan* **39** (1975), 1204–1247.

[155] Wadati, M., *J. Phys. Soc. Japan* **38** (1975), 673–686.

[156] Wilkinson, D. and F.A. Bais, *Phys. Rev. D.* **19** (1979), 2410–2415.

[157] Wilkinson, D. and A.S. Goldhaber, *Phys. Rev. D* **16** (1977), 1221–1231.

[158] Wilson, G., *Ergodic Theory and Dynamical Systems* **1** (1981), 361–380.

[159] Witten, E., *Phys. Rev. Lett.* **38** (1977), 121–125.

[160] Flaschka, H., *Phys. Rev. B* **9** (1974), 1924–1925.

[161] Feigin, B.L. and D.B. Fuchs, *Cohomology of Lie group and Lie algebras*, Encyclopedia of Mathematical sciences, Springer, Berlin, 1991.

[162] Onishchik, A.L. and E.B. Vinberg, *Seminar on algebraic groups and Lie groups*, Springer, Berlin, 1990.

[163] Serganova, V.V., in: *Group-theoretical methods in physics*, Gordon & Breach, New York, 1987.

[164] Serganova, V.V., *Functional Anal. Appl.* **19** (1985), 75–76.

Progress in Physics

— A collection of research-oriented monographs, reports, notes arising from lectures or seminars
— Quickly published concurrent with research
— Easily accessible through international distribution facilities
— Reasonably priced
— Reporting research developments combining original results with an expository treatment of the particular subject area
— A contribution to the international scientific community: for colleagues and for graduate students who are seeking current information and directions in their graduate and post-graduate work

Manuscripts should be no less then 100 and preferably no more than 500 pages in length. We encourage preparation of manuscripts in such forms as LaTeX or AMS TeX for delivery in camera-ready copy which leads to rapid publication, or in electronic form for interfacing with laser printers or typesetters. However, we continue to accept other forms of preparation as well.

Proposals should be sent to:
Birkhäuser Boston, Inc., P.O. Box 2007, Cambridge, MA 02139, USA, or
Birkhäuser Verlag AG, P.O. Box 133, CH-4010 Basel/Switzerland.

12 PIGUET/SIBOLD. Renormalized Supersymmetry: The Perturbation Theory of $N = 1$ Super-symmetric Theories in Flat Space-Time
ISBN 0-8176-3346-4

13 HABA/SOBCZYK. Functional Integration, Geometry and Strings, Proceedings of the XXV Karpacz Winter School of Theoretical Physics
ISBN 3-7643-2387-6

14 SMIRNOV. Renormalization and Asymptotic Expansions
ISBN 3-7643-2640-9